Practical Guide to Neck Dissection

Marco Lucioni

Practical Guide to Neck Dissection

Focusing on the Larynx

Second Edition

Springer

Marco Lucioni, MD
ENT Department
Vittorio Veneto Hospital
Vittorio Veneto
Italy

With a Contribution on Surgical Pathology (chapter 11 and 12) by Duilio Della Libera, Department of Anatomic Pathology, Feltre Hospital, Italy

ISBN 978-3-642-33976-9 ISBN 978-3-642-33977-6 (eBook)
DOI 10.1007/978-3-642-33977-6
Springer Heidelberg New York Dordrecht London

Library of Congress Control Number: 2012954869

Printed on acid-free paper

Springer is part of Springer Science+Business Media (www.springer.com)

Foreword I

It was 3 o'clock in the afternoon: time for anatomy class. A badly lit room, a caretaker to collect tips, and a single lecturer for 30 students. The material on which to study practical anatomy consisted of a humerus, a femur, and an entire decomposing human forearm with skeletised muscles and tendons, reduced to shreds by previous inexperienced dissectors. Then, 2 years later at midday, I found myself in a pathologic anatomy amphitheatre with 300 students. An empty corpse lay on the distant dissection table with various removed organs lined up by its side. The lecturer was giving his last class for the course and gratefully addressed the deceased, "for donating his body to the progress of science". These are my recollections as a student of medicine 35 years ago. Yet Padua was an important university, one of the most ancient, most prestigious universities in Europe! These experiences go back many years, but I do believe the situation has changed very little since then.

These were my thoughts 15 years ago when I was invited to direct a neck dissection course in the corpse at the Vrije Universiteit Brussel. To prepare the course, after 30 years' experience in neck surgery, I went back to, or rather, I found myself for the very first time dissecting a cadaver. I would like to thank all colleagues at the University of Pavia for lending me their dissection theatre. It was a stimulating, highly positive experience, enabling my co-worker and me to broaden and develop our knowledge of neck anatomy and its border areas. While we iconographically documented the various cervical regions and dissection planes, our thoughts turned to past experiences in this type of activity.

Anatomic dissection for research purposes dates back to the Egyptians in Alexandria but was prohibited in the Western world for many centuries by Jewish and Christian religious culture. A decree was passed in the Kingdom of Sicily in 1231 by Frederick II of Swabia, stating that "… all those who studied surgery should become learned in operations and particularly in the anatomy of the human body …". Mondino dei Liuzzi, author in 1316 of the treatise Anathomia, introduced cadaveric dissection into the university teaching curriculum in Bologna. The chief council of the Serenissima Republic in Venice decreed that every year, a number of corpses should be dissected "propter urbis honorem civiumque salutem".

However, the "anatomy century" was undoubtedly the sixteenth century, with its Renaissance anatomists. The most outstanding figure in the scientific revolution of that period was clearly Andreas Vesalius from Brussels (1514–1564) with his De Humani Corporis Fabrica. Prevented from practising dissection at the University of

Leuven, Vesalius came to Padua where, despite his very young age, the Serenissima government appointed him to the chair of anatomy in virtue of his extensive knowledge on the subject and corpse-dissecting skills.

Five and a half centuries later, we were to make the same journey as Vesalius, only in the opposite direction. Prevented by law and custom from holding a course on dissection in Italy, we left the land of the Serenissima in the direction of Brussels, where a modern university organisation provided us with all the necessary technical equipment and 15 cadavers. We armed ourselves with our long and inveterate experience in neck surgery and our more recent know-how in cadaveric neck dissection with related iconography. When, after the second "Andreas Vesalius course", as we call them, Dr. Marco Lucioni, my faithful co-worker in the preparation and conduction of these scientific-teaching ventures, expressed the desire to produce a volume on anatomic neck-dissection techniques, based on our experience; I did not hesitate to encourage him. I then enthusiastically observed the text being drawn up and divided into the various chapters and figures.

Now the volume by Lucioni is complete and ready to go to press. I find it a very carefully prepared, comprehensive, well-illustrated work, constituting an essentially practical, valid reference tool that freshens up notions in normal and topographical neck anatomy and a precious guide for anyone practising anatomic neck dissection in the corpse.

I trust my favourable, but not impartial, judgment will encourage those who wish to browse through, and hopefully read it.

Vittorio Veneto, Italy Italo Serafini

Foreword II

The first report on neck dissection can be traced to Richard Volkmann in 1882; however, it was Franciszek Jawdynski who described the technique of the operation in 1888. Henry Butlin proposed an upper neck dissection for the treatment of tongue cancer at the turn of the nineteenth century; however, George Crile is credited for the first systematic report on classical radical neck dissection over 100 years ago, based on his personal experience of 132 cases. Since then, neck dissection has remained the mainstay of surgical treatment of metastatic cervical lymph nodes from mucosal and cutaneous carcinomas of the head and neck. Increasing experience with this surgical technique and improved understanding of biological progression of metastatic cancer to cervical lymph nodes led to the development of numerous modifications in neck dissection, with the aim of retaining oncologic efficacy but reducing the morbidity of the operation. Thus, Oswaldo Suarez initially proposed a modified neck dissection that was subsequently popularised by Ettore Bocca in English literature. Further modifications in neck dissection were proposed by Allando Ballantyne and others during the latter half of the twentieth century. The systematic classification of various types of neck dissections and its applications have been proposed and popularised by the American Academy of Otolaryngology/Head and Neck Surgery, and these are currently employed in clinical practice worldwide.

Dr. Marco Lucioni is to be complimented on putting together this outstanding piece of work initially stimulated by Italo Serafini. The book is prepared from sequential photographs of cadaver dissections of systematic steps at understanding the topographical anatomy of various layers of tissues in the neck. The author systematically describes anatomic structures in the cadaver under four different headings: the parotid region, submandibular triangle, the lateral neck, and the median cervical region. Each section describes anatomy in the superficial layer as well as the deep layer. Clinical implications of the anatomic structures in therapeutic interventions are highlighted with bullet points indicating "take home messages" and "core messages". Each section begins with a diagram of the anatomic structures important in that region, followed by cadaver dissection, highlighting the salient features of each step of the operation. The book is complemented by a DVD showing video clips of neck dissection in the cadaver, further familiarising the reader with step-by-step anatomic structures encountered during various types of neck dissections. The author has also thus included various modifications in neck dissection, which are currently employed in clinical practice.

For the student of head and neck surgery, this book would be a valuable resource to his or her personal library, since it is a stepwise approach to understanding the anatomy of the neck and its importance in performing a systematic, safe, and effective surgical procedure for excision of cervical lymph nodes, either involved or at risk by metastatic cancer from primary tumours in the head and neck. The photographic reproduction is crisp and clear, both in the cadaver dissections as well as in the DVD. Highly accurate and effective works such as this are crucial to further solidify the surgical prowess of head and neck surgeons of the future.

New York, USA Jatin P. Shah

Foreword III

Through many years of collecting textbooks of anatomy, I have cherished the magnificent descriptions of anatomy provided by the likes of Testut, Latarjet, and Rouviere. As an academic head and neck surgeon practising and teaching in North America, I have frequently struggled not only translating them into English but also making these descriptions intuitively usable by students of head and neck surgery.

A few years ago, I was invited to Italy to lecture on selective neck dissections. As a memento, I was given a copy of foul proof of *Practical Guide to the Neck Dissection* by Marco Lucioni. I was thrilled to encounter in this book the anatomy of the neck depicted in a way that only a surgeon can, when his or her knowledge and expertise are combined with the talents of a good artist and a good photographer.

As I reflect on my reactions when I read the book, I predict that a potential reader, who picks up this book out of curiosity and begins leafing through it, will at first be intrigued, if nothing else, by the exceptional quality of the drawings and by the clarity of the photographs of anatomic dissections. The reader will then feel compelled to study these illustrations and the text that accompanies them and will, shortly thereafter, come to the realisation that this is not just a collection of beautiful illustrations; it is, rather, an insightful documentation of surgical anatomy of the different regions of the neck, the parotid, and the larynx. As such, it would be treasured by medical students of anatomy, who will find in it a clear, almost three-dimensional depiction of the different muscular, vascular, and neural structures of the neck. It would be equally valued by students of otolaryngology and head and neck oncologic surgery for they will find that the complex relationships of these anatomic structures are shown in a manner and sequence similar to what they would encounter during different surgical procedure in the neck, the thyroid, the parotid gland, and the larynx. Teachers of anatomy and of surgery will also find it valuable since it will enable them, as it has often enabled me, to illustrate for students, residents, and fellows important anatomic structures and their relationships in a way that is not always possible in the classroom or in the operating theatre.

This book will find and keep a preferential place in the library of many for it represents what we always hope for in a book of this kind, but rarely get.

Oklahoma City, OK, USA — Jesus E. Medina

Foreword IV

It is my great pleasure to write a preface for this anatomical surgical compendium for head and neck surgery edited by Marco Lucioni.

I have known Lucioni for many years and have had the opportunity in his courses and during visits to Vittorio Veneto to become acquainted with and to appreciate his surgical talents.

Someone with such extensive experience and deftness in head and neck surgery is predestined to edit an anatomically detailed, illustrated presentation of operations of the neck, the larynx, and the salivary glands.

The impressive, excellently photographed intraoperative sites together with the informative schematic drawings will be of help to ENT specialists, laryngologists, and head and neck surgeons in performing anatomically oriented and precise surgical dissections, while preserving structure and function. The excellent illustrations of the complex topographical relationships between muscles, blood vessels, nerves, and lymphatic structures in detailed photographs will allow the surgeon to proceed confidently even in the difficult and risk-fraught dissection of the head and neck region.

This book is a valuable contribution and is to be highly recommended as a guide for head and neck surgeons in the Italian tradition of anatomy and surgery.

Göttingen, Germany Wolfgang Steiner

Foreword V

I met Dr. Lucioni in Milan, about 30 years ago, while he was resident at the Otorhinolaryngology Clinic of the University, then directed by Prof. Bocca. I supported Dr. Lucioni in his thesis on vasomotor rhinitis. This thesis was an excellent one, and Dr. Lucioni entered with top marks the world of Italian otorhinolaryngologists.

He then soon started to get around, looking for a position as an assistant, and at the end of this search, asked my opinion about the chance of joining the group of Prof. Italo Serafini in the hospital of Vittorio Veneto, one of the most outstanding temples of head and neck oncology in Italy. I approved warmly.

Since then, I have had the opportunity to follow Dr. Lucioni in his career at the many meetings organised in Vittorio Veneto by Prof. Serafini. His "learning curve" in head and neck oncologic surgery was reflected in a series of anatomosurgical manuals, of which the present one is the most complete version. Anatomical drawings and beautiful photographs from cadavers are integrated into the schemes of the main surgical procedures, along a teaching path which, through the accuracy of the details and the appealing clarity of the images, achieves a noticeable didactical goal.

This book, which in my opinion is a very useful reminder for any head and neck surgeon, whichever his or her degree of skill, mirrors the talent of Dr. Lucioni and the high quality of the Vittorio Veneto Otorhinolaryngological School.

Brescia, Italy Antonio Antonelli

Preface

Five years have passed since the first edition of the *Practical Guide to Neck Dissection*, and I must say that the book has brought me great satisfaction. Many colleagues have told me that they keep it locked in the cupboard in the operating theatre to be ready whenever needed. The name of an artery, a suturing procedure, the surgical steps of dissection… If it is true that the surgeon's work starts off from structured medical knowledge and multicentral statistical studies from which guidelines are obtained, it is also true that in the end, it is the empirical work of a craftsman, often carried out individually, in a kind of workshop where the operator has brought together good tools, knowledge of materials, and can rely on the precious help of his colleagues and, perhaps, on a handbook like this one.

The second edition is subtitled "focusing on larynx" and involved the collaboration of Duilio Della Libera, an anatomopathologist with whom I have been working for many years. The reader will be able to appreciate the splendid histological macrosections. We shared the passion for laryngeal oncology with our Maestri Giorgio Carlon and Italo Serafini.

Professor Italo Serafini, who inspired this work, passed away on 20 July 2010. Perhaps it was only good luck that we met such a Maestro when making our way in the profession. Proceeding alone, on a teach-yourself basis, requires great effort, and justified doubts often arise concerning the correctness of the professional path you are taking. You don't sleep well. But when the Maestro takes the stage, it's like getting into a carriage. The journey may still be long and tiring, but it is so satisfying to understand where you went wrong and what you can do to improve. Without a Maestro, it's a bit like walking among the clouds…

Vita brevis, ars longa, occasio praeceps, experimentum periculosum, iudicium difficile is a Latin phrase meaning "Life is short, art is long, opportunity fleeting, experiment dangerous, judgement difficult".

It is an aphorism by Hippocrates of Kos in which he stresses the importance of studying and the difficulty of diagnostic analysis. It means substantially that in all the arts, one man's life is not sufficient to achieve perfection, which requires the progressive contribution of several generations.

We hope we may be up to the task.

Vittorio Veneto, August 2012 — Marco Lucioni

Acknowledgments

I thank *Piero Della Libera*, Punto Immagine, for his fine elaboration of the photographs; *Sandra Speirs*, for her English translations; *Andrea Porzionato*, Anatomy Institute of University of Padua, for his information on law rules in cadavers dissection; *Marco Ravanelli* and *Roberto Maroldi*, Department of Radiology of University of Brescia, for their great RMI images of the larynx; and *Sandra Lesny* for her kind aid in editing this book.

Contents

Introduction to Dissection

1

Core Messages

- From the anatomic and surgical point of view, the neck is an extraordinarily interesting place. It is like a bridge where fundamental functional units meet and transit. The operating field is on a convenient scale for the surgeon's hands: not so small that it can be explored only with a microscope (like the brain), nor so large as to require ample movements of the arms (abdomen).

1.1 Prologue

> Sul parquet, tra la tavola e la credenza piccola, a terra … quella cosa orribile … Un profondo, un terribile taglio rosso le apriva la gola, ferocemente. Aveva preso metà il collo, dal davanti verso destra, cioè verso sinistra, per lei, destra per loro che guardavano: sfrangiato ai due margini come da un reiterarsi dei colpi, lama o punta: un orrore! da nun potesse vede. Palesava come delle filacce rosse, all'interno, tra quella spumiccia nera der sangue, già raggrumato, a momenti; un pasticcio! con delle bollicine rimaste a mezzo. Curiose forme, agli agenti: parevano buchi, al novizio, come dei maccheroncini color rosso, o rosa. "La trachea", mormorò Ingravallo chinandosi, la carotide! la iugulare … Dio!
>
> ("On the floor, between the table and the sideboard, lay a horrible sight … A ferociously deep red cut opened her throat. Half her neck had been removed, from front to right, that is front to left for her or front to right for any onlookers. It was frayed at the edges as though she had been struck again and again, by a point or blade: what horror! A sight for no eyes! Red strips were showing on the inside, between the blackened, coagulating blood. What a mess! Strange shapes emerged: to the police they looked like holes, to the novice like red or pink macaroni. "The trachea", murmured Ingravallo, bending over her. "The carotid! The jugular … Oh my God!") [1].

This piece from a high school novel presents a dramatically curious, subtly humorous approach to the neck. Other cervical images that come to mind are the pale, lunar necks in Bram Stocker's original black-and-white screenplay versions of Dracula; the "long", ethereal simplicity of Modigliani's necks; or photographs of the ringed necks of Burmese women depicted in National Geographic. During a

M. Lucioni, *Practical Guide to Neck Dissection*,
DOI 10.1007/978-3-642-33977-6_1, © Springer-Verlag Berlin Heidelberg 2013

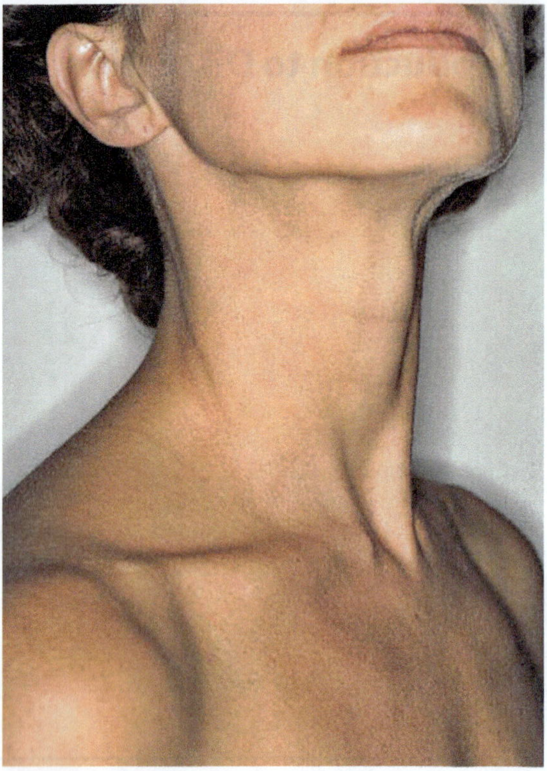

Fig. 1.1 Carolina's neck

school trip to Castello del Buon Consiglio in Trento, I vividly recollect feeling very uneasy when I saw the unnatural posture of Cesare Battisti's head that had been photographed after execution by hanging.

The neck does indeed conjure up more images than any other part of the body, depending on mode of reproduction. It can inspire the maternal sweetness of sixteenth-century images of Madonna and Child, erotic fantasies of long-legged models on metropolitan catwalks, or anxiety as the strangler's hands close around it in a horror movie. Its versatility probably stems from its being anatomically and conceptually hard to define, and the lack of a material or symbolic identity of its own, compared, say, to the eye or liver. It presents virtual anatomic boundaries, with arbitrary lines rather than natural limits of its own. Its main function of supporting the head has nothing special or exclusive about it. Its true essence seems instead to be its function as a linking structure, a sort of bridge between head and body, transporting blood, air, emotions, and information on movement and sensitivity, that is, it is the point where the "breath of life" converges and is conveyed. We use the neck of a classic ballerina, like Carolina (Fig. 1.1), as a graceful introduction to our dissection class (Figs. 1.2 and 1.3). Let us start by getting to know the superficial landmarks.

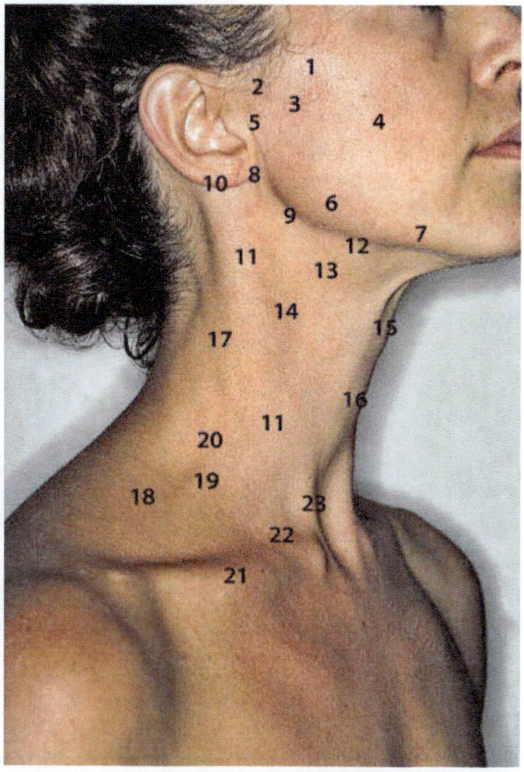

Fig. 1.2 Superficial landmarks: lateral view. *1* Zygomatic process of the temporal bone, *2* auriculotemporal nerve and superficial temporal pedicle, *3* caput mandibulae, *4* parotid duct, *5* external auditory canal, *6* angle of mandible, *7* facial pedicle, *8* transverse process of atlas, *9* inferior parotid pole, *10* apex of mastoid, *11* sternocleidomastoid muscle, *12* submandibular gland, *13* apex of greater cornu of hyoid bone, *14* carotid bifurcation, *15* laryngeal prominence, *16* cricoid cartilage, *17* emergence of spinal accessory nerve (peripheral branch), *18* trapezius and entrance of spinal accessory nerve (peripheral branch), *19* inferior belly of omohyoid muscle, *20* external jugular vein, *21* clavicle, *22* sternocleidomastoid muscle (clavicular head), *23* sternocleidomastoid muscle (sternal head)

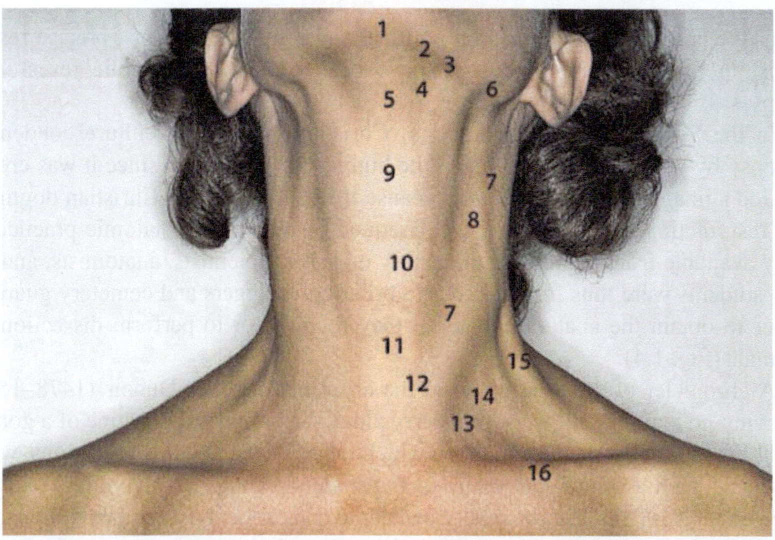

Fig. 1.3 Superficial landmarks: anterior view. *1* Mental eminence, *2* inferior border of mandible, *3* facial pedicle, *4* submandibular gland, *5* hyoid bone, *6* angle of mandible, *7* sternocleidomastoid muscle, *8* external jugular vein, *9* laryngeal prominence, *10* cricoid, *11* isthmus of thyroid gland, *12* sternocleidomastoid muscle (sternal head), *13* sternocleidomastoid muscle (clavicular head), *14* inferior belly of omohyoid muscle, *15* anterior border of trapezius muscle, *16* clavicle

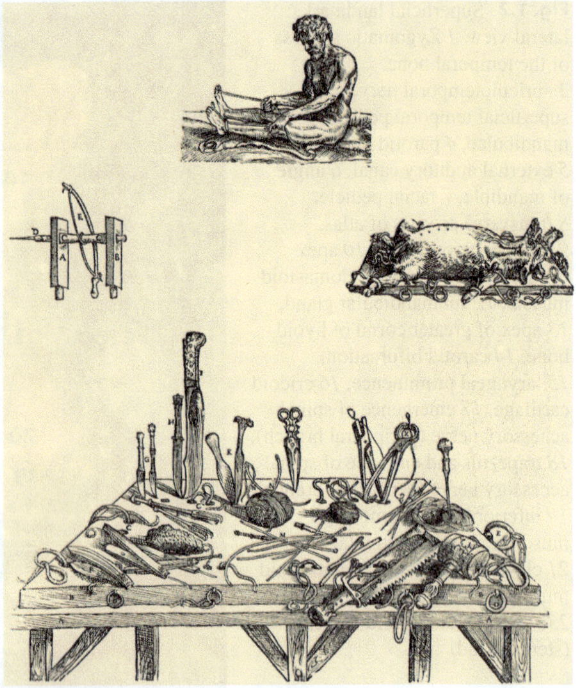

Fig. 1.4 Sixteenth-century dissection instrumentation

1.2 Releasing a Corpse for Research Purposes

Over the eras, in accordance with political and religious precepts, precise restrictions, in many cases prohibitions, have been placed on scientific research on corpses.

In the Western world in particular, Christian and Jewish culture condemned autopsy by virtue of the belief that "the human body is sacred since it was created in God's image and likeness," and because it was "contrary to Christian dogma on the resurrection of the flesh" [2]. Consequently, records on anatomic practice are only available from the thirteenth century onwards. Scientists, anatomists, and fine arts students were thus forced either to bribe gravediggers and cemetery guards in order to obtain the anatomic material they required or to perform dissections on animals (Fig. 1.4).

A chronicler of the time wrote of the anatomist Jacques Dubois (1478–1555): "Having no manservant, I saw him carry alone the uterus and intestine of a goat, or the thigh or arm of a hanged man, on which to perform anatomic dissections, which produced such a stench that many of his students would have vomited, had they been able" [3]. Even the University of Padua, one of the most famous in Europe in the early sixteenth century, was allowed a quota of two corpses, one male and one female, on which to practise dissection, thanks to a specific privilege granted by the

Fig. 1.5 Anatomy Theatre, Palazzo del Bo, Padua

Church. However, the chronicles of the period speak of the secret conveyance of the bodies of hangman's victims through an underground river passage leading directly to the Anatomy Theatre of Palazzo del Bo, where Andreas Vesalius taught for 5 years (Figs. 1.5 and 1.6).

The sixteenth century was the century of the great anatomists, and Vesalius stands head and shoulders above them all. With the Renaissance, anatomy moved away from the religious and dogmatic doctrines that had dominated the Middle Ages and was subordinate to the neutral observation of natural phenomena. Vesalius was therefore the successor of Galen, just as in physics Copernicus took over from Ptolemy. With Vesalius, anatomical science officially became an essential part of the experimental method. In teaching, "Vesalius's reform" meant the replacement of a method of teaching anatomy based on books and dogma with another, revolutionary method, based on the practice of direct and systematic dissection, and therefore more "faithful to anatomical reality". In 1543, Vesalius published the first great modern treatise on anatomy, *De Humani Corporis Fabrica*, an educational text with very clear text and illustrations. He was helped by painters such as Jan Stephan van Calcar, a student of Titian, and the drawings were transferred into woodcuts by Valverde. The frontispiece of the *Fabrica* is in the Academy of Medicine in New York; it shows a lesson held by Vesalius in the Anatomy Theatre of Padua University (Fig. 1.7).

Anatomic dissection has always been considered a fundamental subject for the teaching of medicine. Nevertheless, in European degree courses in medicine and

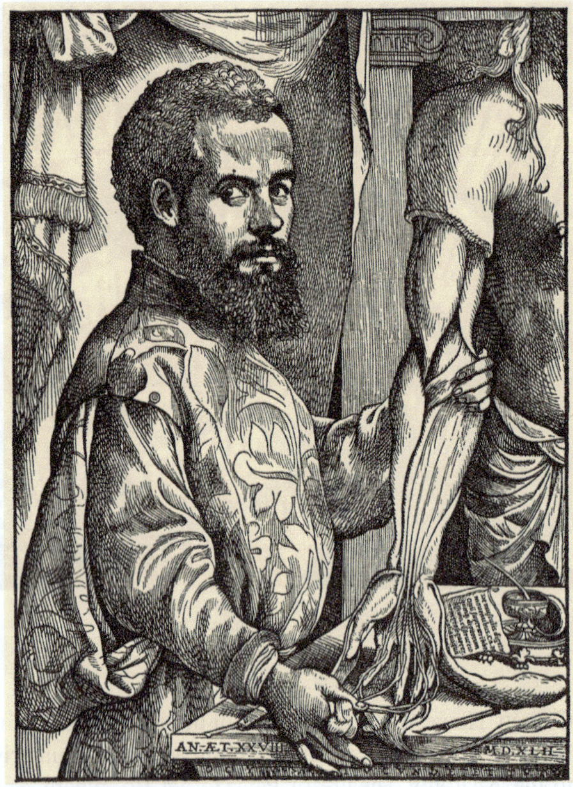

Fig. 1.6 Andreas Vesalius

surgery, in recent decades there has been a drastic reduction in the hours, methods, and contents of the teaching of human anatomy and in particular of the hours of practical lessons. However, there has recently been a renewed interest in the subject, and it is usually specialists in surgery who want to perfect their surgical techniques on cadavers or learn new ones. For this reason, there is a growing offer of courses in surgical anatomy on cadavers.

In Italy, the use of corpses for research purposes is considered a legitimate practice, albeit governed by specific state legislation; reference should be made in particular to the Consolidation Act on Higher Education Legislation (1933) and the Mortuary Police Regulations (1990).

First, the place of dissection is established, that is, at a university institution. Theoretically, the law permits hospitals to request parts of corpses from university institutions, but, in practice, the excessive bureaucracy involved makes such requests prohibitive (suffice it to consider the transportation of corpses or parts of them).

Regarding the selection procedure for cadavers for teaching and research purposes, Italian legislation allows only the following: corpses admitted to forensic investigation (through the courts) but not requested by family members (excluding

Fig. 1.7 Frontispiece of *De Humani Corporis Fabrica*, 1543

suicides) and corpses for whom transportation has not been paid by the respective family but has been provided free of charge by the local authorities.

Anyone during his or her lifetime can donate by a living will the entire body for teaching and research purposes. This is not, however, a customary practice in Italy. Indeed, in order to have several corpses simultaneously, the three editions of the Practical Course in Neck Dissection (1991, 1992, and 1994), edited by the ENT team of Vittorio Veneto, were carried out in Brussels, Belgium, where the decision to leave one's own body to medical science is a far more common practice. This probably derives from the fact that in other European countries and in the United States, the law has already approved and regulated this possibility for several years now.

Our hypothetical dissection class therefore takes place in a university institution of normal human anatomy or pathologic anatomy. A diagnosis has recently been formulated for the corpse before us; hence, at least 24 h have passed since time of death, and rigor mortis is resolving. We have already ascertained the absence of disease and previous surgical operations on the neck in the structures to be dissected. We are very fortunate if the person in question was fairly tall as this will greatly aid dissection.

1.3 Instrumentarium

Anatomic dissection is a contemplative manual activity. It requires silence and above all should be subject to no time restrictions, as its value is depreciated by hurried performance. Very good lighting conditions are needed and are best provided by scialytic operating lamps. Alternatively, two revolving cold light lamps can be adopted. As a last resort, environmental light focally reinforced by a Clar forehead mirror can be used. Figure 1.8 illustrates the operating instruments that we consider necessary for neck dissection, in addition to a few helpful tools.

Neck dissection may be conducted by a lone surgeon, but this makes it a very awkward task. Two surgeons should instead be involved, alternating with each other in the roles of chief and assistant, thereby promoting efficacy and cultural exchange. The classic error to avoid is to have two surgeons acting separately at opposite sides of the neck.

Last, at the end of dissection, the body should be carefully recomposed. Where possible, unnecessary deforming manoeuvres should be avoided. Consideration and respect should reign at all times towards those who have willingly or unknowingly donated their bodies to science.

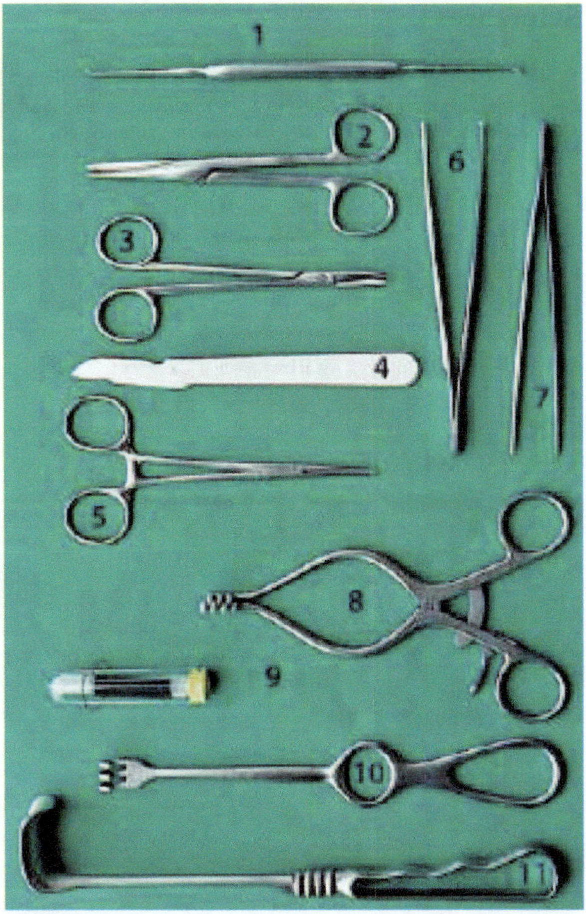

Fig. 1.8 Instrumentarium. *1* Septum-type separator, *2* medium surgical scissors, *3* small surgical scissors, *4* disposable scalpel, *5* cocker, *6* surgical forceps, *7* anatomic forceps, *8* self-retaining retractor, *9* silk, *10* three-point hook, *11* medium-sized Farabeuf

Take-Home Messages

- Anatomic dissection is a contemplative manual activity. It requires silence, and haste should be avoided at all costs. It is best to have two surgeons working on the neck dissection, because one has to help the other expose the field and possibly discuss the concepts learned.
- My professor used to say that on the learning scale, it is one thing for a surgeon to find a structure and know how to recognise it, while it is quite a different thing to look for that structure in the precise place where he is sure to find it.

References

1. Gadda CE (1957) Quer pasticciaccio brutto de via Merulana. Garzanti, Milan
2. Giusti G, Malannino S (1988) Legislazione Sanitaria Tanatologica. Cedam, Padua
3. Guerrier Y, Mounier P (1989) La gola. In: Kuhn F (ed) Storia delle malattie dell' orecchio, del naso e della gola. Editiemme, Milan

General Anatomic Layout

2

Core Messages

- At the start of the dissection exercise, we must take a panoramic look for orientation. We then establish the limits of the area of operation and the main landmarks.

2.1 Anatomic Layout

The neck is the part of the trunk that joins the head and the chest and constitutes its most mobile part. It is essentially cylindrical in shape; length is constant while diameter varies. The expression "long neck/short neck" is incorrect, because the length of the neck, understood to be the cervical portion of the vertebral column, does not present significant variations. Conversely, neck width, determined by the development of muscular and adipose masses is extremely variable [1].

Significant Anatomical Structures: nuchal region, trachelian region, superficial cervical fascia, middle cervical fascia, deep cervical fascia, superficial cervical lymph node system, deep lymph node system, Delphian lymph node.

Landmarks: mandible, external auditory canal, mastoid apex, superior nuchal line, external occipital protuberance, clavicle, spinous process of the seventh cervical vertebra, linea alba.

2.2 Neck Boundaries

Its upper limits run along the inferior and posterior borders of the mandible, the extreme posterior of the zygomatic arches, the anteroinferior borders of the external auditory canals, the profiles of the mastoid apex, the superior nuchal line, and the external occipital protuberance. Its lower boundaries lie along the superior border of

M. Lucioni, *Practical Guide to Neck Dissection*,
DOI 10.1007/978-3-642-33977-6_2, © Springer-Verlag Berlin Heidelberg 2013

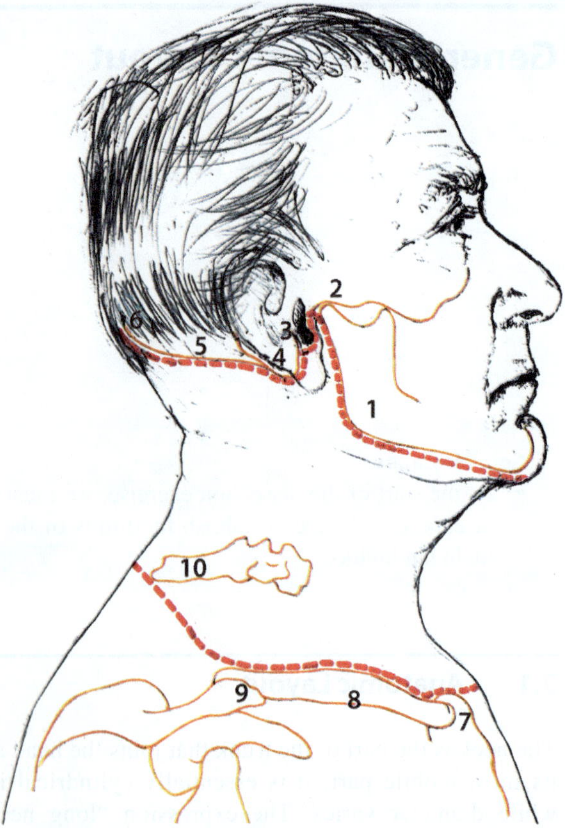

Fig. 2.1 Neck boundaries. *1* Mandible, *2* zygomatic process of the temporal bone, *3* external auditory canal, *4* mastoid, *5* superior nuchal line, *6* external occipital protuberance, *7* manubrium sterni, *8* clavicle, *9* acromioclavicular joint, *10* spinous process of seventh cervical vertebra

the sternum and clavicles, the acromioclavicular joints, and an imaginary line joining the acromioclavicular joints to the spinous process of the seventh cervical vertebra (Fig. 2.1).

2.3 Structure

On transverse section, the neck appears to be roughly divided into two parts, a posterior or nuchal region (osteo-muscular) and an anterior or trachelian region (muscular–fascial). The conventional dividing line extends from the transverse vertebral processes to the anterior edges of the trapezius muscles (Fig. 2.2).

The function of the posterior region is essentially static and dynamic – powerful, articulated muscles support a bone framework with the head at the top. This structure functions as an articulated joint since the two interapophyseal joints between one vertebra and the next permit head movement; it also functions as a shock absorber for intravertebral disk compressibility in addition to being a fastening point for the muscles of mastication, swallowing, and speech. The cervical portion of the vertebral column is curved with anterior convexity (cervical lordosis). In contrast, the anterior region, which is the object of this dissection, holds the internal organs.

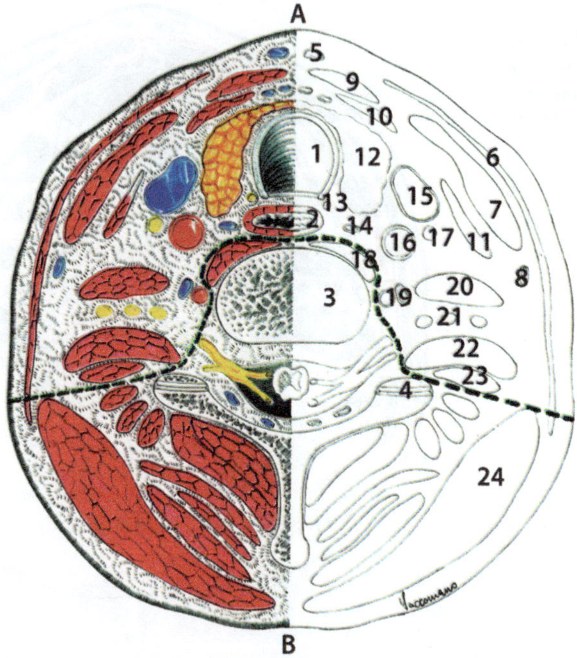

Fig. 2.2 Transverse cervical section: trachelian region and nuchal region. (**a**) Trachelian region. (**b**) Nuchal region. *1* Trachea, *2* oesophagus, *3* vertebral body of seventh cervical vertebra, *4* interapophyseal articulation, *5* anterior jugular vein, *6* platysma muscle, *7* sternocleidomastoid muscle, *8* external jugular vein, *9* sternohyoid muscle, *10* sternothyroid muscle, *11* omohyoid muscle, *12* thyroid gland, *13* recurrent nerve, *14* inferior thyroid vein, *15* internal jugular vein, *16* common carotid artery, *17* vagus nerve, *18* prevertebral muscles, *19* vertebral artery and vein, *20* anterior scalene muscle, *21* brachial plexus, *22* medial scalene muscle, *23* posterior scalene muscle, *24* trapezius muscle

It contains the parotid and submandibular glands, the thyroid gland, and several lymph nodes and is crossed by important blood and lymphatic vessels and nerves and by the respiratory and digestive tracts.

In addition to being prevalently a structure of transit and union, the neck is an important point of autonomous physiological activity, linked to the presence of exocrine glands (parotid and submandibular), endocrine glands (thyroid, parathyroid, and thymus), muscle and tendon neuroreceptors, visceral receptors, vascular chemopressoreceptors, and lymph nodes.

2.4 Cervical Fasciae

Almost all cervical viscera originate from or lead to the thorax or upper extremities; the loose connective tissue surrounding them is in direct, continuous contact with the loose connective tissue of the mediastinum and axillary regions. In some points, the loose connective tissue thickens to form fibrous sheaths (around neurovascular bundles,

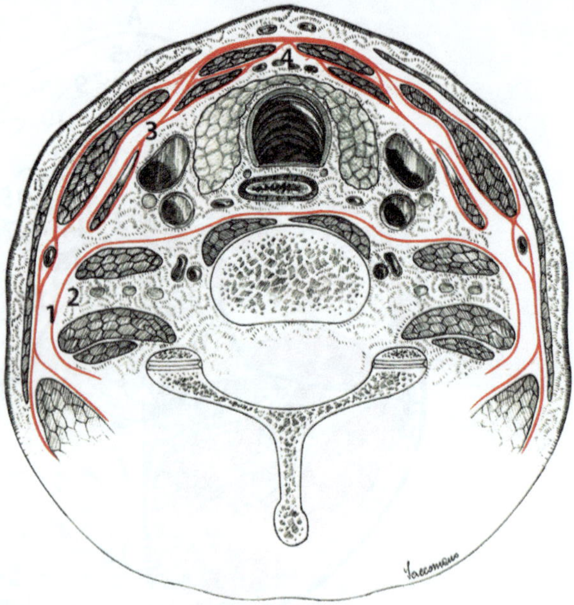

Fig. 2.3 Transverse cervical section: cervical fasciae. *1* Superficial cervical fascia, *2* deep cervical fascia, *3* middle cervical fascia, *4* infrahyoid linea alba

the laryngotracheal canal, and the thyroid) and perimuscular aponeuroses. These latter define important dissection planes, particularly:

1. The superficial cervical fascia, extending from the anterior edge of the trapezius and splenus capitis muscles on both sides, which divides into two to enclose the sternocleidomastoid muscles, parotid gland, and submandibular gland; it fuses with the middle cervical fascia on the midline (linea alba).
2. The middle cervical fascia, lying between the omohyoid muscles on both sides; as a whole, it forms a triangle with the hyoid bone at its apex and the clavicles at the base; it divides in two to contain the infrahyoid muscles.
3. The deep (or prevertebral) cervical fascia, investing the prevertebral muscles and dividing laterally to contain the scalene and levator scapulae muscles (Fig. 2.3).

2.5 Lymph Node Stations

The cervical lymphatic system forms a three-dimensional network into whose nodal points the lymph nodes are intercalated. Although they vary in number and dimensions, they do keep a relatively constant position, and they can thus be considered topographically grouped into lymph gland stations (Fig. 2.4).
These are divided in the neck as follows:

1. A subfascial, superficial lymph node system with a circular arrangement between chin and occiput (occipital, mastoid, parotid, submandibular, and submental lymph nodes) and along the course of the external jugular vein

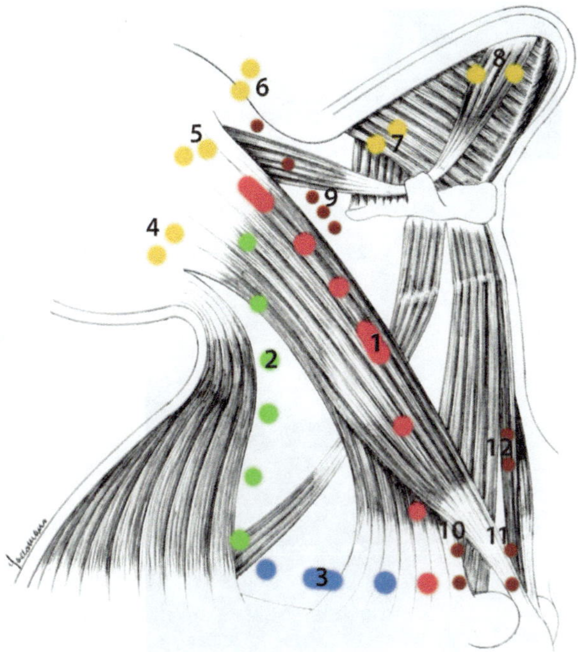

Fig. 2.4 Lymph node stations. *1* Jugular chain, *2* spinal chain, *3* supraclavicular chain, *4* occipital lymph nodes, *5* mastoid lymph nodes, *6* parotid lymph nodes, *7* submandibular lymph nodes, *8* submental lymph nodes, *9* retropharyngeal lymph nodes, *10* recurrent lymph nodes, *11* pretracheal lymph nodes, *12* prelaryngeal lymph nodes

2. A more consistent, deep lymph node system in a bilateral triangular arrangement, bounded anteriorly by lymph nodes adjacent to the internal jugular vein and posteriorly by the spinal lymph node chain, with a supraclavicular lymph node
3. A perivisceral lymph node system close to the median viscera (prethyroidean, pretracheal, retropharyngeal, recurrent, and finally prelaryngeal lymph nodes, the more defined of which, called Delphian lymph node, is situated between the cricothyroideal muscles)

Remarks: The relationships between the lymph nodes/lymphatic vessels and the muscles/vessels/nerves and glands in the neck are of a contiguous nature, always in normal conditions and nearly always in pathological ones. Thanks to the removal of the fascia, they may be separated from the contiguous structures and moved away easily. There may be an interruption of the fascia and colonisation of the contiguous structures only if the lymph node capsule gives way as a result of carcinomatous invasion [2].

2.6 Before Beginning the Dissection

Anatomists divide the neck into two major regions:

1. The anterior region, situated between the two sternocleidomastoid muscles, encompassing the suprahyoid, infrahyoid, and prevertebral regions
2. The lateral regions, comprising the parotid, sternocleidomastoid or carotid, and supraclavicular regions

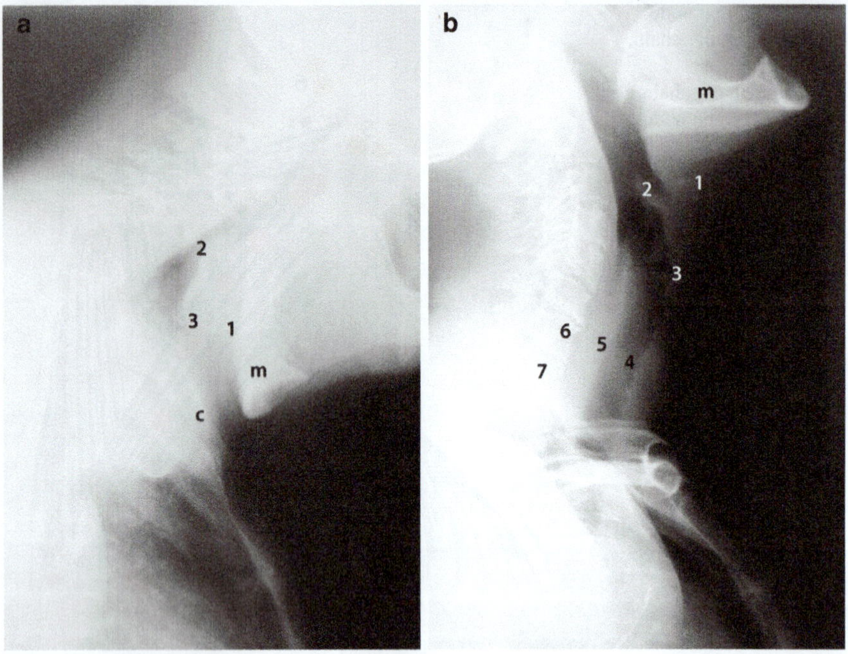

Fig. 2.5 Neck mobility. (**a**) Flexion, (**b**) Extension. *m* Mandible, *c* clavicle, *1* hyoid bone, *2* epiglottis, *3* Morgagni's ventricle, *4* trachea, *5* cervical oesophagus, *6* seventh cervical vertebra, *7* first thoracic vertebra

For the sake of simplicity and for dissection purposes, we instead divide the neck into three lateral regions (parotid, submandibular, and laterocervical) and three median regions (inferior median, superior median, and prevertebral).

The anatomic arrangement of the neck organs varies considerably with neck movements, especially flexing–extending movements. For example, at maximum flexion, the hyoid bone, one of the more cranial structures, can almost reach the thorax (Fig. 2.5a, b). Surgeons should bear this in mind since they can take advantage of great cervical mobility to achieve the widest possible dissection areas.

Remarks: We stress that the symmetrical posture of the neck is commonly defined the normal position. The surgical manoeuvres will be carried out on a neck which we shall try to hyperextend as much as possible. To obtain this position, a thickness of at least 10 cm must be placed under the scapulae. That is as far as the anterior regions are concerned. For the lateral regions, the head must be turned contralaterally with respect to the operator; this is defined the operating position. Instead, when the head is bent and slightly inclined towards the explored side, the structures relax and this allows deep exploration of the neck. This is defined as the clinical exploration position.

Take-Home Messages

- The correct position of the head (extended as far as possible) is crucial both in anatomic specimens and when operating in vivo.

References

1. Testut L, Jacob O (1977) Trattato di Anatomia Topografica. UTET, Torino
2. Bocca E (1972) Chirurgia dei linfonodi cervicali. In: Naumann HH (ed) Chirurgia della testa e del collo. Casa Editrice Ambrosiana, Milano, pp 153–187

Superficial Dissection

3

Core Messages

- A large area of operation makes dissection easier. The cutaneous flap is raised between the platysma and the superficial cervical fascia, as in vivo. The superficial cervical fascia is interrupted as little as possible: it contains the vessels and lymph nodes which in neck dissection would be removed with the specimen.

3.1 Anatomic Layout

The neck is placed in a normal position, hyperextended. The incision is very low and posterior, also to allow reconstitution of the cadaver at the end of dissection without scars that disfigure the uncovered cutaneous areas. Our references are the mastoid and the inferior margin of the mandible superiorly and the clavicles and the sternal manubrium inferiorly.

Significant Anatomical Structures: platysma muscle, sternocleidomastoid muscle, anterior triangle, digastric muscle, omohyoid muscle, posterior triangle, Robbins level system, selective neck dissection, Robbins level VII.

Landmarks: mastoid apex, inferior margin of the mandible, clavicle, mental eminence, hyoid bone, trapezius muscle, cricoid ring, laryngeal prominence, mandible angle, suprasternal notch.

3.2 Cutaneous Flap

A large cutaneous flap is raised with an incision approximately 3 cm beneath the inferior margin of the clavicle, extending along the acromioclavicular joint and ascending laterally by approximately 3 cm behind the trapezius margin and posterosuperiorly to the posterior profile of the mastoid apex, beyond the level of the external auditory canal (Fig. 3.1).

M. Lucioni, *Practical Guide to Neck Dissection*,
DOI 10.1007/978-3-642-33977-6_3, © Springer-Verlag Berlin Heidelberg 2013

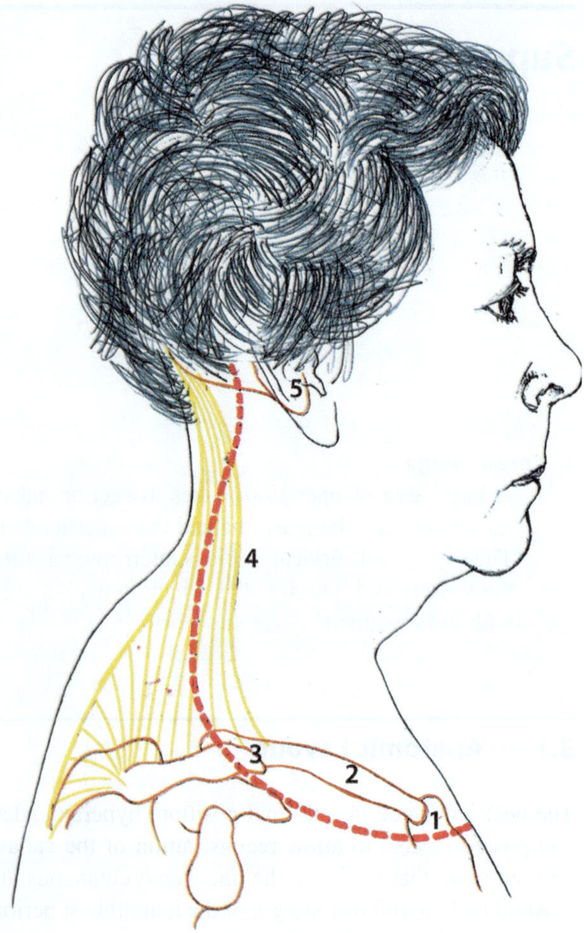

Fig. 3.1 Cutaneous line of incision. *1* Manubrium sterni, *2* clavicle, *3* acromioclavicular joint, *4* anterior margin of trapezius muscle, *5* mastoid

The flap may be raised above the platysma, which thus becomes fully exposed (Fig. 3.2).

The platysma muscle extends from the corpus mandibulae to the outer surface of the clavicle. Its lateral margin crosses the sternocleidomastoid muscle between its third median and third superior and then descends towards the acromioclavicular joint; from the mental eminence, its medial margin deviates from the midline in an inferior direction; its outer surface is more or less rectangular and invested with subcutaneous tissue, and its inner surface is contiguous with the superficial cervical fascia. The platysma is innervated by a branch of the facial nerve (Fig. 3.3).

Remarks: This anatomic cutdown, which permits excellent platysma exposure, is not always easy to perform in preserved cadavers, owing to the muscle's slenderness and fragility. Accordingly, a flap incorporating the platysma is often required, and it is indeed more useful for teaching purposes. In routine surgical practice, preparation of a flap formed by skin, subcutaneous tissue, and the platysma is in fact

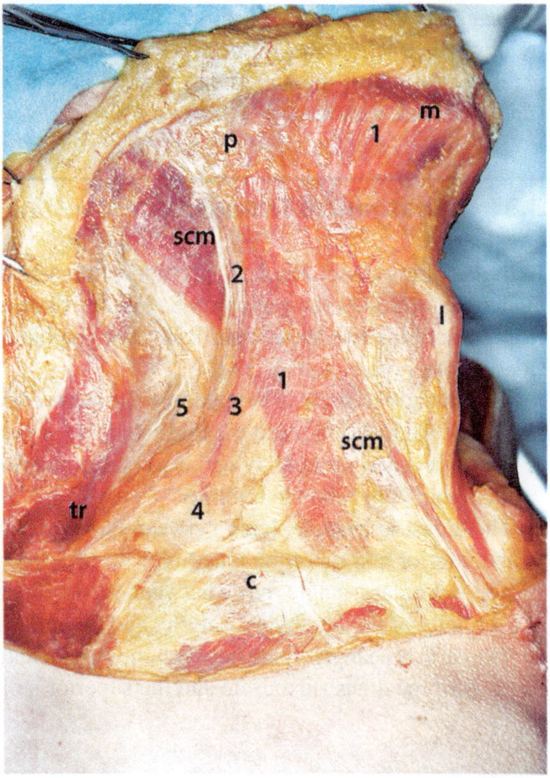

Fig. 3.2 Platysma muscle plane. *m* Mandible, *p* parotid, *scm* sternocleidomastoid muscle, *tr* trapezius muscle, *c* clavicle, *l* larynx, *1* platysma muscle, *2* great auricular nerve, *3* external jugular vein, *4* superficial cervical fascia, *5* spinal accessory nerve (peripheral branch)

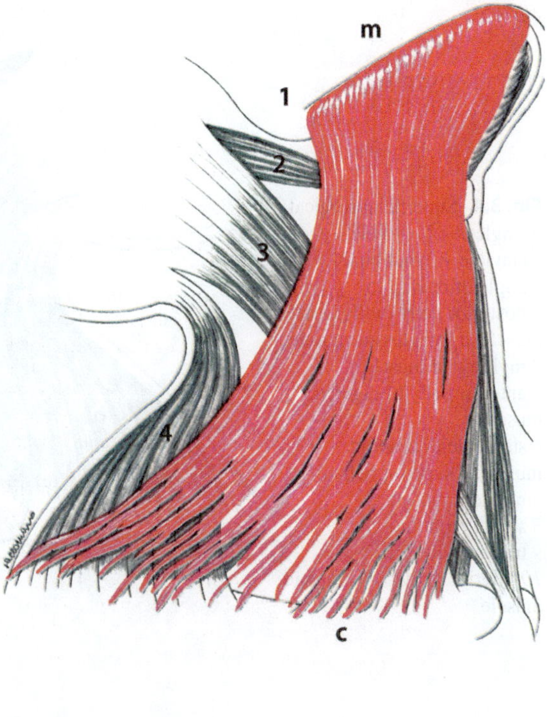

Fig. 3.3 Platysma muscle. *m* Mandible, *c* clavicle, *1* angle of mandible, *2* posterior belly of digastric muscle, *3* sternocleidomastoid muscle, *4* trapezius muscle

envisaged in all cervical operations. It is raised from the superficial cervical fascia by upwards traction and cut with a scalpel at a tangent to the flap; if this plane is carefully followed, the superficial vessels and nerves in the fascia are not interrupted because they remain below.

3.3 Anatomical Triangles

In the resulting dissection field, sternocleidomastoid muscle prominence is clearly evident as it crosses the region on both sides from top to bottom and from back to front, describing two large superficial, topographic triangles on each side, one anterior and one posterior (Fig. 3.4).

The anterior triangle is bounded by the sternocleidomastoid muscle, the inferior margin of the mandible, and the midline. It is further divided into:

1. The submental triangle, lying between the anterior belly of the digastric muscle, the hyoid bone, and the midline
2. The digastric triangle, lying between the two bellies of the digastric muscle and the inferior margin of the mandible
3. The muscular triangle, lying between the sternocleidomastoid muscle, the superior belly of the omohyoid muscle, and the midline
4. The carotid triangle, lying between the sternocleidomastoid muscle, the posterior belly of the digastric muscle and the superior belly of the omohyoid muscle

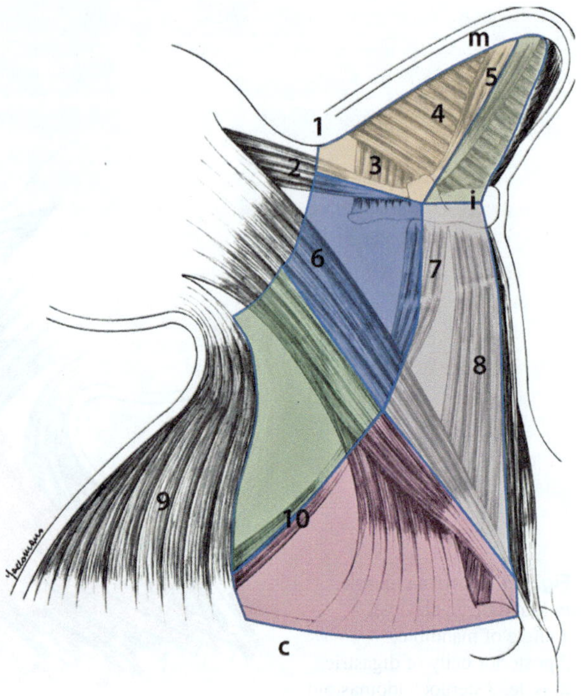

Fig. 3.4 Superficial surgical triangles. *m* Mandible, *c* clavicle, *i* hyoid bone, *1* angle of mandible, *2* posterior belly of digastric muscle, *3* hyoglossus muscle, *4* mylohyoid muscle, *5* anterior belly of digastric muscle, *6* sternocleidomastoid muscle, *7* superior belly of omohyoid muscle, *8* sternohyoid muscle, *9* trapezius muscle, *10* inferior belly of omohyoid muscle

The posterior triangle is bounded by the sternocleidomastoid muscle, trapezius muscle, and clavicle. It is further divided into:

1. The spinal triangle, lying between the sternocleidomastoid muscle, the trapezius, and the inferior belly of the omohyoid muscle
2. The supraclavicular triangle, lying between the sternocleidomastoid muscle, the inferior belly of the omohyoid muscle, and the clavicle

The above topographic division of the neck is the one used by anatomists and is certainly a helpful method of orienting general anatomy.

3.4 Robbins Levels

In routine oncological practice, importance is laid on an additional, internationally accepted topographical subdivision, introduced by K. Thomas Robbins in 1991 [1]; it was updated by him in 2002 [2] and finally in 2008 [3]. It is now internationally accepted. Its aim is to achieve uniformity in the nomenclature of various types of cervical lymph node neck dissection, which it does by classifying the various topographical regions involved in the excision and any sacrificed anatomic structures. The neck is therefore divided into a total of seven levels (four plus sublevel IB on each side and sublevel IA, sixth, and seventh anterior median levels). Levels I, II, and V have two sublevels each (Fig. 3.5).

The use of the Robbins level system aims to delineate the location of lymph node disease in the neck and it is well known and easy to remember. It serves as the basis for describing various selective neck dissections, with specific notation of the levels or sublevels removed, and relies less on historical terms that were less precise. Lymph nodes involving regions not located within the six levels would be referred to by the name of their specific nodal group. Examples include the retropharyngeal lymph nodes, the periparotid lymph nodes, the buccinator nodes, the postauricular, and the occipital nodes.

The last Consensus Statement on the Classification and Terminology of Neck Dissection (American Head and Neck Society- AHNS, 2008) introduces new recommendations regarding the following:

1. The boundaries between levels I and II and between levels III/IV and VI.
   To separate the sublevel IB from sublevel IIA, the stylohyoid muscle was indicated. Since it is not a very practical marker during clinical examination, nor can it be identified by the radiologist on imaging studies, the vertical plane defined by the posterior edge of the submandibular gland is now preferred. So, for the surgeon who is interested in removing level II and preserving level I, the dissection plane is along the fascia overlying the posterior aspect of the submandibular gland. A radiologic landmark was also incorporated – the medial aspect of the common carotid artery – as a boundary between levels III/IV and VI.
2. The terminology of the superior mediastinal nodes.
   We have a new entry, Robbins level VII, to indicate the extension of the chain of paratracheal nodes below the suprasternal notch (the dividing line between level VI and VII) to the level of the innominate artery.

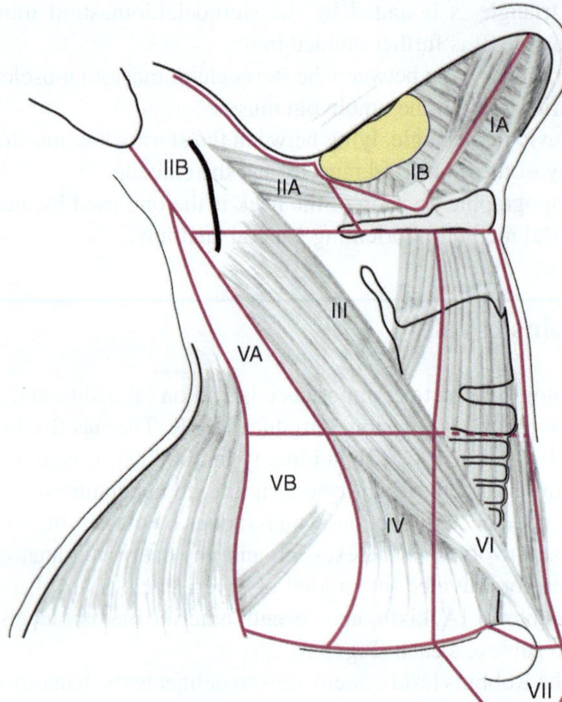

Fig. 3.5 Cervical levels according to Robbins (2002) revised in 2008: lymph node groups. *Submental* (*sublevel IA*) anterior bellies of digastric muscles and the hyoid bone, *submandibular* (*sublevel IB*) anterior belly of digastric muscle, posterior aspect of the submandibular gland and mandible, *upper jugular* (*sublevel IIA/IIB*) skull base and inferior border of hyoid bone; posterior aspect of the submandibular gland, posterior border of sternocleidomastoid muscle; spinal nerve divides the two sublevels, *middle jugular* (*level III*) inferior border of hyoid bone and inferior border of cricoid cartilage; sternohyoid muscle, posterior border of sternocleidomastoid muscle, *lower jugular* (*level IV*) inferior border of cricoid cartilage and clavicle; sternohyoid muscle, posterior border of sternocleidomastoid muscle, *posterior triangle group* (*sublevel VA/VB*) apex formed by convergence of sternocleidomastoid and trapezius muscles, clavicle; posterior border of sternocleidomastoid muscle, anterior border of trapezius muscle; a horizontal plane marking the inferior border of cricoid cartilage divides the two sublevels, *anterior compartment* (*level VI*) hyoid bone and suprasternal notch; medial aspect of common carotid artery bilaterally, *superior mediastinal nodes* (*level VII*) suprasternal notch and level of innominate artery

3. The method of submitting surgical specimens for pathological analysis.
   The surgeon should divide the neck dissection specimen into levels and sublevels and submit them to the laboratory in separate containers. This is because in current surgical practice, most neck dissections are of the selective type, and specimens typically lack anatomic landmarks that historically allowed pathologists to orient specimens.

Remarks: The concept of neck dissection as an indispensable complement to the treatment of tumours of the upper aerodigestive tract began with George Crile

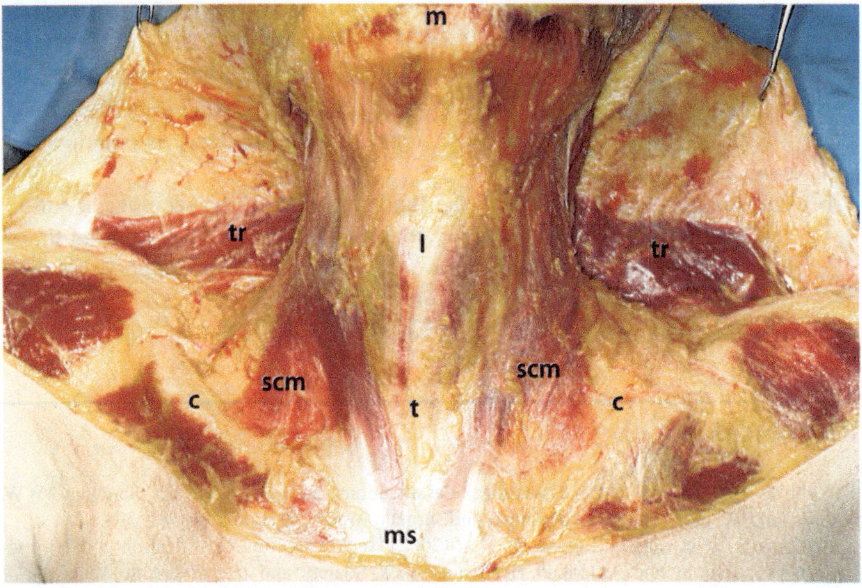

Fig. 3.6 Cutaneous flap. *m* Mandible, *l* larynx, *t* thyroid gland, *ms* manubrium sterni, *c* clavicle, *scm* sternocleidomastoid muscle, *tr* trapezius muscle

more than a century ago [4]. Neck dissection was always carried out with the demolitive technique. In the 1960s, Ettore Bocca introduced the so-called functional neck dissection in Europe [5]. It is based on Osvaldo Suarez's assertion that there are no lymph node formations outside the fascial investments of the neck. So the surgeon can be just as radical as in the neck dissection proposed by Crile while preserving important structures such as the sternocleidomastoid muscle, the internal jugular vein, and the spinal accessory nerve. This applies as long as the lymph node capsule is intact. This new method has led to an appreciable reduction of morbidity.

In recent years, the study of the pattern of metastatic diffusion of tumours of the head and neck has led surgeons performing prophylactic neck dissection (i.e. in N0 necks), to neglect the lymphatic areas that are statistically less exposed to metastatic colonisation. Selective neck dissections were therefore introduced in routine surgery (Robbins, 1991). The reason behind this evolution is to reduce as far as possible the functional sequelae of cervical neck dissections.

At the end of this surgical phase, the very vast dissection field extends inferiorly from the trapezius muscles to the clavicles and superiorly to encompass the mandible and external auditory canal (Fig. 3.6).

We shall now try to establish the limits of the Robbins levels conceptually and by palpation. At the top, we identify the mastoid and the hyoid bone; farther down, the inferior margin of the cricoid ring, larynx prominence, mandible angle, hyoid bone, and then the suprasternal notch and the clavicle; and posteriorly, the anterior margin of the trapezius.

Take-Home Messages

- Neck dissection is the most complete and fascinating surgical procedure as regards the anatomical knowledge of the neck. Succeeding in performing it with methodological exactness, sureness, and confidence is one of the goals of the excellent surgeon.
- The Robbins levels (2008) are the fundamental map for oncological surgery of the neck. Cervical adenopathies should always be located in the Robbins levels, both in the objective examination prior to surgery and in the description of the neck dissection.

References

1. Robbins KT (1994) Neck dissection: classifications and incisions. In: Shockley WW, Pillsbury HC (eds) The neck: diagnosis and surgery. Mosby, St. Louis, pp 381–391
2. Robbins KT, Clayman G, Levine PA et al (2002) Neck dissection classification update: revision proposed by the American Head and Neck Society and the American Academy of Otolaryngology- Head and Neck Surgery. Arch Otolaryngol Head Neck Surg 128(7):751–758
3. Robbins KT, Shaha AR, Medina JE et al (2008) Consensus statement on the classification and terminology of neck dissection. Arch Otolaryngol Head Neck Surg 134(5):536–538
4. Crile G (1906) Excision of cancer of head and neck with special reference to the plan of dissection based on one hundred and thirty two operation. JAMA 47:1780
5. Bocca E, Pignataro O (1967) A conservation technique in radical neck dissection. Ann Otol (St Louis) 76:975–987

Parotid Region

4

Core Messages

- The essence of parotid surgery consists of removing the gland without harming the facial nerve and its branches. The first surgical stage always consists of identifying the common trunk of the facial nerve.
- The identification of the facial nerve and the isolation of its branches may be carried out using the operating microscope, with a magnifying prismatic loop (enlargement between 2× and 4×) or even with the naked eye, depending on what the surgeon is accustomed to.

4.1 Anatomic Layout

The parotid region is bounded anteriorly by the ramus of the mandible with the masseter muscle laterally and the medial pterygoid muscle medially; posteriorly, by the mastoid, sternocleidomastoid muscle, and posterior belly of the digastric muscle; medially by the jugular carotid tract, the styloid process with the stylienus muscles (Riolan's bundle), and the pharyngeal wall (superior constrictor muscle of the pharynx); superiorly, by the external auditory canal and the extreme posterior of the zygomatic arch; and inferiorly by the imaginary horizontal line between the angle of the mandible and the anterior margin of the sternocleidomastoid muscle.

The superficial and deep parotid fasciae invest the gland and are formed by the division of the superficial cervical fascia into two. The parotid lymph nodes are concentrated in two sites, one superficial, immediately below the fascia, and one deep, intraparotid site, adjacent to the external carotid artery (Fig. 4.1).

Significant Anatomical Structures: superficial cervical fascia, great auricular nerve, external jugular vein, retromandibular vein, facial nerve branch serving the platysma, buccal branches of facial nerve, Stensen's duct, transverse artery of the face, zygomatic branches of facial nerve, temporal branches of facial nerve, superficial temporal artery, auriculotemporal nerve, posterior auricular artery, facial

M. Lucioni, *Practical Guide to Neck Dissection*,
DOI 10.1007/978-3-642-33977-6_4, © Springer-Verlag Berlin Heidelberg 2013

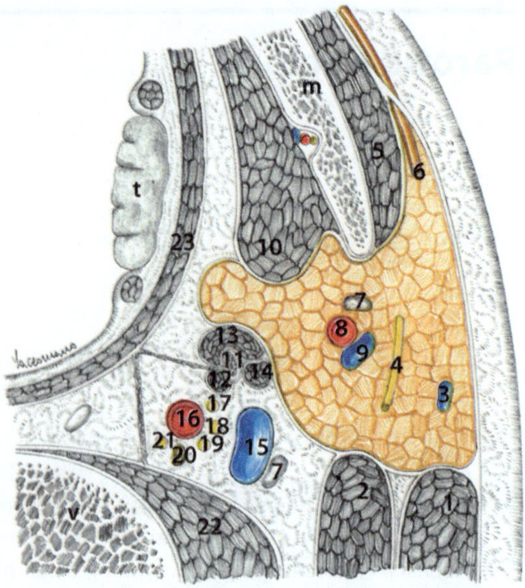

Fig. 4.1 Parotid region: cross section. *m* Mandible, *t* palatine tonsil, *v* vertebral body, *1* sternocleidomastoid muscle, *2* posterior belly of digastric muscle, *3* external jugular vein, *4* facial nerve, *5* masseter muscle, *6* Stensen's duct, *7* lymph node, *8* external carotid artery, *9* retromandibular vein (or posterior facial vein), *10* internal pterygoid muscle, *11* styloid process, *12* stylopharyngeus muscle, *13* styloglossus muscle, *14* stylohyoid muscle, *15* internal jugular vein, *16* internal carotid artery, *17* glossopharyngeal nerve, *18* spinal accessory nerve, *19* vagus nerve, *20* cervical sympathetic chain, *21* hypoglossal nerve, *22* prevertebral muscles, *23* superior constrictor muscle of the pharynx

nerve, stylomastoid artery, Riolan's bundle, internal jugular vein, external carotid artery, glossopharyngeal nerve, Hering's nerve, goose's foot, temporofacial trunk, cervicofacial trunk, Ponce–Tortella's loop, superior constrictor muscle of the pharynx, ascending palatine artery, ascending pharyngeal artery, Frey's syndrome, and masseter muscle

Landmarks: corner of the mandible, external auditory canal, posterior belly of the digastric muscle, mastoid apex, pointer, superficial plane of the digastric muscle, styloid process, sentinel artery, facial valley, stylomastoid foramen

4.2 Superficial Fascial Plane

Elevation of the cutaneous flap must extend superiorly beyond the caput mandibulae, after dissection of the external auditory canal, and ascend anteriorly to the zygomatic arch (posterior portion). At this point, we can recognise the limits of the parotid gland. We can also find our way by identifying a few landmarks, such as the corner of the mandible, the external auditory canal, and the sternocleidomastoid muscle (Fig. 4.2).

On removal of the superficial cervical fascia, the superior superficial pedicles of the parotid cavity are immediately visible. Now we look for and isolate the superficial temporal artery, which in vivo can be felt pulsating just in front of the tragus (Fig. 4.3).

Inferiorly, the platysma (unless already removed) and superficial cervical fascia are dissected and everted, exposing the inferior portion of the parotid cavity (Fig. 4.4).

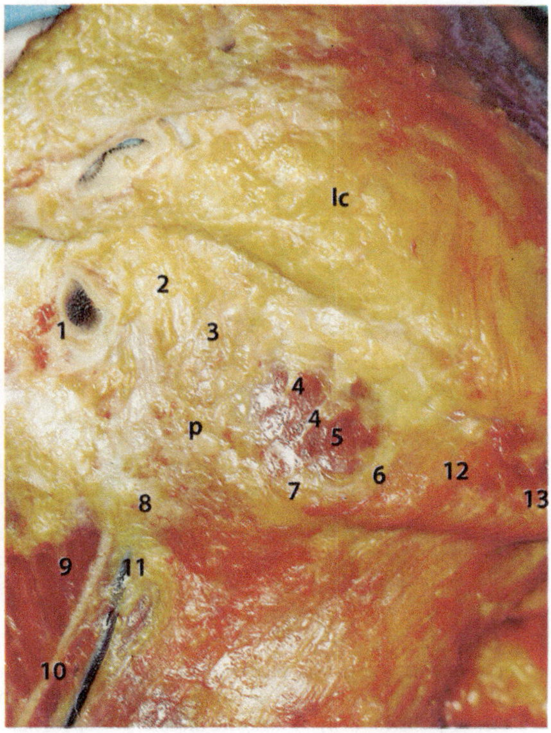

Fig. 4.2 Superficial fascial plane. *p* Parotid, *lc* everted cutaneous flap, *1* external auditory canal cartilage, *2* mandibular caput mandibulae, *3* ramus of the mandibulae, *4* buccal branches (facial nerve), *5* masseter muscle, *6* marginal branch (facial nerve), *7* angle of mandible, *8* superficial cervical fascia, *9* sternocleidomastoid muscle, *10* great auricular nerve, *11* external jugular vein, *12* platysma muscle, *13* basis mandibulae

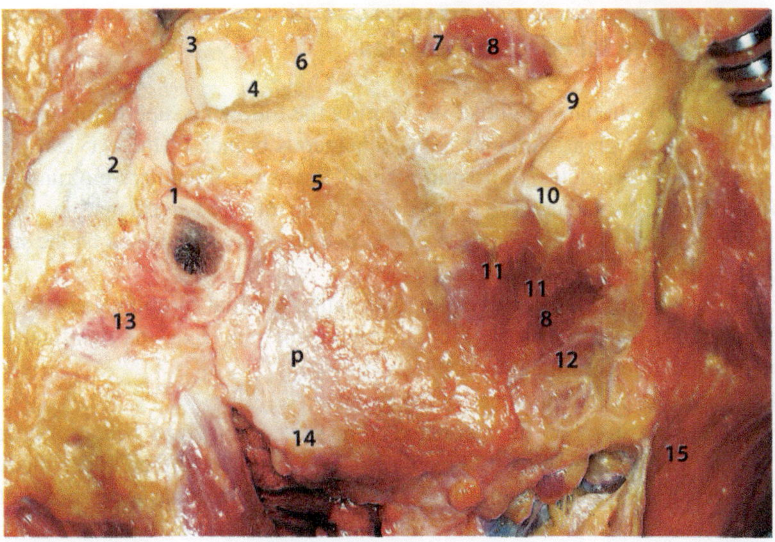

Fig. 4.3 Subfascial plane (I). *p* Parotid, *1* external auditory canal cartilage, *2* fascia temporalis, *3* superficial temporal artery, *4* auriculotemporal nerve, *5* caput mandibulae, *6* temporal branches (facial nerve), *7* zygomatic branches (facial nerve), *8* masseter muscle, *9* transverse facial artery, *10* Stensen's duct, *11* buccal branches (facial nerve), *12* marginal branch (facial nerve), *13* mastoid, *14* angle of mandible, *15* platysma muscle

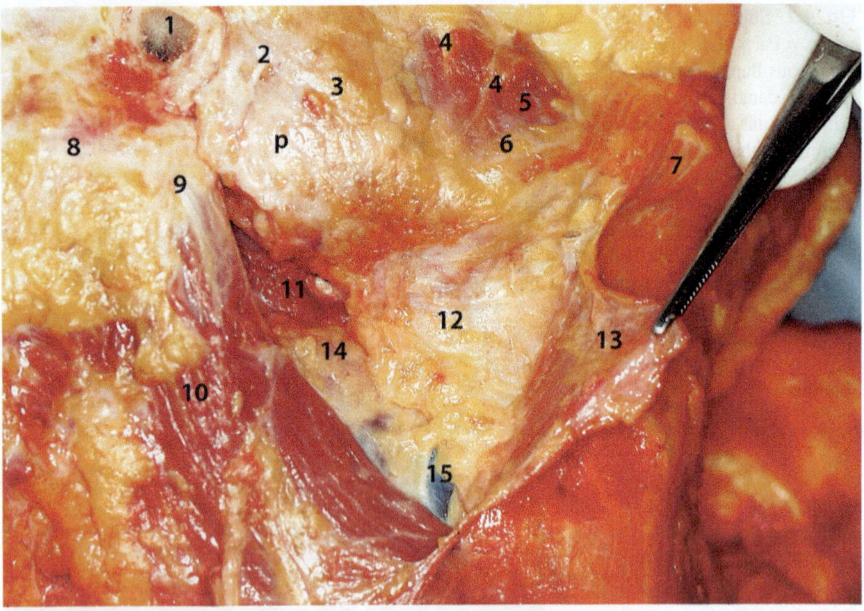

Fig. 4.4 Subfascial plane (II). *p* Parotid, *1* external auditory canal, *2* caput mandibulae, *3* ramus of the mandible, *4* buccal branches (facial nerve), *5* masseter muscle, *6* marginal branch (facial nerve), *7* basis mandibulae, *8* mastoid, *9* sternocleidomastoid tendon, *10* sternocleidomastoid muscle, *11* posterior belly of digastric muscle, *12* superficial cervical fascia, *13* platysma muscle, *14* lymph node, *15* thyrolinguofacial trunk

4.3 Superficial Parotid Pedicles

Examining the right parotid gland, we identify the following superficial structures:

- 7 o' clock: the great auricular nerve (cutaneous branch of the cervical plexus, innervating the auricle and parotid region); the external jugular vein runs alongside the great auricular nerve in proximity to the inferior parotid margin and exits the region. The two subfascial structures can be easily recognised on the surface of the sternocleidomastoid muscle [1].
- 5 o'clock: the retromandibular vein (also called external carotid vein or posterior facial vein), the facial nerve branch serving the platysma, the marginal nerve serving the inferior mimetic muscles.
- 4 o'clock: the buccal branches of the facial nerve.
- 3 o'clock: the Stensen's duct, situated at the apex of the gland's anterior process; it passes horizontally forward beyond Bichat's fat pad and then bends medially, embedding itself deep within the buccinator fibres; the transverse artery of the face, branch of the internal arteria maxillaris.
- 2 o'clock: the zygomatic branches of the facial nerve.
- 1 o'clock: the temporal branches of the facial nerve.
- 12 o'clock: the superficial temporal artery (and vein), a branch of the carotid artery arising in the parotid gland; the auriculotemporal nerve, arising in the mandibular branch of the trigeminal nerve, emerging anteriorly to the external

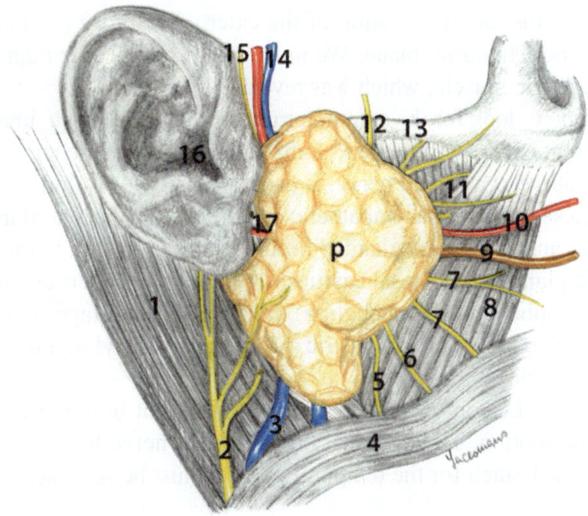

Fig. 4.5 Superficial parotid pedicles. *p* Parotid, *1* sternocleidomastoid muscle, *2* great auricular nerve, *3* external jugular vein, *4* platysma muscle, *5* platysma branch (facial nerve), *6* marginal branch (facial nerve), *7* buccal branches (facial nerve), *8* masseter muscle, *9* Stensen's duct, *10* transverse facial artery, *11* zygomatic branches (facial nerve), *12* temporal branches (facial nerve), *13* zygomatic arch, *14* superficial temporal artery and vein, *15* auriculotemporal nerve, *16* external auditory canal, *17* posterior auricular artery

auditory canal and accompanying the ascent of the superficial temporal artery. It also sends secretory parasympathetic fibres to the parotid gland (through glossopharyngeal nerve, tympanic nerve, lesser petrosal nerve, otic ganglion, auriculotemporal nerve to parotid); and the caput mandibulae.

- 10 o'clock: the external auditory canal.
- 9 o'clock: the posterior auricular artery (and vein), a branch of the external carotid artery arising in the parotid gland, passing over the sternocleidomastoid tendon (Fig. 4.5).

4.4 Parotidectomy

We begin the parotidectomy by freeing the superficial portion of the posteroinferior aspect of the gland and dissect the posterior auricular artery, great auricular nerve, and external jugular vein.

The posteroinferior portion of the parotid gland is elevated from the anterior margin of the sternocleidomastoid muscle. More deeply, we uncover the posterior belly of the digastric muscle and free its anterior margin. In this phase, we advise the use of a self-retaining retractor clamped between the parotid gland and sterno-cleidomastoid tendon. Superiorly, dissection should not exceed the horizontal plane crossing the mastoid apex, to avoid encountering the facial nerve. Digital elevation is effective and avoids damage.

Now we free the anterior portion of the external auditory canal, taking care to remain on the perichondrial plane. We must not go any deeper than the plane tangent to the digastric muscle, which was revealed previously.

It is now time to look for the common trunk of the facial nerve, immediately after the point where it emerges from the stylomastoid foramen of the temporal bone.

The facial nerve is a mixed nerve. It carries sensitivity from the isthmus of the fauces; it has a secretory parasympathetic component for the tear glands and for the submandibular and sublingual glands (through chorda tympani to lingual nerve), as well as for the glands of the nasal cavities (through great superficial petrosal nerve, Vidian nerve to sphenopalatine ganglion). It innervates the stapes muscles, the platysma, the posterior belly of the digastric muscle, and the stylohyoid muscle, as well as the mimic muscles of the face.

Complications: Lesion of the facial nerve may result in important asymmetries of facial mimic motion. The marginal branch of the nerve for the cervical portion and the orbicular branch for the temporal portion must be accurately identified and preserved.

4.5 In the Search of Facial Nerve

In parotidectomy, the search for the common trunk of the facial nerve is carried out by identifying the inferior end of the cartilaginous external auditory canal that inferoposteriorly ends with a pointed triangular appendix [2]. Rather like a thick compass needle, it indicates the facial nerve trunk (pointer). In regard to depth, reference is made to the superficial plane of the digastric muscle. It is less advisable to use the styloid process as a landmark because its dimensions vary; moreover, the facial nerve runs anterolaterally to the styloid process, and therefore, on finding the styloid process, a medial position has already been reached in relation to the nerve. Normally just above the facial nerve and following the same direction, we can see the stylomastoid artery, branch of posterior auricular artery. On account of its position, it is also called the sentinel artery because the nerve is to be found immediately beneath it (Fig. 4.6).

Exercise 1: Facial Nerve (Fig. 4.7)

To find the nerve, we must have a clear idea of the landmarks of approach to the facial nerve, which are (1) the anterior margin of the external auditory canal, (2) the anterior margin of the sternocleidomastoid muscle, and (3) the posterior belly of the digastric muscle.

Next, we must remember the landmarks of interception of the facial nerve, which are (1) for the direction in which to search, the pointer, and (2) for the depth, the plane tangent to the lateral surface of the posterior belly of the digastric muscle.

Once the facial common trunk has been identified, more distally we shall uncover the goose's foot, which gives us further confirmation that we are indeed dissecting the facial nerve.

Next, we shall free the facial nerve tree, following the same dissection methods that are adopted in vivo. With Pean forceps inserted in a parallel position above each nerve trunk, the overlying parenchyma will be widened into a "bridge" and cut between the divaricated branches, lifting it up from the nerve.

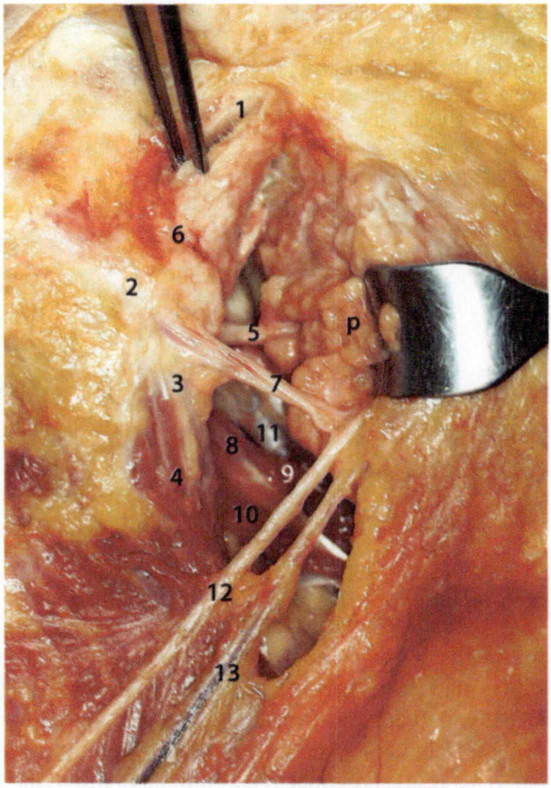

Fig. 4.6 Locating the facial trunk (I). *p* Parotid, *1* external auditory canal, *2* mastoid, *3* sternocleidomastoid tendon, *4* sternocleidomastoid muscle, *5* facial nerve, *6* petrotympanic suture (in depth), *7* posterior auricular artery and vein, *8* stylohyoid muscle, *9* styloglossus muscle, *10* posterior belly of digastric muscle, *11* internal jugular vein, *12* great auricular nerve, *13* external jugular vein

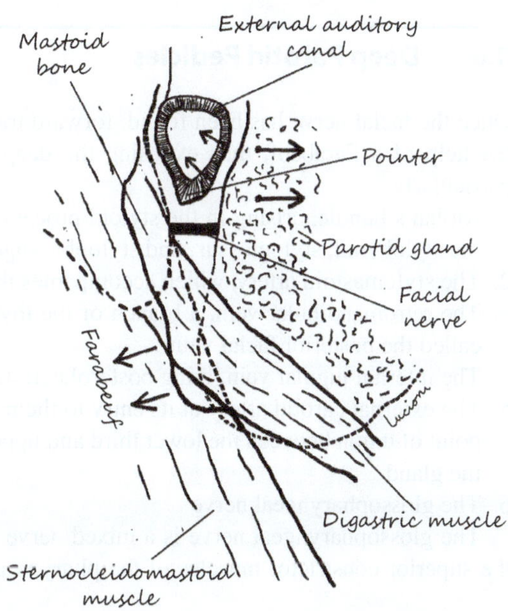

Fig. 4.7 Exercise 1: facial nerve

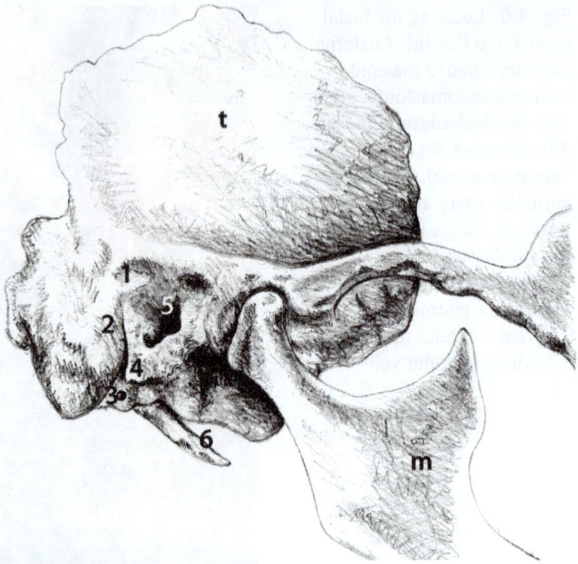

Fig. 4.8 Locating the facial trunk (II). *t* Temporal bone, *m* mandible, *1* suprameatal spine, *2* petrotympanic suture, *3* stylomastoid for amen, *4* pointer, *5* external auditory canal, *6* styloid process

A second possibility for identifying the common trunk of the facial nerve is to follow the petrotympanic suture, also called the facial valley. A retroauricular incision is made in the periosteum, which is elevated anteriorly to expose the suprameatal spine, which continues caudally with the petrotympanic suture. The entire descent of this structure is then followed: the stylomastoid foramen, where the facial nerve emerges, is always situated 6–8 mm medially to the point where the petrotympanic suture ends (Fig. 4.8).

4.6 Deep Parotid Pedicles

Once the facial nerve has been found, forward traction is applied to the gland with the help of a Farabeuf, thus exposing the deep structures of the parotid cavity, particularly:

1. Riolan's bundle, arising in the styloid process of the temporal bone, formed by the stylohyoid, styloglossus, and stylopharyngeus muscles
2. The stylomastoid artery, which accompanies the facial nerve trunk
3. The retromandibular vein, a branch of the thyrolinguofacial venous trunk, also called the posterior facial vein
4. The internal jugular vein, lying posterolaterally to the styloid process
5. The external carotid artery at its entry to the parotid gland, at approximately the point of union between the lower third and upper two thirds of the medial wall of the gland
6. The glossopharyngeal nerve

The glossopharyngeal nerve is a mixed nerve since it contains motor fibres for the superior constrictor muscles of the pharynx and stylopharyngeus, parasympa-

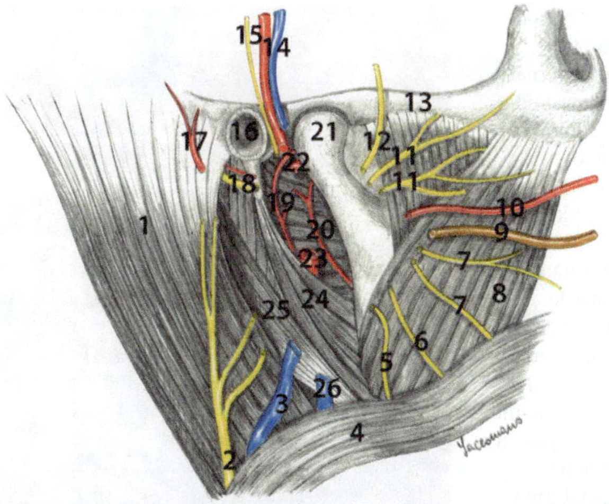

Fig. 4.9 Deep parotid pedicles. *1* Sternocleidomastoid muscle, *2* great auricular nerve, *3* external jugular vein, *4* platysma muscle, *5* platysma branch (facial nerve), *6* marginal branch (facial nerve), *7* buccal branches (facial nerve), *8* masseter muscle, *9* Stensen's duct, *10* transverse facial artery, *11* zygomatic branches (facial nerve), *12* temporal branches (facial nerve), *13* zygomatic arch, *14* superficial temporal artery and vein, *15* auriculotemporal nerve, *16* external auditory canal, *17* posterior auricular artery, *18* facial nerve and stylomastoid artery, *19* ascending pharyngeal artery, *20* ascending palatine artery, *21* caput mandibulae, *22* internal maxillary artery, *23* external carotid artery, *24* stylienus muscles, *25* posterior belly of digastric muscle, *26* retromandibular vein

thetic pregangliar fibres for secretory innervation of the parotid (through tympanic nerve, otic ganglion, mandibular nerve to auriculotemporal nerve), and sensory fibres (sensory innervation of the middle ear and pharynx; sensory innervation of taste buds in the area immediately anteriorly and posteriorly to the lingual "V"). Together with the vagus nerve, it also governs circulatory and respiratory homeostasis. It emerges from the cranial cavity through the posterior foramen lacerum, extending anteriorly and describing an anterosuperiorly concave curve; it initially runs between the internal jugular vein and the internal carotid artery, then between the styloglossus and stylopharyngeus muscles and, laterally accompanying the pharynx, reaches the base of the tongue. It transmits impulses generated by stretch receptors located in the arterial wall of the bifurcation through one of its peripheral branches. This is called the Hering's nerve and runs from the bifurcation of the carotid artery along the anterolateral surface of the internal carotid artery until it joins the main branch. Information on the partial pressure of oxygen in the blood perfusing the carotid bodies travels along the same pathway. In turn, the central nervous system transmits to the periphery impulses that tend to modify arterial pressure (reduction in peripheral vascular resistance, heart rate, and contractility and increase in venous system capacity) and impulses that increase breathing capacity (increase in respiratory rate and tidal volume). A similar reflex arc is supported by the vagus, but in this case, afferent impulses arise from receptors located in the aortic arch (Figs. 4.9 and 4.10).

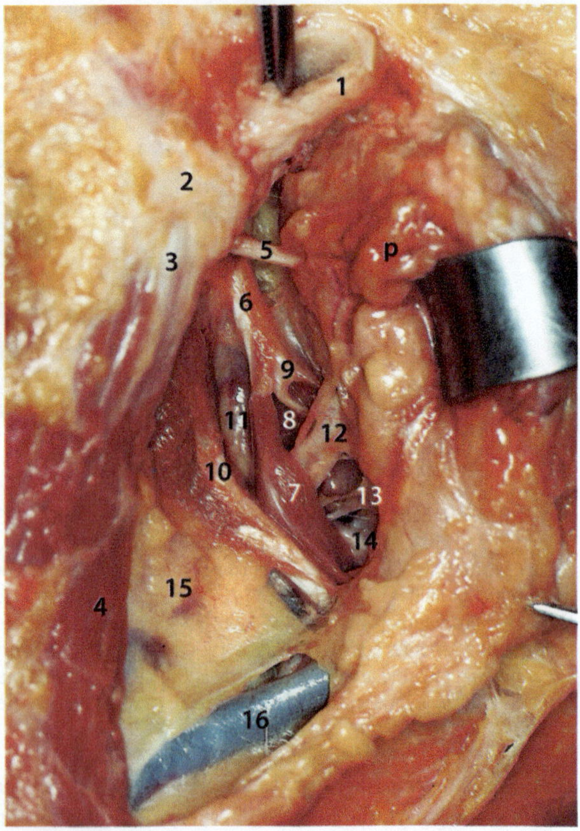

Fig. 4.10 Parotid region: deep plane. *p* Parotid, *1* anterior wall of external auditory canal, *2* mastoid, *3* sternocleidomastoid tendon, *4* anterior margin of sternocleidomastoid muscle, *5* facial nerve, *6* styloid process, *7* stylohyoid muscle, *8* stylopharyngeus muscle, *9* styloglossus muscle, *10* posterior belly of digastric muscle, *11* internal jugular vein, *12* external carotid artery, *13* ascending palatine artery, *14* glossopharyngeal nerve, *15* lymph node, *16* thyrolinguofacial trunk

Complications: Interruption of the glossopharyngeal nerve is usually manifested by slight difficulty in swallowing and alterations in taste; however, since the regions innervated by the vagus and glossopharyngeal nerves bilaterally cover a similar area, effective control is maintained over circulation and respiration. This nerve is rarely identified in cervical surgery. Iatrogenic lesions of the glossopharyngeal nerve may occur in the course of otoneurosurgery of the cerebellopontine angle or base of the skull, during cervical excision (dissection of voluminous jugulodigastric adenopathies), in lateral pharyngotomies and in ablation of parapharyngeal tumours. Surgical operations on tumours of the pharynx, tonsils, and tongue base may also cause injury to the glossopharyngeal nerve, with functional sequelae of dysphagia and dysgeusia secondary to surgical excision. In tonsillectomy, the glossopharyngeal nerve, running in deep proximity to the inferior tonsil pole, may be injured during dissection or electrocoagulation; however, damage is usually reversible.

Last, it should be borne in mind that intraoperative stimulation through manipulation of either the glossopharyngeal or vagus nerve may induce transitory bradycardia and hypotension.

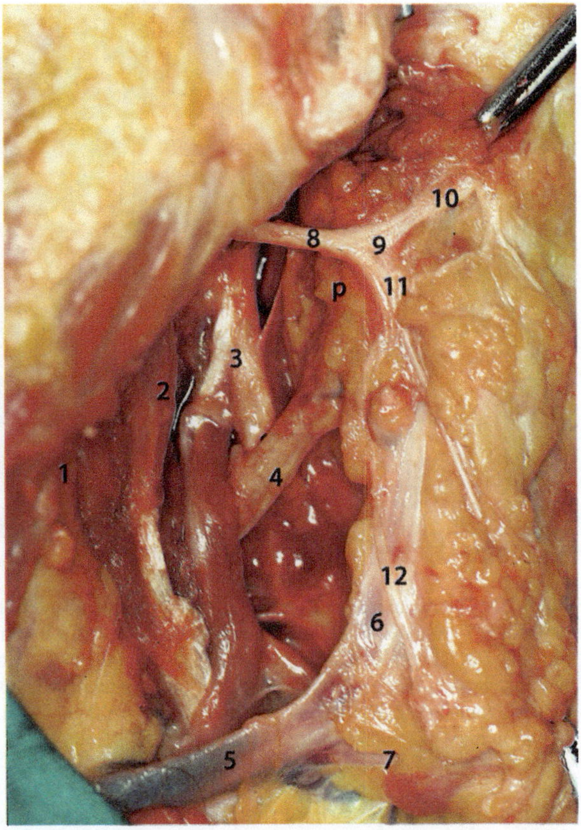

Fig. 4.11 The goose's foot. *p* Parotid, *1* anterior margin of sternocleidomastoid muscle, *2* posterior belly of digastric muscle, *3* styloid process and stylienus muscles, *4* external carotid artery, *5* thyrolinguofacial trunk, *6* retromandibular vein, *7* facial vein, *8* facial nerve, *9* goose's foot of facial nerve, *10* temporofacial trunk (facial nerve), *11* cervicofacial trunk (facial nerve), *12* marginal branch (facial nerve)

We now expose the intraglandular tract of the facial nerve. There is some debate about the existence of a superficial and deep parotid lobe. Indeed, there is no real cleavage plane between the two so-called lobes, and the superficial parotid portion is far more voluminous than the deep portion, comprising about 90 % of the whole glandular parenchyma.

Following the facial trunk from its emergence at the periphery, we find the goose's foot, that is, the subdivision of the nerve into its two terminal trunks, the temporofacial trunk, and the cervicofacial trunk. The first is appreciably more voluminous than is the second and has more collateral branches. An imaginary horizontal line crossing the labial commissure roughly divides the areas of musculocutaneous innervation of the two trunks. In particular, it can be seen how the most important of these, the marginal branch, is situated laterally to the retromandibular vein. Remember that the conformation of the facial trunk is rather inconstant. Anastomoses occur frequently between the two main trunks (Ponce–Tortella's loop), and this may explain the functional recovery of iatrogenic mediofacial lesions. Instead, the absence of collaterals in the front and mandibular branches would explain the non-reversibility of the deficits caused by the interruption of the nerve branches in these locations (Fig. 4.11).

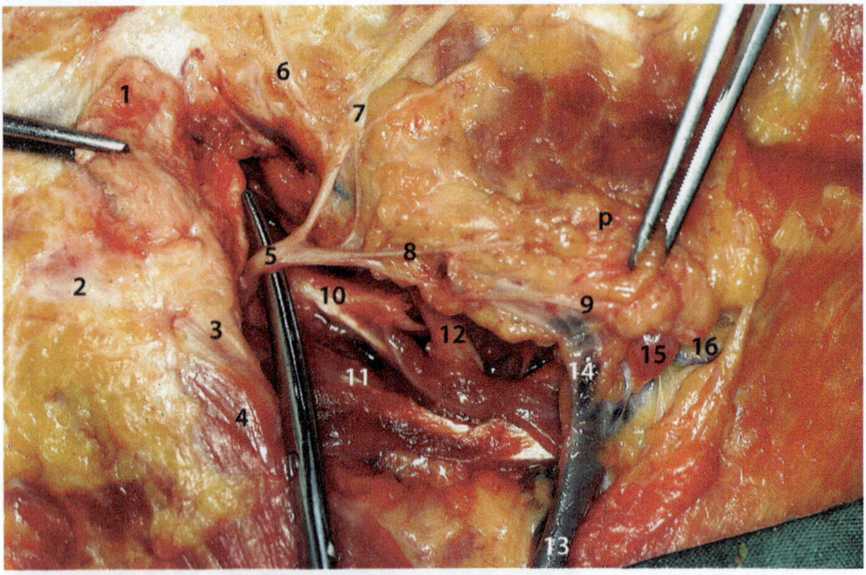

Fig. 4.12 The facial tree. *p* Anterior parotid remnants, *1* anterior wall of external auditory canal, *2* mastoid, *3* sternocleidomastoid tendon, *4* sternocleidomastoid muscle, *5* facial nerve, *6* temporal branches (facial nerve), *7* zygomatic branches (facial nerve), *8* buccal branches (facial nerve), *9* marginal branch (facial nerve), *10* styloid process and stylienus muscles, *11* posterior belly of digastric muscle, *12* external carotid artery, *13* thyrolinguofacial trunk, *14* retromandibular vein, *15* lymph node of facial pedicle, *16* facial vein

The superficial portion of the parotid is stretched anterosuperiorly, thus isolating the terminal branches of the facial nerve. The Stensen's duct and the superficial temporal artery are identified and sectioned. The transverse facial artery, which comes at depth from the internal arteria maxillaries and rises to the surface anteriorly on the masseter muscle, is left intact (Fig. 4.12).

4.7 Deep Parotid Lobe

After having removed the superficial portion, another dissection exercise is ablation of the deep portion of the gland, posteroanteriorly exposing the styloid process, cervical vasculonervous bundle, cervical sympathetic nerve trunk, and glossopharyngeal, accessory, and hypoglossal nerves (Fig. 4.13).

Dissection may be extensive, elevating the pharyngeal process of the parotid as far as the superior constrictor muscle of the pharynx, whose surface reveals the ascending palatine artery, branch of the facial and, posteriorly to the latter, the ascending pharyngeal artery, branch of the external carotid. The following elements are then dissected: (1) the retromandibular vein, (2) the external carotid artery at the entrance to the gland, and (3) the internal maxillary artery and vein, anteriorly, at 2 o' clock.

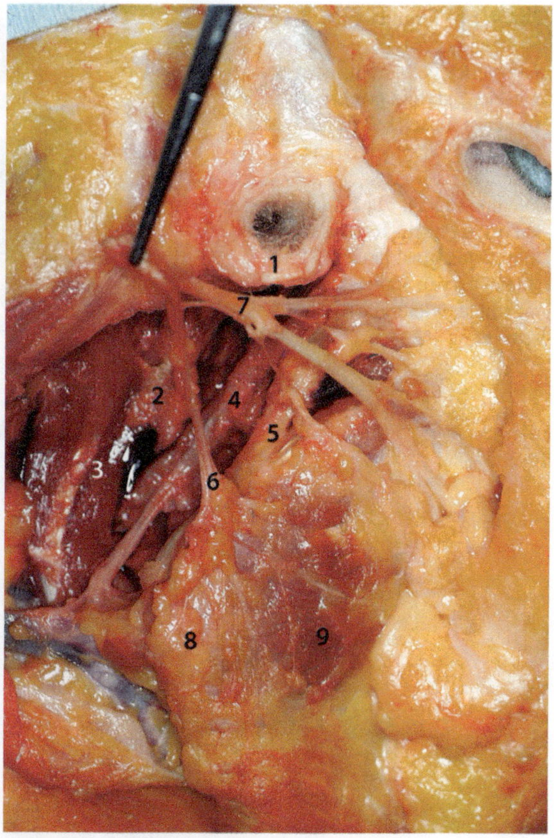

Fig. 4.13 Terminal branches of facial nerve (I). *1* External auditory canal, *2* styloid process and stylienus muscles, *3* posterior belly of digastric muscle, *4* retromandibular vein, *5* external carotid artery, *6* cervicofacial trunk (facial nerve), *7* temporofacial trunk (facial nerve), *8* angle of mandible, *9* masseter muscle

Following ablation of the deep portion of the parotid gland, the parotid cavity is completely cleared of its contents. The various components of the facial nerve can now be examined (Fig. 4.14).

Complications: Periprandial symptomatology may occasionally manifest itself after parotidectomy and is characterised by hyperhidrosis and reddening of the cutis around the area served by the auriculotemporal nerve (Frey's syndrome). This phenomenon is due to abnormal innervation by auriculotemporal parasympathetic fibres that, after interruption by gland ablation, communicate with the sympathetic nervous system directed towards the skin glands and vessels. In some cases, symptoms regress spontaneously. Where this is not the case, the syndrome can only be cured by resection of the tympanic nerve, which runs along the medial wall of the middle ear.

At this point, the anatomical "minus" that remains after the complete removal of the gland can be clearly seen. A further dissection exercise may be to cut away a small flap from the anterior edge of the sternocleidomastoid muscle, hinged at the top. The anterior rotation and suture at the cranial end of the masseter muscle can fill the space and partially make up for the unaesthetic appearance, besides reducing the incidence of Frey's syndrome [3].

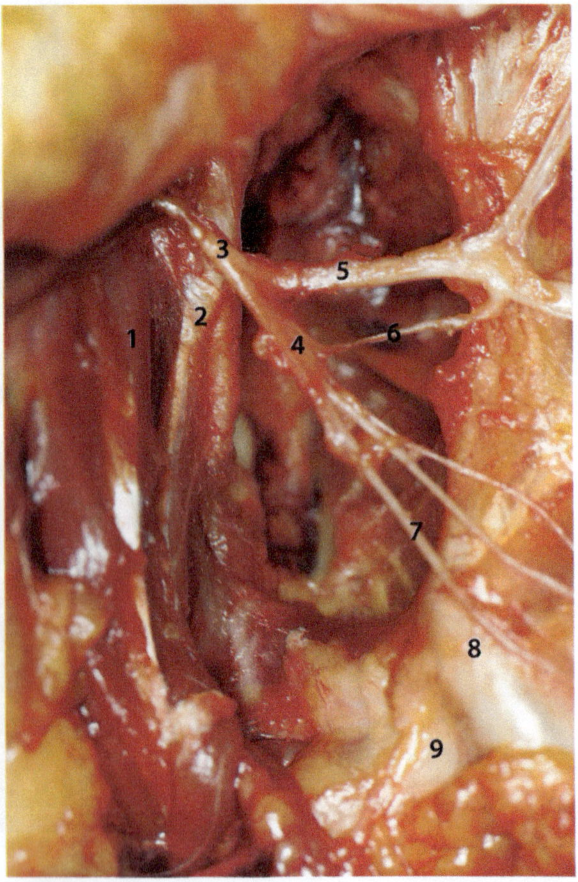

Fig. 4.14 Terminal branches of facial nerve (II).
1 Posterior belly of digastric muscle, *2* styloid process and stylienus muscles, *3* facial trunk, *4* cervicofacial trunk (facial nerve),
5 temporofacial trunk (facial nerve), *6* Ponce–Tortella's loop, *7* marginal branch (facial nerve), *8* angle of mandible, *9* interglandular septum

Take-Home Messages

- To identify the common trunk of the facial nerve, we must constantly remember the landmarks of approach and the landmarks of interception.
- We must bear in mind that the marginal edge of the facial nerve usually crosses the retromandibular vein laterally; consequently, the ligating and sectioning of this vein, which we encounter early at the inferior pole of the gland, are superfluous in the exeresis of the superficial lobe. Indeed, it may be of assistance in identifying the common trunk of the facial nerve with a retrograde approach, starting from the vein at the inferior pole, identifying the marginal branch at this level, and coming up along the nerve to the goose's foot.
- We must consider that the great auricular nerve should not be completely sectioned in the phase of isolating the anterior margin of the sternocleido-mastoid muscle. Intervention may be limited to the cutaneous anaesthesia

of the auricle and of the neighbouring zones, sectioning only the branches that enter the gland while leaving intact the posterior branches that go up along the mastoid region.

- Last, the flap of skin over the parotid gland should be cut in an arbitrary intra-adipose plane, more superficial than the cervical fascia that covers the gland. This guards against any lesions of the terminal branches of the goose's foot which, anteriorly, rise to the surface on the masseter muscle.

References

1. Pernkoff E (1986) Atlante di anatomia umana, vol 1. Piccin Nuova Libraria, Padova
2. Testut L, Jacob O (1977) Trattato di anatomia topografica. UTET, Torino
3. Osborne RF, Jesse W, Hamilton JS, Calcaterra TC (2004) Bipedicle sternocleidomastoid muscle flap for reconstruction of tail of parotid defects. Laryngoscope 114(11):2045–2047

5 Submandibular–Submental Region (Robbins Level I)

Core Messages

- Submandibular surgery essentially consists of gland ablation or complete excision of the region; some important structures must, however, be preserved, such as the marginal branch of the facial nerve and the lingual and hypoglossal nerves. The most significant surgical stage is to succeed in revealing, on the plane of the hyoglossal, the lingual nerve, Wharton's duct, and the hypoglossal nerve.

5.1 Anatomic Layout

The region we are going to dissect corresponds to Robbins level I. The sublevel IA coincides with the submental region, while the sublevel IB coincides with the submandibular region. The two sublevels are separated by the fascia overlying the posterior aspect of the submandibular gland [1].

The almond-shaped submandibular gland is located in the cavity of the same name and invested by a layer of superficial cervical fascia. The cavity has a superomedial wall contiguous with the mylohyoid and hyoglossal muscles, a lateral wall contiguous with the body of the mandible. The inferolateral wall invested with split-open superficial cervical fascia, subcutaneous tissue, and skin. The anterior end of the gland is inserted between the mylohyoid and hyoglossal muscles and communicates with the sublingual cavity. The posterior end of the gland is separated from the parotid by the interglandular septum, which marks a thickening in the superficial cervical fascia, and is in close contact with the origin of the facial artery. The submandibular lymph nodes are prevalently subfascial and are situated by the superolateral margin of the gland. The submandibular cavity is bounded caudally by the digastric muscle. The anterior belly bounds the submental region with its median line (Fig. 5.1).

Significant Anatomical Structures: superficial cervical fascia, marginal nerve, facial vein, submental vein, retromandibular vein, interglandular septum, distal facial pedicle, masseter muscle, Hayes Martin's manoeuvre, proximal facial pedicle,

M. Lucioni, *Practical Guide to Neck Dissection*,
DOI 10.1007/978-3-642-33977-6_5, © Springer-Verlag Berlin Heidelberg 2013

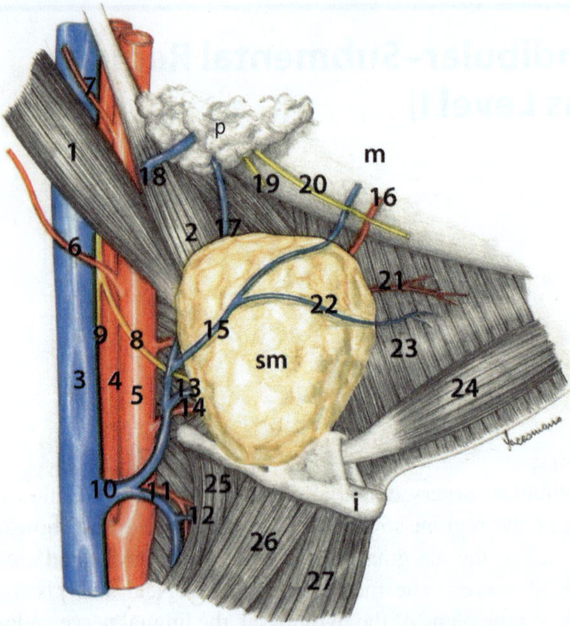

Fig. 5.1 Ablation of the submandibular gland (I). *sm* Submandibular gland, *p* parotid, *m* mandible, *i* hyoid bone, *1* posterior belly of digastric muscle, *2* stylohyoid muscle, *3* internal jugular vein, *4* external carotid artery, *5* internal carotid artery, *6* occipital artery, *7* posterior auricular artery, *8* hypoglossal nerve, *9* descending branch of hypoglossal nerve, *10* thyrolinguofacial venous trunk, *11* superior thyroid artery and vein, *12* superior laryngeal artery and vein, *13* lingual vein, *14* lingual artery, *15* facial vein, *16* facial artery, *17* retromandibular vein, *18* external jugular vein, *19* platysma branch (facial nerve), *20* marginal branch (facial nerve), *21* submental artery, *22* submental vein, *23* mylohyoid muscle, *24* anterior belly of digastric muscle, *25* thyrohyoid muscle, *26* omohyoid muscle, *27* sternohyoid muscle

digastric muscle, submental artery, hyoglossal muscle, mylohyoid muscle, lingual nerve, Wharton's duct, hypoglossal nerve, lingual artery, Beclard's triangle, Pirogoff's triangle, submental region, suprahyoid linea alba.

Landmarks: inferior margin of mandible, posterior margin of mylohyoid muscle, greater cornu of the hyoid bone.

5.2 Cervical Fascia and Distal Facial Pedicle

Below the platysma, the region is invested with superficial cervical fascia, which divides into two at this level to envelop the gland. In the thickness of the fascia, we can identify two of the inferior branches of the facial nerve, that is, the marginal nerve, and nerve serving the platysma muscle. The former runs 1 cm above the inferior margin of the mandible; the latter, which is more difficult to find, runs through the posterosuperior angle of the region, descending to innervate the platysma (Fig. 5.2).

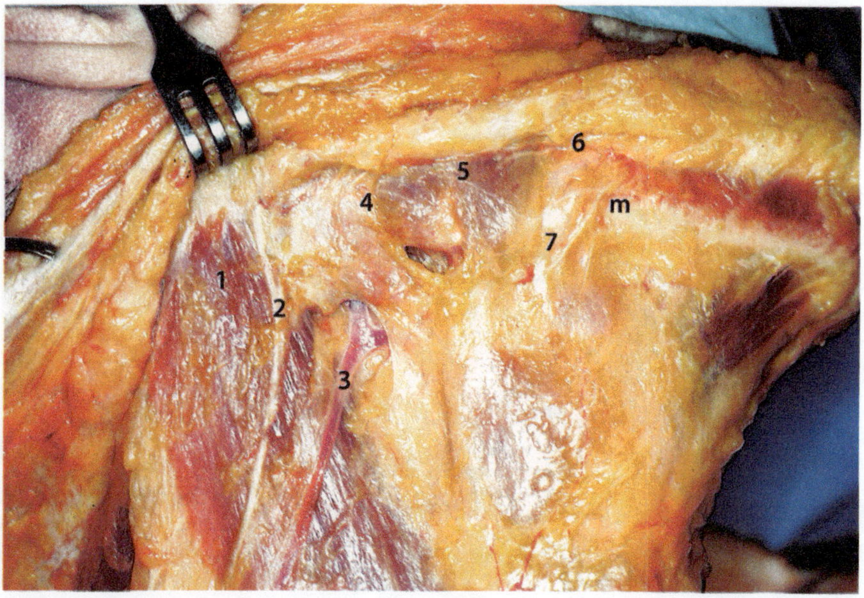

Fig. 5.2 Fascial plane. *m* Mandible, *1* sternocleidomastoid muscle, *2* great auricular nerve, *3* external jugular vein, *4* angle of mandible, *5* masseter muscle, *6* marginal branch (facial nerve), *7* facial pedicle

After dissecting the superficial cervical fascia, the submandibular gland is exposed. On the surface of its posterior pole, we look for the facial vein, which in its downward course unites anteriorly with the submental vein and posteriorly with the retromandibular vein (or external carotid vein) to form the facial venous trunk. It should be borne in mind that venous circulation in this region is somewhat variable, and the situation described is the most frequent one. The interglandular septum can be viewed further behind, which is a thickening of the superficial cervical fascia separating the submandibular gland from the parotid (Fig. 5.3).

Dissection then proceeds by elevating the superficial cervical fascia from the contents of the cavity, exposing at the top the distal part of the distal facial pedicle. At the bottom, the two bellies are uncovered and the intermediate tendon of the digastric muscle which bounds the submandibular cavity at the bottom (Fig. 5.4).

The facial pedicle can be found straddling the inferior margin of the mandible, by the anterior border of the masseter muscle. The marginal branch of the facial nerve crosses the facial pedicle at the top and innervates the mimetic muscles of the lower lip. We ligate the distal facial pedicle 1–2 cm from the inferior margin of the mandible (Fig. 5.5).

Complications: Traumatisation of the marginal nerve causes temporary paresis of the depressor labii inferioris. It is therefore good practice to maintain a caudal position with respect to the cutaneous incision, to avoid exerting excessive traction on the flap in proximity to the mandibular margin, and, where necessary, to dissect the facial pedicle as close as possible to the gland. In the latter case, we are sure to

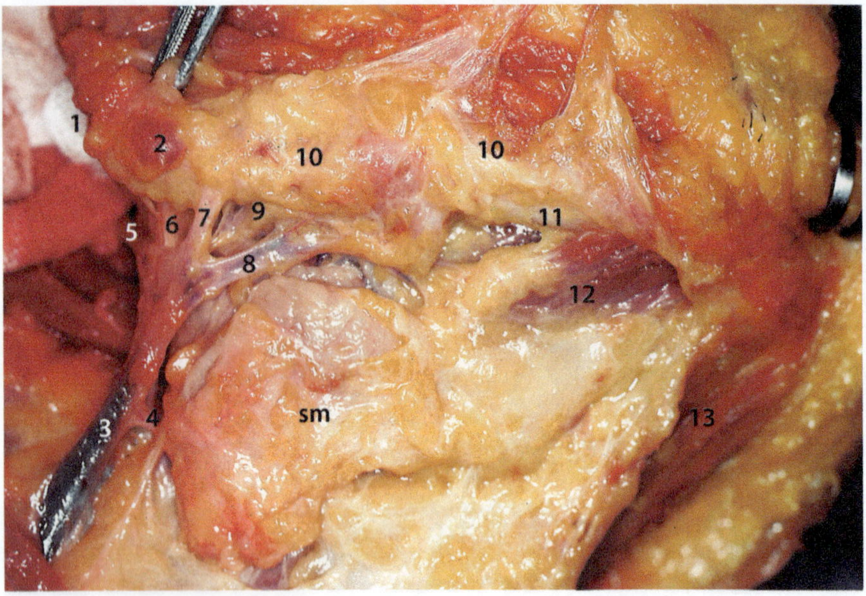

Fig. 5.3 Subfascial plane. *sm* Submandibular gland, *1* angle of mandible, *2* lymph node, *3* linguofacial venous trunk, *4* lingual vein, *5* retromandibular vein, *6* interglandular septum, *7* facial vein, *8* submental vein, *9* facial artery, *10* marginal branch (facial nerve), *11* mandibular inferior margin, *12* mylohyoid muscle, *13* anterior belly of digastric muscle

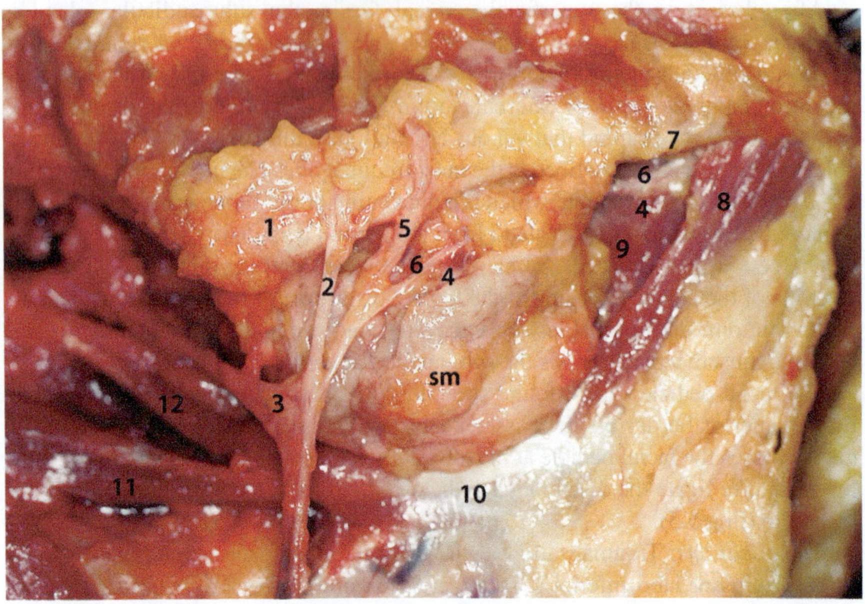

Fig. 5.4 Facial pedicle plane. *sm* Submandibular gland, *1* angle of mandible, *2* facial vein, *3* retromandibular vein, *4* submental vein, *5* facial artery, *6* submental artery, *7* mandibular inferior margin, *8* anterior belly of digastric muscle, *9* mylohyoid muscle, *10* intermediate tendon of digastric muscle, *11* posterior belly of digastric muscle, *12* stylohyoid muscle

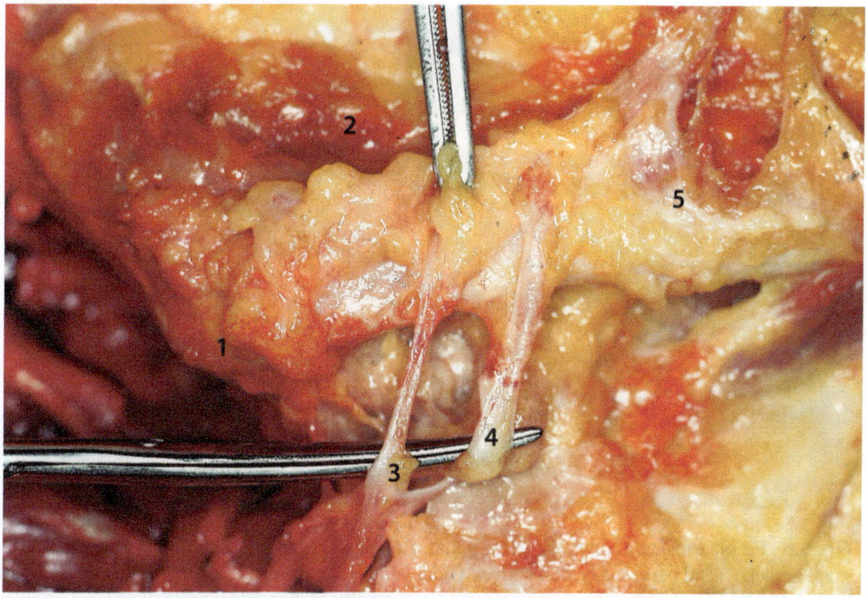

Fig. 5.5 Facial pedicle. *1* Angle of mandible, *2* masseter muscle, *3* facial vein, *4* facial artery, *5* marginal branch (facial nerve)

preserve it by turning the sectioned pedicle upwards: the nerve, which always passes over the pedicle, is thus stretched upwards, away from the surgical field (Hayes Martin's manoeuvre).

5.3 Gland Dissection

Gland ablation begins from the posterior pole, demonstrating the course of the proximal facial pedicle. The facial artery is a branch of the external carotid artery. It emerges behind the posterior belly of the digastric muscle, posteriorly skimming the submandibular gland; running backwards and forwards and upwards and downwards, it surfaces to surround the inferior margin of the mandible, immediately anterior to the facial vein. We ligate the proximal facial pedicle where it appears behind the digastric muscle. In the benign pathology of the submandibular gland, the facial artery is preserved as a rule (Fig. 5.6).

Its anterior branch, the submental artery, thrusts itself in an anteromedial direction, towards the submental region, and is the only important vessel above the mylohyoid muscle. Once we arrive at this plane, we reveal the posterior margin of the muscle (Fig. 5.7).

The submental artery is dissected together with the previously isolated venous collectors of the facial trunk (Fig. 5.8).

The gland is then raised from the deep muscle plane (hyoglossal muscle) and intermediate muscle plane (mylohyoid muscle) and everted (Fig. 5.9).

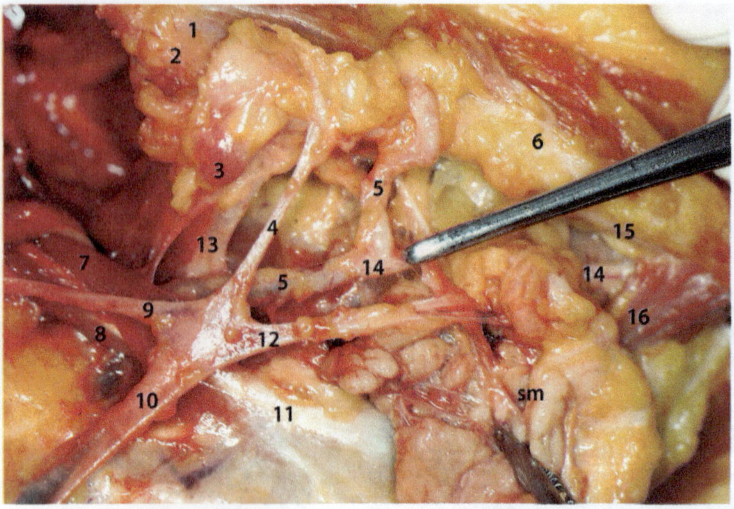

Fig. 5.6 Facial artery. *sm* Submandibular gland, *1* angle of mandible, *2* proximal portion of marginal branch (facial nerve), *3* lymph node, *4* facial vein, *5* facial artery, *6* distal portion of marginal branch (facial nerve), *7* stylohyoid muscle, *8* posterior belly of digastric muscle, *9* retromandibular vein, *10* facial venous trunk, *11* intermediate tendon of digastric muscle, *12* submental vein, *13* interglandular septum, *14* origin of submental artery, *15* mandibular inferior margin, *16* anterior belly of digastric muscle

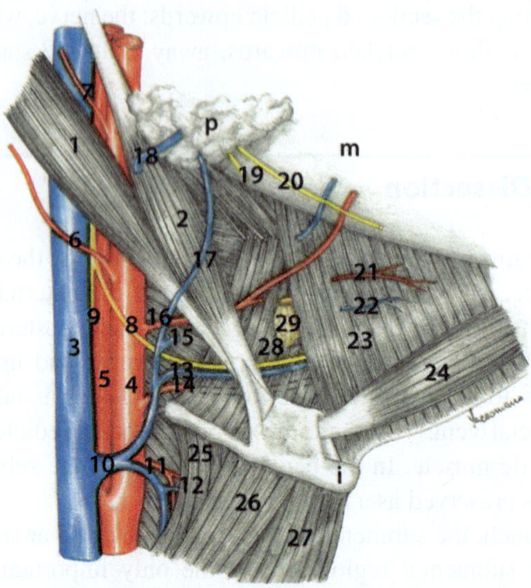

Fig. 5.7 Ablation of the submandibular gland (II). *p* Parotid, *m* mandible, *i* hyoid bone, *1* posterior belly of digastric muscle, *2* stylohyoid muscle, *3* internal jugular vein, *4* external carotid artery, *5* internal carotid artery, *6* occipital artery, *7* posterior auricular artery, *8* hypoglossal nerve, *9* descending branch of hypoglossal nerve, *10* thyrolinguofacial venous trunk, *11* superior thyroid artery and vein, *12* superior laryngeal artery and vein, *13* lingual vein, *14* lingual artery, *15* facial vein, *16* facial artery, *17* retromandibular vein, *18* external jugular vein, *19* platysma branch (facial nerve), *20* marginal branch (facial nerve), *21* submental artery, *22* submental vein, *23* mylohyoid muscle, *24* anterior belly of digastric muscle, *25* thyrohyoid muscle, *26* omohyoid muscle, *27* sternohyoid muscle, *28* hyoglossal muscle, *29* anterior process of submandibular gland

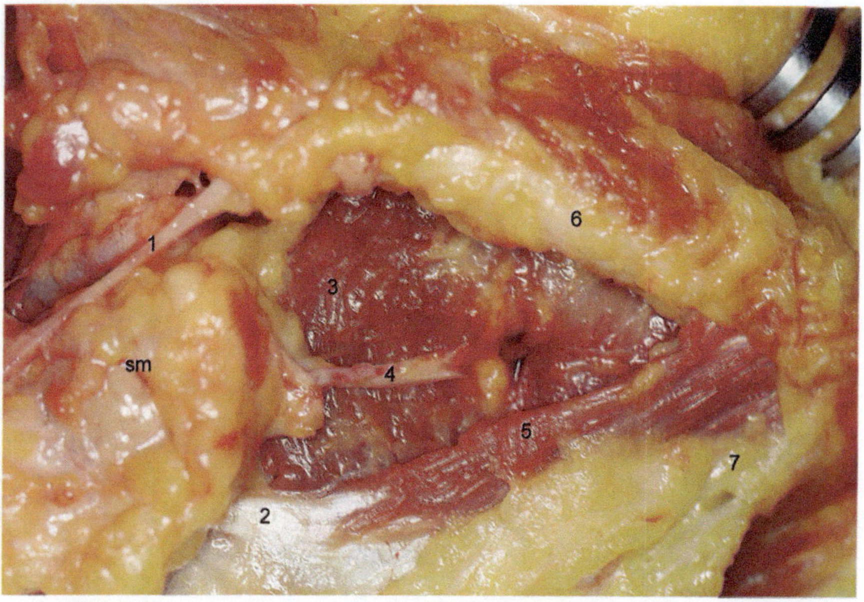

Fig. 5.8 Submental pedicle. *sm* Submandibular gland, *1* facial artery, *2* intermediate tendon of digastric muscle, *3* mylohyoid muscle, *4* submental pedicle, *5* anterior belly of digastric muscle, *6* mandibular inferior margin, *7* submental region

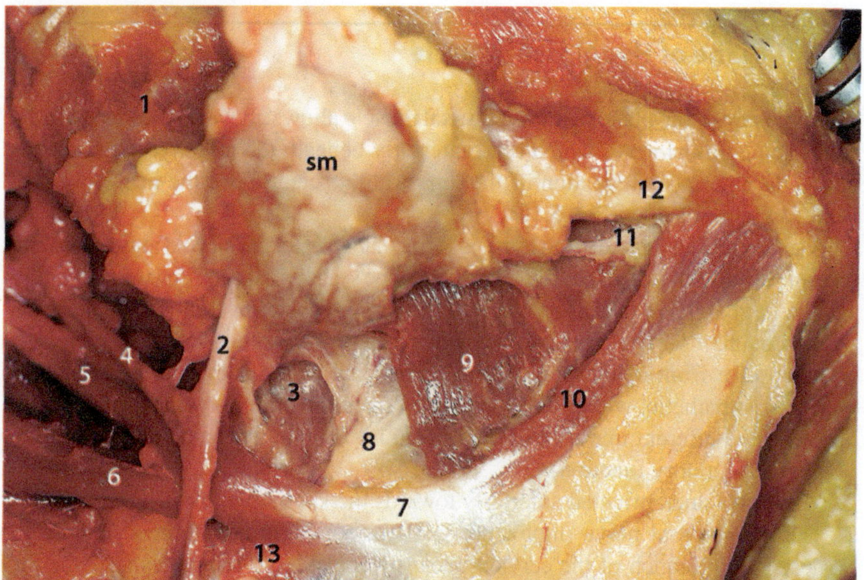

Fig. 5.9 Mylohyoid muscle plane (I). *sm* Submandibular gland, *1* masseter muscle, *2* facial vein, *3* hyoglossal muscle, *4* retromandibular vein, *5* stylohyoid muscle, *6* posterior belly of digastric muscle, *7* intermediate tendon of digastric muscle, *8* hypoglossal nerve, *9* mylohyoid muscle, *10* anterior belly of digastric muscle, *11* submental artery, *12* mandibular inferior margin, *13* apex of greater cornu of hyoid bone

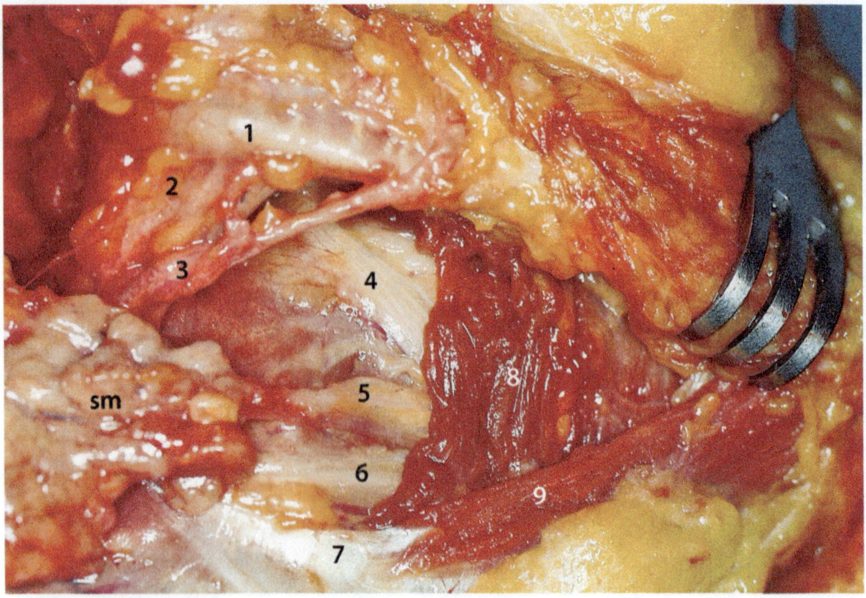

Fig. 5.10 Mylohyoid muscle plane (II). *sm* Submandibular gland, *1* angle of mandible, *2* interglandular septum, *3* facial artery, *4* lingual nerve, *5* Wharton's duct, *6* hypoglossal nerve, *7* intermediate tendon of digastric muscle, *8* anterior belly of digastric muscle

5.4 Plane of Hyoglossal Muscle

The exposure of the plane of the hyoglossal muscle allows above all the identification of the hypoglossal nerve which runs anteriorly beneath the mylohyoid muscle and above the intermediate tendon of the digastric muscle. Above the nerve, we shall isolate Wharton's duct (Fig. 5.10).

A small Farabeuf is used to move the posterior margin of the mylohyoid muscle forwards, revealing the hyoglossal plane. The following can be seen from the top downwards:

1. The lingual nerve (a sensory nerve arising in the posterior trunk of the mandibular branch of the trigeminal nerve; it provides sensory and taste innervation of the mucosa in front of the lingual "V") connected to the submandibular ganglion (parasympathetic, with afferent impulses from the chorda tympani of the facial nerve and efferent impulses to the lingual nerve with a submandibular and sublingual secretory function)
2. Wharton's duct, oriented anteriorly towards the sublingual gland
3. The hypoglossal nerve (motor nerve of the tongue and – in concert with the descending branch of the cervical plexus – the subhyoid muscles save the thyrohyoid muscle, which it innervates separately) (Fig. 5.11)

Complications: On reaching the hyoglossal muscle plane, it is essential when ligating Wharton's duct to avoid injuring the lingual nerve or, worse still, the hypoglossal nerve, as by rush cautery. Lesion of the hypoglossal nerve causes dysphagia, and the tongue, when protruded, deviates towards the paretic side.

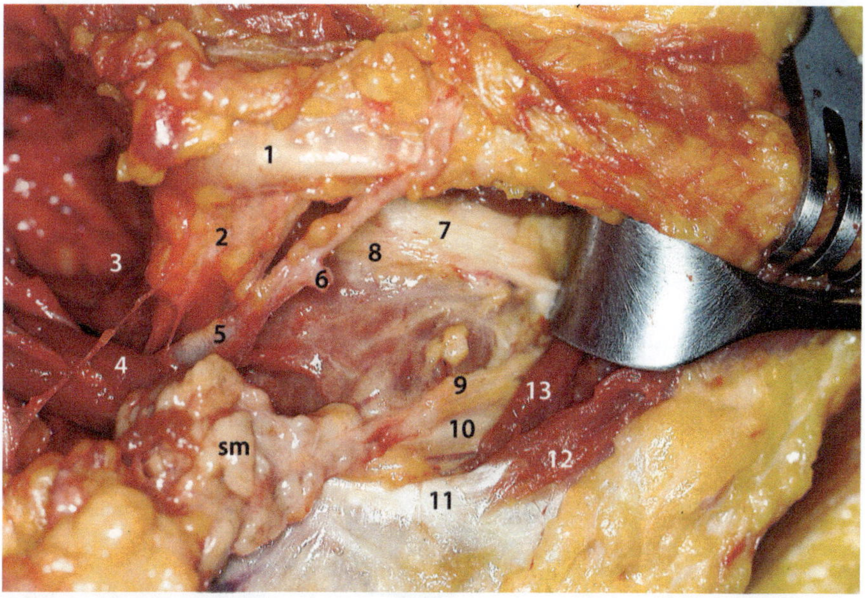

Fig. 5.11 Hyoglossal muscle plane. *sm* Submandibular gland, *1* angle of mandible, *2* interglandular septum, *3* parotid region, *4* stylohyoid muscle and posterior belly of digastric muscle, *5* facial artery, *6* origin of submental artery, *7* lingual nerve, *8* submandibular ganglion, *9* Wharton's duct, *10* hypoglossal nerve, *11* intermediate tendon of digastric muscle, *12* anterior belly of digastric muscle, *13* mylohyoid muscle

The lingual artery, which is the second branch of the external carotid artery, is sought and bound. Almost immediately after its origin, accompanying the middle constrictor of the pharynx, it comes into contact with the posterior margin of the hyoglossal muscle, which takes a horizontal, parallel route to the greater cornu of the hyoid bone, approximately half a centimetre above it (Fig. 5.12).

Exercise 2: Lingual Artery (Fig. 5.13)

In clinical practice, the seeking and binding of the lingual artery are indicated as the preliminary stage of surgery of the oropharynx and of the oral cavity and are carried out at the point of origin. In dissection classes, it is nonetheless interesting to isolate it behind and in front of the posterior belly of the digastric muscle, where anatomists locate Beclard's triangle and Pirogoff's triangle, respectively. The former is bounded by the posterior belly of the digastric muscle, the greater cornu of the hyoid bone, and the posterior margin of the hyoglossal muscle. Dissection in this space involves the hyoglossal fibres, just below the hypoglossal nerve and the lingual vein. The latter triangle is formed by the intermediate tendon of the digastric muscle, the hypoglossal nerve, and the posterior margin of the mylohyoid muscle. In this case too, the lingual artery is isolated by dissecting the hyoglossal muscle fibres. Such well-defined anatomic details enable the lingual artery to be identified and ligated with extreme precision.

To conclude the exercise, dissection is extended anteriorly to the submental region which lies between the two anterior bellies of the digastric muscles. We shall remove the

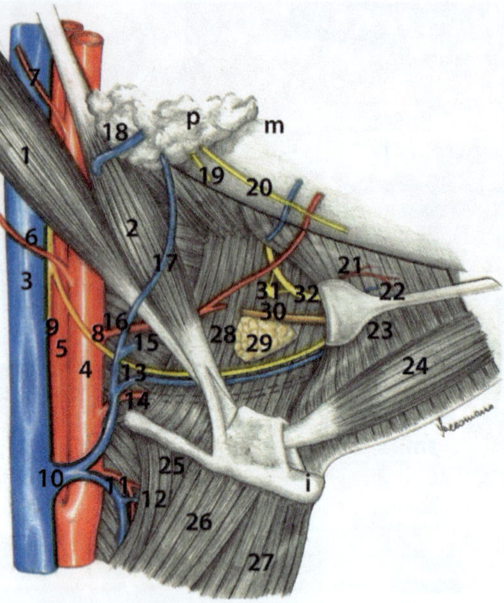

Fig. 5.12 Ablation of the submandibular gland (III). *p* Parotid, *m* mandible, *i* hyoid bone, *1* posterior belly of digastric muscle, *2* stylohyoid muscle, *3* internal jugular vein, *4* external carotid artery, *5* internal carotid artery, *6* occipital artery, *7* posterior auricular artery, *8* hypoglossal nerve, *9* descending branch of hypoglossal nerve, *10* thyrolinguofacial venous trunk, *11* superior thyroid artery and vein, *12* superior laryngeal artery and vein, *13* lingual vein, *14* lingual artery, *15* facial vein, *16* facial artery, *17* retromandibular vein, *18* external jugular vein, *19* platysma branch (facial nerve), *20* marginal branch (facial nerve), *21* submental artery, *22* submental vein, *23* mylohyoid muscle, *24* anterior belly of digastric muscle, *25* thyrohyoid muscle, *26* omohyoid muscle, *27* sternohyoid muscle, *28* hyoglossal muscle, *29* anterior process of submandibular gland, *30* Wharton's duct, *31* submandibular ganglion, *32* lingual nerve

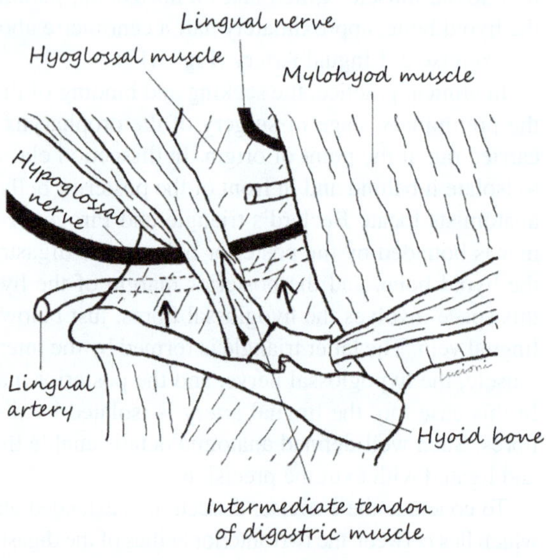

Fig. 5.13 Exercise 2: lingual artery

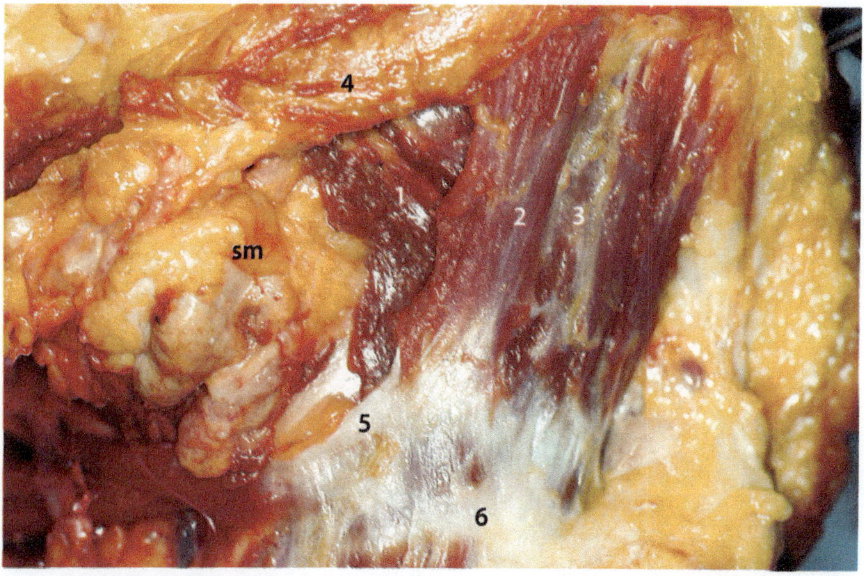

Fig. 5.14 Submental region. *sm* Submandibular gland, *1* mylohyoid muscle, *2* anterior belly of digastric muscle, *3* suprahyoid linea alba, *4* mandibular inferior margin, *5* intermediate tendon of digastric muscle, *6* hyoid bone

adipose tissue which fills this space until we expose the plane of the mylohyoid muscles which, uniting on the median line, form a fibrous raphe extending from the hyoid bone to the mental protuberance, known as the suprahyoid linea alba (Fig. 5.14)

Take-Home Messages

- In submandibular surgery in benign pathology, we must remember that, after repeated phlogosis, for example, in sialolithiasis, the removal of the gland may be more exacting due to scars and to more intense bleeding. In these cases, there is an increased risk of lesion of the lingual and hypoglossal nerves. In the case of calculosis, it is necessary to check that the section of Wharton's duct does not let any calculi and parenchyma pass into the distal stump.
- In submandibular surgery in malignant pathology, the ablation includes the gland and the adipose and fascial tissue of the region; when required, exeresis may extend to the deep muscles, to the lingual artery, and, if infiltrated by neoplasm, also to the hypoglossal nerve. The excision of this region is required for the rare primitive tumours of the gland or as a stage of laterocervical excisions (Robbins level I), especially for tumours of the oropharynx, of the oral cavity and of the lower lip. It may also be a transit surgical stage for access to the parapharyngeal space, after having dissected the digastric and stylienus muscles, as an alternative to transmandibular access.

Reference

1. Robbins KT, Shaha AR, Medina JE et al (2008) Consensus statement on the classification and terminology of neck dissection. Arch Otolaryngol Head Neck Surg 134(5):536–538

Laterocervical Region (Supraclavicular Region: Robbins Level V)

6

Core Messages

- The surgery of this region has a specific oncological significance for the treatment of lymph nodal metastases of tumours of the rhinopharynx and oropharynx and of the posterior cutaneous tumours of the head and neck. It may also be considered for tumours of the larynx or of the hypopharynx if the presence of metastases at Robbins levels II or III has been ascertained. In the surgical exploration of this region, the peripheral branch of the spinal accessory nerve must be identified and preserved.

6.1 Anatomic Layout

The laterocervical region is bounded posteriorly by the anterior margin of the trapezius and by the splenius capitis muscle, anteriorly by the lesser cornu of the hyoid bone and lateral margins of the sternothyroid and thyrohyoid muscles, inferiorly by the superior margin of the clavicle, and superiorly by the inferior margin of the digastric muscle. The deep boundary of the region corresponds to the scalene, levator scapulae, and prevertebral muscle plane (Fig. 6.1).

Dissecting from bottom to top and from rear to front, we will adhere closely to the correct technique used for neck dissection in oncological patients, performing it here at least theoretically, to avoid the spread of any metastatic emboli.

We shall start from the supraclavicular region and then move on to the jugulocarotid region. Translating the anatomic nomenclature of the Robbins levels, our dissection will start with level V and then proceed, in the following chapter, with levels II, III, and IV.

The supraclavicular region corresponds to Robbins level V. It is bounded superiorly by the apex formed by the convergence of the trapezius and sternocleidomastoid muscles, inferiorly by the clavicle, anteriorly by the posterior margin of the sternocleidomastoid muscle, and posteriorly by the anterior margin of the trapezius.

M. Lucioni, *Practical Guide to Neck Dissection*,
DOI 10.1007/978-3-642-33977-6_6, © Springer-Verlag Berlin Heidelberg 2013

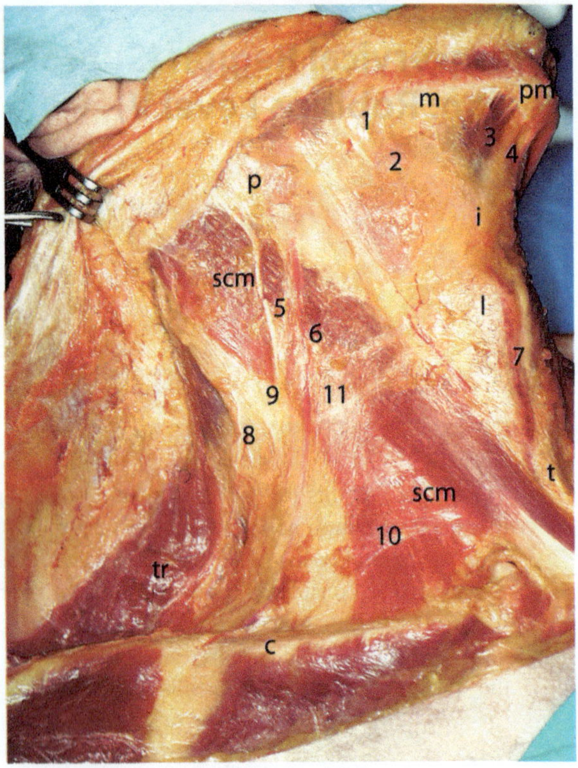

Fig. 6.1 Laterocervical region. *p* Parotid, *m* mandible, *pm* mental protrusion, *scm* sternocleidomastoid muscle, *i* hyoid bone, *l* larynx, *tr* trapezius muscle, *t* thyroid gland, *c* clavicle, *1* facial pedicle, *2* submandibular gland, *3* anterior belly of digastric muscle, *4* interdigastric (submental) area, *5* great auricular nerve, *6* external jugular vein, *7* anterior jugular vein, *8* spinal accessory nerve (peripheral branch), *9* Erb's point, *10* superficial cervical fascia, *11* cutaneous cervical nerve

This level has the shape of a pyramid with the base at the bottom, where the first rib separates it from the pulmonary apex. In depth, the emerging of the cervical and brachial plexi separates level V from levels II, III, and IV. An imaginary horizontal line, inferiorly at a tangent to the cricoid cartilage, divides level V into VA (upper, lymph nodes of the spinal chain) and VB (lower, supraclavicular lymph nodes).

The celluloadipose content of this region is superiorly and medially in continuity with that of the jugulocarotid region, inferiorly and medially with that of the superior mediastinum, and inferiorly and laterally with that of the axilla.

The significant groups of lymph nodes are those adjacent to the peripheral portion of the spinal accessory nerve and those of the transverse cervical artery.

Significant Anatomical Structures: platysma, great auricular nerve, external jugular vein, cutaneous cervical nerve, superficial cervical fascia, spinal accessory nerve, sternocleidomastoid muscle, shoulder syndrome, lesser occipital nerve, cervical plexus, omohyoid muscle, brachial plexus, scalene muscles, Pancoast syndrome, transverse cervical artery, transverse scapular artery, phrenic nerve, subclavian artery, anterior scalene muscle syndrome, Troisier's sign

Landmarks: clavicle, Erb's point, anterior margin of the trapezius, Lisfranc's tubercle

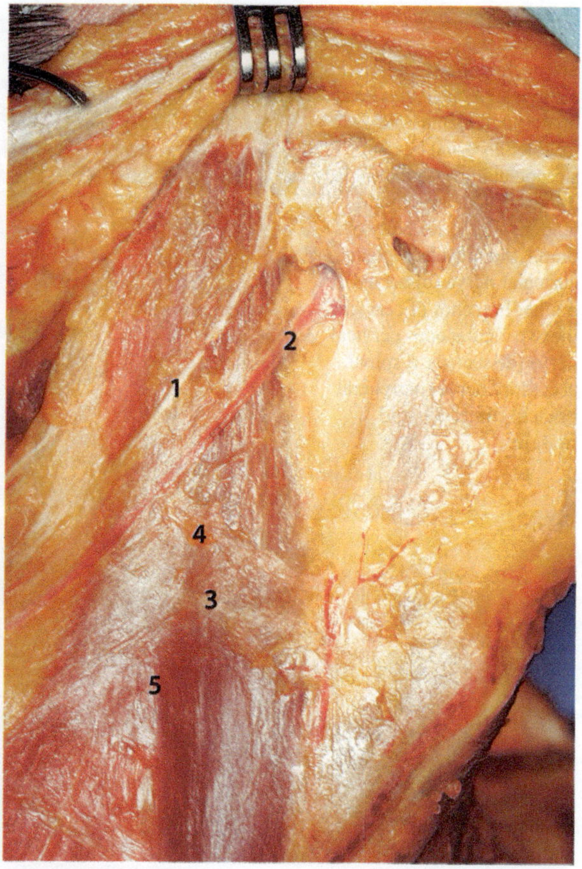

Fig. 6.2 Superficial cervical fascia plane. *1* Great auricular nerve, *2* external jugular vein, *3* cutaneous cervical nerve, *4* superficial cervical fascia, *5* sternocleidomastoid muscle

6.2 Superficial Cervical Fascia and Spinal Accessory Nerve

The neck is extended and rotated as far as possible in the opposite direction to the operator. If still present, the platysma is now completely resected, leaving the superficial cervical fascia in place. On the surface of the sternocleidomastoid muscle, under the superficial cervical fascia, three structures can clearly be seen which cross the muscle: (1) the great auricular nerve, (2) the external jugular vein with its branches, and (3) the cutaneous cervical nerve; both nerves are cutaneous (sensory) branches of the cervical plexus (Fig. 6.2).

The superficial cervical fascia is dissected along the external surface of the sternocleidomastoid muscle, in the centre, following a craniocaudal direction, and so the abovementioned structures are interrupted. The fascia is raised from the muscle fibres by holding the scalpel at a tangent to the muscle along its entire length (Fig. 6.3).

The dissection of level V begins with the identification and isolation of the spinal accessory nerve. The accessory nerve originates in the cranium from the union of

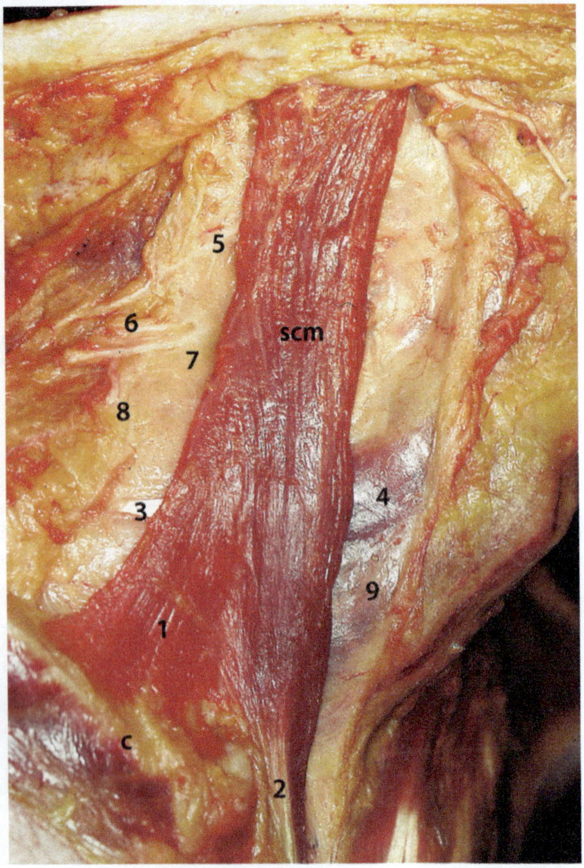

Fig. 6.3 Sternocleidomastoid muscle. *scm* Sternocleidomastoid muscle, *c* clavicle, *1* clavicular head of sternocleidomastoid muscle, *2* sternal head of sternocleidomastoid muscle, *3* intermediate omohyoid tendon, *4* superior belly of omohyoid muscle, *5* great auricular nerve (dissected), *6* other branches of cervical plexus, *7* cutaneous cervical nerve (dissected), *8* spinal accessory nerve (peripheral branch), *9* sternohyoid muscle

the vagal accessory nerve (parasympathetic fibres/visceral effector) and spinal accessory nerve (somatic motor); it exits from the posterior foramen lacerum and divides once again: the vagal portion (internal or medial branch), joins the vagus nerve, and participates in innervating the larynx. The spinal portion (external or lateral branch) passes anteriorly to the internal jugular vein, enters the sternocleidomastoid muscle (which it innervates), and exits in proximity to the posterior margin of the muscle. Running from top to bottom and from front to rear, the peripheral portion of the nerve then enters the trapezius, which it innervates.

Exercise 3: Spinal Accessory Nerve (Fig. 6.4)

We shall look for the peripheral portion of the spinal accessory nerve in two points:

1. At the exit from the posterior margin of the sternocleidomastoid muscle, about 1 cm superiorly to Erb's point, that is, where the great auricular nerve, which is part of the cervical plexus, surrounds the muscle and surfaces
2. On entry to the trapezius, about 2 cm above the point where this muscle and the inferior belly of the omohyoid muscle cross

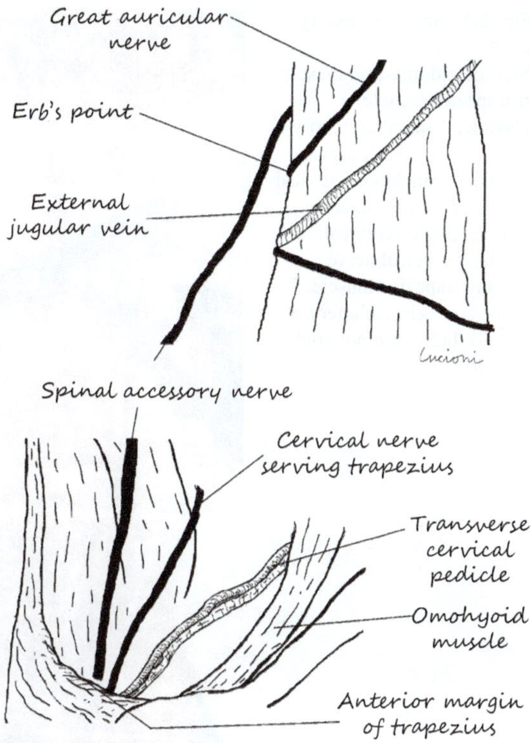

Fig. 6.4 Exercise 3: spinal accessory nerve

The second approach is the more practical because neck dissection is normally performed from bottom upwards and from back to front. First, we must identify the anterior margin of the trapezius muscle just beneath the skin. The nerve, which penetrates the muscle medially at its anterior margin, is thus more easily protected. Here we shall identify the cervical branch for the trapezius and, after that, the distal portion of the transverse pedicle of the neck. Once identified, the spinal accessory nerve is isolated along its entire course from the trapezius to the sternocleidomastoid muscle (Fig. 6.5). During this procedure, some spinal chain lymph nodes may be found, which follow the course of the nerve.

Complications: The trapezius and the sternocleidomastoid muscle have a double innervation, one coming from the spinal accessory nerve and another pertaining to the roots C2 and C3 of the cervical plexus. The sectioning of both afferents leads to what is defined shoulder syndrome and consists of the lowering and anterolateral rotation of the shoulder and of pain associated with the movements of lifting the limb. In some cases, this may be followed by marked hypertrophy of the sternoclavicular articulation due to microfractures or capsular distortions from lifting and anteriorisation of the medial section of the clavicle. Clinically speaking, a clavicular "pseudotumour" is presented, which, at first sight, may lead to the suspicion of metastases at level IV or secondary bone localisation.

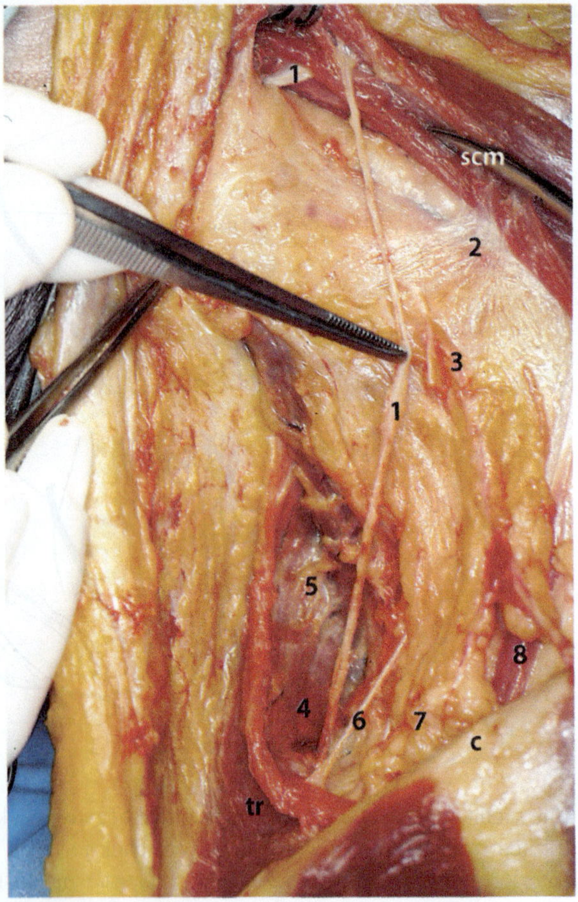

Fig. 6.5 Spinal accessory nerve. *scm* Sternocleidomastoid muscle, *tr* trapezius muscle, *c* clavicle, *1* spinal accessory nerve, *2* superficial cervical fascia, *3* branches of cervical plexus, *4* levator scapulae muscle, *5* deep cervical fascia, *6* cervical nerve serving trapezius muscle, *7* transverse cervical artery, *8* inferior belly of omohyoid muscle

The medial surface of the trapezius is freed from the overlying loose connective tissue until, at the top and on a deeper plane, the levator scapulae muscle and the scalene muscles are revealed, covered by the deep cervical fascia (level VA). On the levator scapulae muscle, the lesser occipital nerve can be identified, another cutaneous branch of the cervical plexus (Fig. 6.6).

6.3 Cervical Plexus and Transverse Pedicle

Dissection will encounter other posterior branches of the plexus, and will stop medially at the level where the anastomotic loops of the cervical plexus emerge, medially to which there are the Robbins levels II and III.

The following structures are sought and isolated below the spinal accessory nerve (level VB): (1) the distal portion of the transverse cervical artery and (2) the cervical plexus branch serving the trapezius. These structures are exposed by

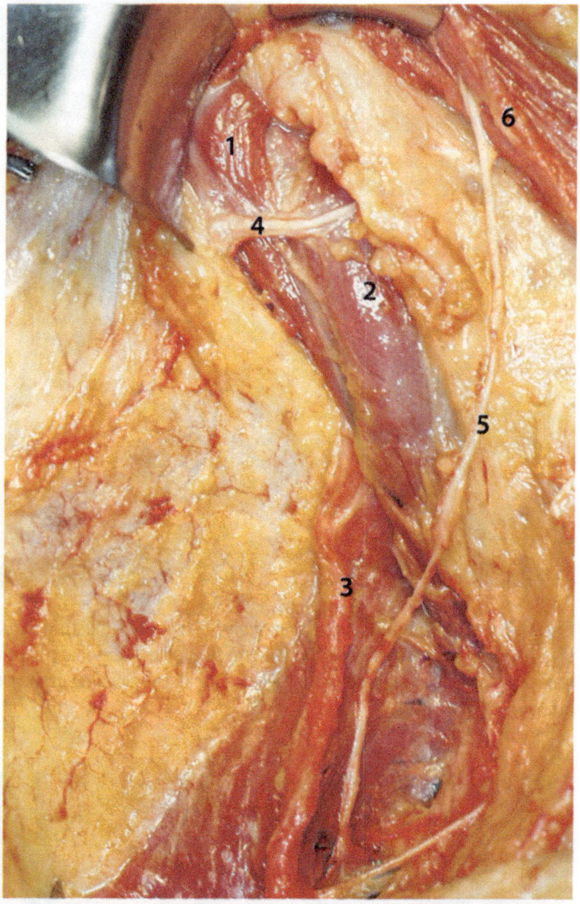

Fig. 6.6 Robbins level V. *1* Levator scapulae muscle, *2* scalene muscles, *3* trapezius muscle, *4* lesser occipital nerve, *5* spinal accessory nerve, *6* sternocleidomastoid muscle

medially lifting the loose connective tissue from the supraclavicular fossa with the scissors (Fig. 6.7).

The omohyoid muscle is identified in the superficial portion of the supraclavicular triangle. The external jugular vein is evident in the immediate subfascial plane, thus above the plane of the omohyoid muscle. It arises from the external surface of the sternocleidomastoid muscle, lateralises and descends towards the clavicle, and then meets the subclavian vein. It is served laterally by a single significant venous branch, that is, the transverse cervical vein. These vessels are isolated and dissected at their ends (Fig. 6.8).

The next step is to isolate the inferior belly of the omohyoid muscle, which is invested in the more lateral portion of the midcervical fascia divided into two (Fig. 6.9).

We section the omohyoid muscle distally and evert it. Any hypertrophic lymph nodes of the supraclavicular chain lying on the posterosuperior margin of the clavicle are identified. With the aid of dry gauze, the adipose tissue is lifted medially,

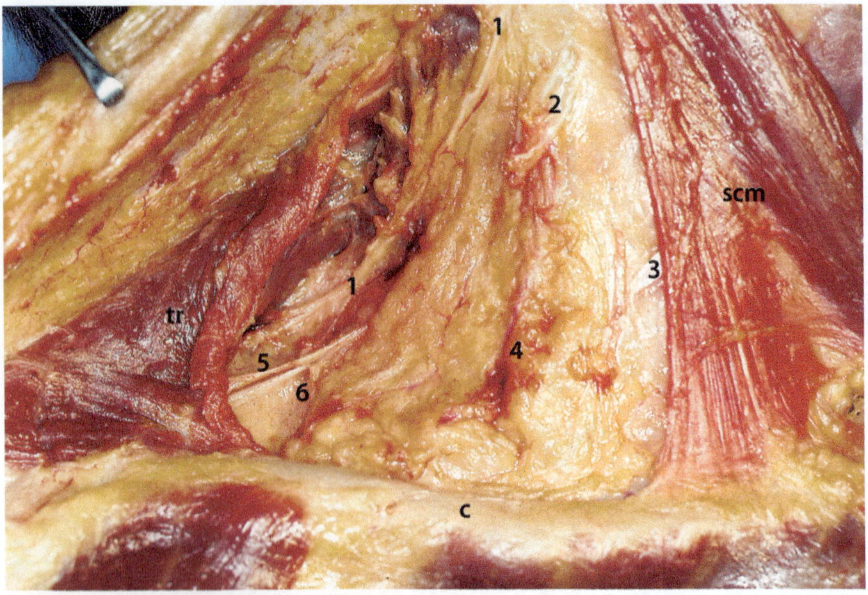

Fig. 6.7 Inferior subfascial plane. *scm* Sternocleidomastoid muscle, *tr* trapezius muscle, *c* clavicle, *1* spinal accessory nerve (peripheral branch), *2* cervical plexus nerve, *3* intermediate omohyoid tendon, *4* external jugular vein, *5* cervical nerve serving trapezius muscle, *6* transverse cervical artery

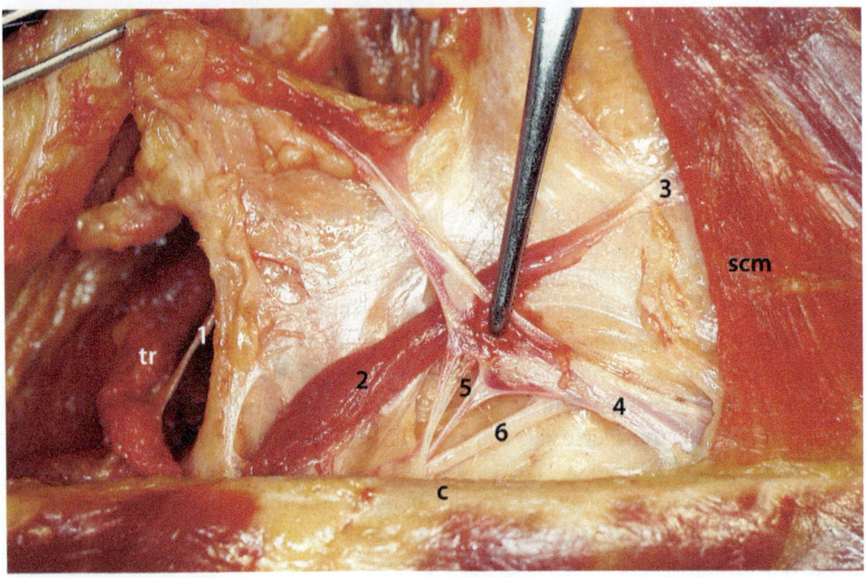

Fig. 6.8 Omohyoid muscle plane. *scm* Sternocleidomastoid muscle, *tr* trapezius muscle, *c* clavicle, *1* spinal accessory nerve (peripheral branches), *2* inferior belly of omohyoid muscle, *3* intermediate omohyoid tendon, *4* external jugular vein, *5* transverse cervical vein, *6* transverse scapular artery

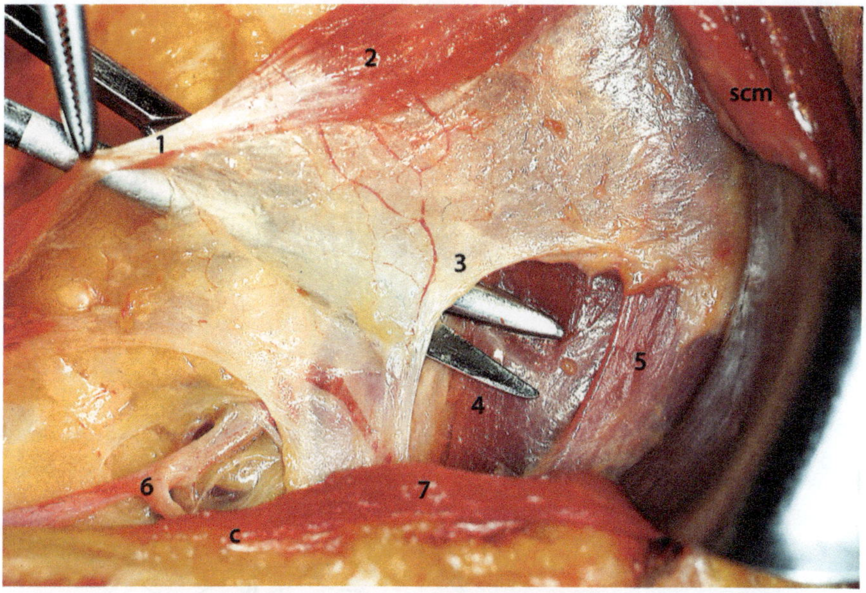

Fig. 6.9 Middle cervical fascia. *scm* Sternocleidomastoid muscle, *c* clavicle, *1* intermediate omohyoid tendon, *2* superior belly of omohyoid muscle, *3* middle cervical fascia, *4* sternothyroid muscle, *5* sternohyoid muscle, *6* transverse scapular artery and vein, *7* clavicular insertion or sternocleidomastoid muscle

thus revealing the deep plane where we identify the plane of the scalene muscles, the brachial plexus, and the overlying transverse cervical artery.

6.4 Scalene Muscles Plane and Brachial Plexus

There are three scalene muscles: the anterior, medial, and posterior. They descend from the cervical column, diverging laterally and inserting in the first and second ribs. They are invested by the deep cervical fascia, which continues medially on the prevertebral muscles (Fig. 6.10).

The brachial plexus is formed by the anterior branches of the fifth through eighth cervical nerves and of the first thoracic nerve. Three primary nerve trunks exit between the anterior scalene muscle and the median scalene muscle. One branch of the brachial plexus, the dorsal scapular nerve, exits between the median scalene and the posterior scalene muscles. The brachial plexus innervates the upper limb.

Remarks: Pancoast's syndrome is the painful symptom complex propagated to the arm due to compression of the brachial plexus by laterocervical metastasis or a primary tumour of the apex of the lung.

Complications: In neck surgery, particularly neck dissection, lesions of the brachial plexus are very rare. The plexus is readily identifiable as a white, fibrous, triangular-shaped cord with an inferior base, forming the space between the scalene muscles. The plexus and muscles are invested by the deep cervical fascia (Fig. 6.11).

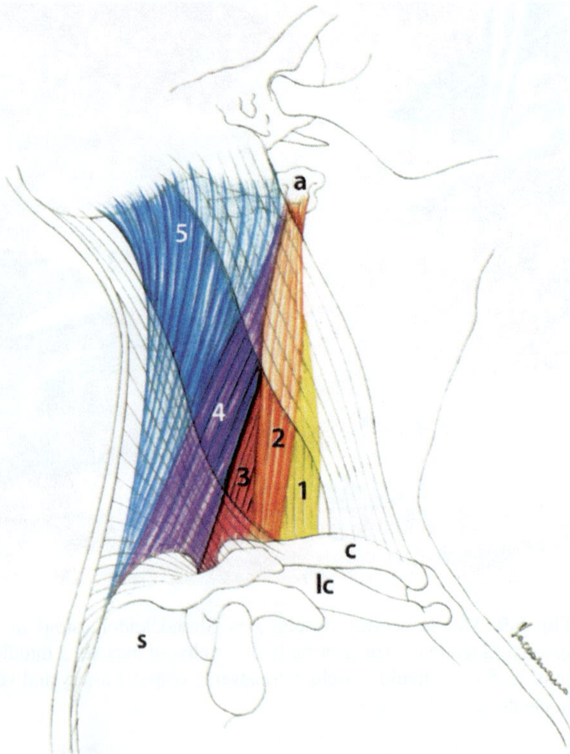

Fig. 6.10 Deep cervical muscles. *a* Transverse process of atlas, *c* clavicle, *Ic* first rib, *s* scapula, *1* anterior scalene muscle, *2* medial scalene muscle, *3* posterior scalene muscle, *4* levator scapulae muscle, *5* splenius capitis muscle

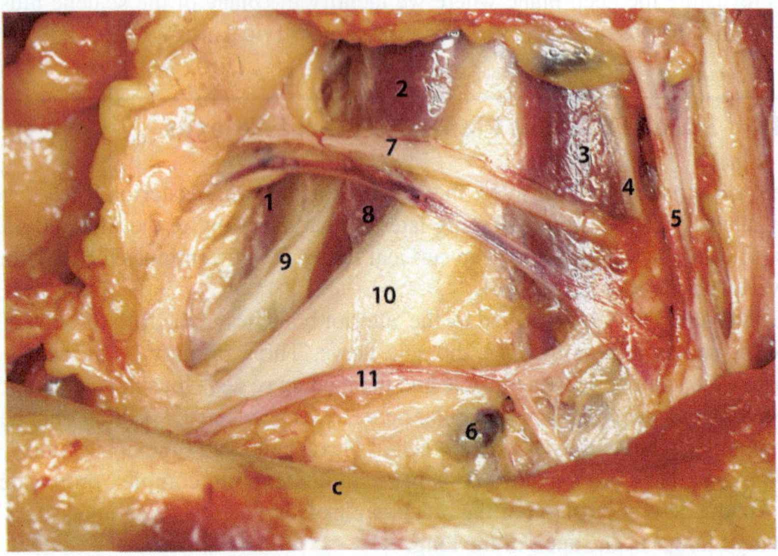

Fig. 6.11 Brachial plexus. *c* Clavicle, *1* posterior scalene muscle, *2* medial scalene muscle, *3* anterior scalene muscle, *4* phrenic nerve, *5* internal jugular vein, *6* anthracotic lymph node, *7* transverse cervical artery and vein, *8* deep cervical fascia, *9* dorsal scapular nerve, *10* brachial plexus, *11* transverse artery of the scapula

It is generally easy to elevate the supraclavicular celluloadipose tissue from the scalene plane with gauze since the surface of the deep cervical fascia is an excellent cleavage plane. Since the superior primary nerve trunk (C5–C6) is in a more superficial position than are the medial and inferior trunks, it is more exposed to trauma or lesions. Anatomic variants are also possible: in the loose supraclavicular cellular tissue, I personally witnessed the C5–C6 trunk running superficially and consequently, accidentally sectioned. This iatrogenic lesion induces motor impairment in the shoulder, which becomes lowered, with frequent dislocation of the head of the humerus; the arm droops on the trunk, exhibiting internal rotation and pronation. There is abduction paralysis of the arm and flexion paralysis of the forearm; 2–3 weeks later, atrophy appears in the muscles concerned.

The transverse cervical artery (and vein) (or superficial cervical artery) and transverse scapular artery (and vein) originate from the thyrocervical trunk. They enter the region medially and diverge laterally, crossing at two different levels of the brachial plexus. They must be isolated and their course followed to the region boundaries.

6.5 Phrenic Nerve

The phrenic nerve is a ramus muscularis of the four of the cervical plexus. It induces movement of the diaphragm and contains sensory fibres for the pulmonary pleura and pericardium. It rests on the surface of the anterior scalene muscle, taking a slightly diverging latero-medial course with respect to the brachial plexus (as a memory aid, the phrenic nerve can be thought of as the thumb of a hand, while the other four fingers represent the branches of the brachial plexus).

The phrenic nerve can be easily identified by continuing digital elevation medially along the cleavage plane formed by the deep cervical prescalene fascia. It appears medially to the brachial plexus, invested by fascia on the external surface of the anterior scalene muscle. Dissection of the cutaneous branches of the cervical plexus, with the scissor point cranial directed, must be performed on a more superficial plane to the course of the phrenic nerve, which must always be identified beforehand.

Complications: Injury to or dissection of a phrenic nerve presents as paralysis of a hemidiaphragm and its elevation. Patients with monolateral phrenoplegia are generally asymptomatic; however, they may complain of dyspnea when lying down since the contents of the abdomen tend to raise the flaccid hemidiaphragm. The resulting compensatory expansion of the rib cage forces the intercostal and accessory muscles to work hard to produce an effective inspiratory volume. Spirometry in patients with monolateral phrenoplegia exhibits a 25 % decrease in total lung capacity (TLC), vital capacity (VC), inspiratory capacity, and maximum inspiratory pressure (MIP), while the reduction in forced expiratory volume at the first second (FEVI) may be as high as 40 %. These values are not normally associated with important clinical consequences, except in the presence of previous pulmonary pathologies with reduced respiratory functioning.

Bilateral phrenoplegia, which is very rare in cervical surgery, is more commonly related to central or systemic neurological pathologies. Transitory bilateral paralysis may in some cases result from heart surgery-related hypothermia. Assisted ventilation is required in such cases. Reparatory operations, requiring optimum physical performance, may also be conducted in patients with monolateral phrenoplegia and consist of "folding" the flaccid hemidiaphragm to reduce compliance.

6.6 Subclavian Artery

To conclude dissection of this region, it may be worth seeking and isolating the subclavian artery. It lies immediately inferomedially to the brachial plexus in the tract where the artery, straddling the first rib beneath the scalene muscles and passing below the clavicle, becomes the axillary artery. Its passage on the first rib occurs immediately laterally to Lisfranc's tubercle, which is a bony prominence where the anterior scalene muscle is attached. This is an excellent landmark for ligating the interscalene portion of the subclavian artery. To reveal it, it is advisable to dissect the lateral portion of the anterior scalene muscle, of course after having identified and preserved the phrenic nerve.

Exercise 4: Subclavian Artery (Fig. 6.12)

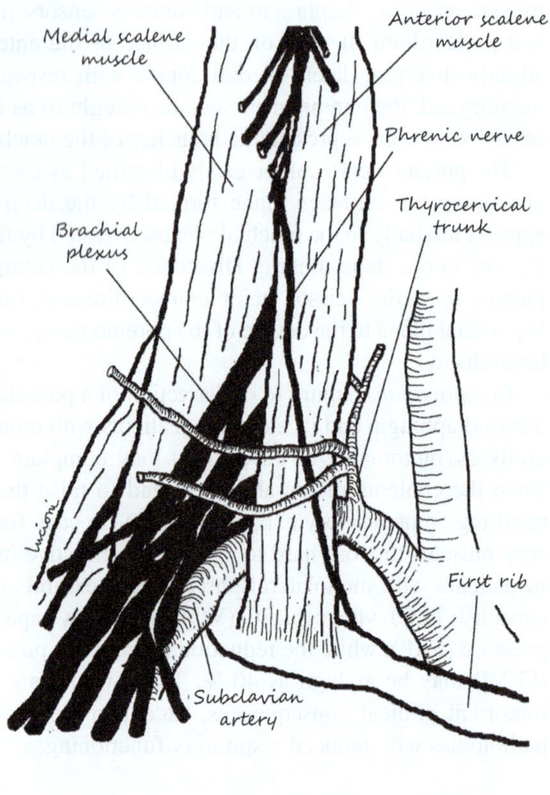

Fig. 6.12 Exercise 4: subclavian artery

We identify the attachment of the anterior scalene muscle on the first rib and seek Lisfranc's tubercle by palpation. We look for the artery at the superior margin of the first rib under the emergence of the brachial plexus. We can reveal it better by dissecting the lateral portion of the anterior scalene muscle always after having revealed and preserved the phrenic nerve.

Remarks: At this point, we recall the rare anterior scalene muscle syndrome, which consists of ischemic disturbances of the upper limb and of the hand and of ulnar neuralgia; these disturbances are accentuated with the limb hanging down and are alleviated when it is raised; they are due to compression of the subclavian artery and of the brachial plexus in the fissure between the median and anterior scalene muscles. The disturbances are cured by sectioning the anterior scalene muscle.

Take-Home Messages

- Regarding the spinal accessory nerve, recall that at the bottom, it may be confused with the branch of the cervical plexus for the trapezius, so before cutting the nerve, it should be isolated to ascertain its identity. It may present anomalies in its course: for example, in approximately 6 % of cases, it ends in the sternocleidomastoid muscle [1], and in 30 % of cases, it does not enter the muscle but remains posterior to it [2, 3]. It has also been demonstrated that the part of the spinal nerve with by far the greatest risk of iatrogenic lesion is its peripheral portion [4].
- In regard to the peripheral branches of the cervical plexus, it must be considered that their sectioning (considered in both radical and modified radical neck dissection by Calearo and Teatini [5], but not in Bocca's functional neck dissection [6]) involves a hypoanaesthesia of the skin, which may extend from the auricle to the skin of the thorax adjacent to the clavicle.
- Solitary metastases on level VB must point to suspect neoplasias coming from the lung, the oesophagus, the breast, and the stomach (on the left, Troisier's sign).

References

1. Krause HR, Bremevic A, Herrmann M (1993) The innervation of the trapezius muscle in connection with radical neck dissection: an anatomical study. J Craniomaxillofac Surg 21:102–106
2. Kierner AC, Burian M, Bentzien S et al (2002) Intraoperative electromyography for identification of the trapezius muscle innervation: clinical proof of a new anatomical concept. Laryngoscope 112:1853–1856
3. Kierner AC, Zelenka I, Burian M (2001) How do the cervical plexus and the spinal accessory nerve contribute to innervation of the trapezius muscle? As seen from within using Sihler's stain. Arch Otolaryngol Head Neck Surg 127:1230–1232
4. Cappiello J, Piazza C, Giudice M et al (2005) Shoulder disability after different selective neck dissection (level II–IV versus level II–V): a comparative study. Laryngoscope 115:259–263
5. Calearo C, Teatini G (1983) Functional neck dissection: anatomical grounds, surgical techniques, clinical observations. Ann Otol Rhinol Laryngol 92:215–222
6. Bocca E, Pignataro O (1967) A conservation technique in radical neck dissection. Ann Otol 76:975–987

Laterocervical Region (Sternocleidomastoid Region: Robbins Levels II, III, and IV)

7

Core Messages

- The surgery of this region has a specific oncological significance for the treatment of lymph node metastases of tumours of the upper respiratory and digestive tracts. The surgical exploration of this region will concern mainly the jugulocarotid axis and its adnexa. It is the area that the oncological surgeon of the head and neck explores most frequently.

7.1 Anatomic Layout

The sternocleidomastoid region as defined by anatomists corresponds approximately to Robbins levels II, III, and IV. It comprises roughly the sternocleidomastoid muscle and all that lies below it, considering the head in a normal position.

Robbins's classification [1] gives its precise limits, which are the base of the skull and the stylohyoid muscle at the top, the clavicle at the bottom, the posterior margin of the sternocleidomastoid muscle at the side, and anteriorly, the lateral edge of the sternocleidomastoid muscle.

The three levels are divided in the craniocaudal direction by the inferior edge of the hyoid bone and the inferior edge of the cricoid cartilage.

The significant lymph node groups are above all those of level II and III, which represent the principal stations of lymphatic drainage of the neck.

Significant Anatomical Structures: splenio capitis muscle, elevator scapulae muscle, scalene muscles, spinal accessory nerve, internal jugular vein, occipital artery, sternocleidomastoid muscle, omohyoid muscle, cervical plexus, hypoglossal loop, phrenic nerve, thyrolinguofacial trunk, perijugular lymph nodes, common carotid artery, carotid glomus, external carotid artery, superior laryngeal artery, hyoglossal muscle, facial artery, ascending pharyngeal artery, posterior auricular artery, hypoglossal nerve, greater cornu of hyoid bone, Farabcuf's triangle, vagus nerve, thoracic duct, thyrocervical trunk, transverse scapular artery, transverse cervical

M. Lucioni, *Practical Guide to Neck Dissection*,
DOI 10.1007/978-3-642-33977-6_7, © Springer-Verlag Berlin Heidelberg 2013

artery, ascending cervical artery, inferior thyroid artery, internal thoracic artery, thoracoacromial artery, subclavian artery, recurrent nerve

Landmarks: transverse process of atlas, digastric muscle, carotid tubercle

7.2 Spinal Nerve and Robbins Level IIB

We begin the dissection of this region from the most cranial part. We identify the deep musculofascial plane, which is formed, latero-medially, by the splenius capitis, levator scapulae, and scalene muscles. After having applied traction medially on the sternocleidomastoid muscle with a Farabeuf, we seek by palpation the transverse process of the atlas, which is an important landmark and the upper limit of lymph node drainage of the neck, close by running the occipital artery or some of its branches. The overlying loose cellular connective tissue is stretched medially, and the lesser occipital nerve, a cutaneous branch of the cervical plexus, can be demonstrated in the deep muscle plane (Fig. 7.1).

It is important at this point to find the common trunk of the spinal accessory nerve, which runs anteriorly to the internal jugular vein in a mediolateral direction and penetrates the sternocleidomastoid muscle. A reliable method of identifying it amid the celluloadipose tissue is to grasp the muscle with the hand, after isolating it from bottom to top, and to palpate with the index finger the stretched nerve within this soft tissue setting (Fig. 7.2).

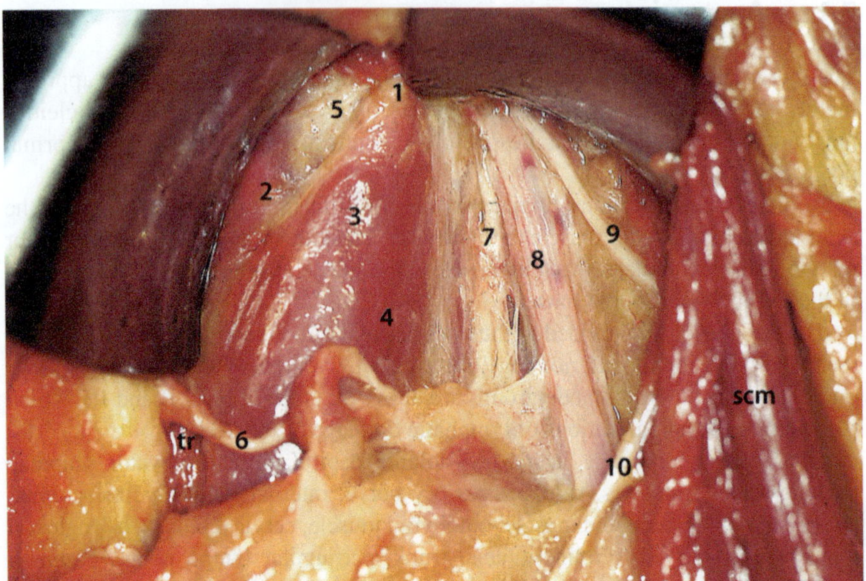

Fig. 7.1 Cervical deep plane and atlas. *scm* Sternocleidomastoid muscle, *tr* trapezius muscle, *1* transverse process of atlas, *2* splenius capitis muscle, *3* levator scapulae muscle, *4* medial scalene muscle, *5* occipital artery, *6* lesser occipital nerve, *7* cervical plexus, *8* internal jugular vein, *9* spinal accessory nerve (common trunk), *10* spinal accessory nerve (peripheral branch)

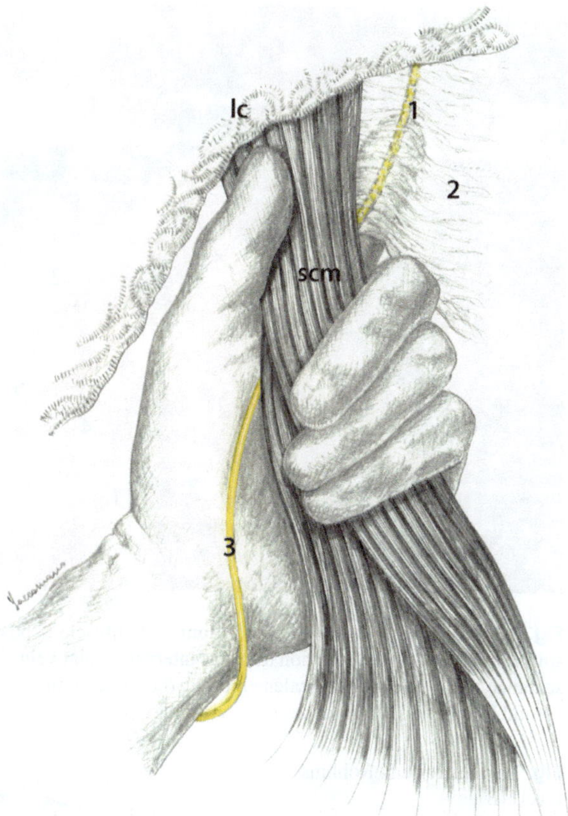

Fig. 7.2 Common trunk of the spinal accessory nerve. *lc* Cutaneous flap, *scm* sternocleidomastoid muscle, *1* spinal accessory nerve (common trunk), *2* loose and fascial connective tissue, *3* spinal accessory nerve (peripheral branch)

Instead, the most practical method is direct access from level II. The anterior margin of the sternocleidomastoid muscle and the posterior belly of the digastric muscle are freed; the latter is another important landmark, on the surface of which there are no dangerous structures for dissection. Divaricating these two muscles allows access to Robbins level II. The common trunk of the spinal accessory nerve, which normally (though not always) runs above the internal jugular vein, divides into an anterior sector (level IIA) and a posterior one (level IIB) (Fig. 7.3).

Remarks: In tumours of the larynx, the presence of metastases at level IIB when level IIA is intact is a negligible occurrence. So, once the intraoperative histological examination has ascertained that level IIA is free from metastases, level IIB is not removed.

Exercise 5: Robbins Level IIB (Fig. 7.4)

It may be interesting to proceed with the surgical technique of approach to level IIB as used in modified radical and selective dissections. Once the common trunk of the spinal accessory nerve has been identified and the sternocleidomastoid and digastric muscles have been well divaricated, we first identify the internal jugular vein, which must be left medially. The nerve is delicately raised and separated from the adipose tissue below. We dissect the adipose tissue along an arched line with upper convexity until we reach the deep muscle plane (levator scapulae and splenius

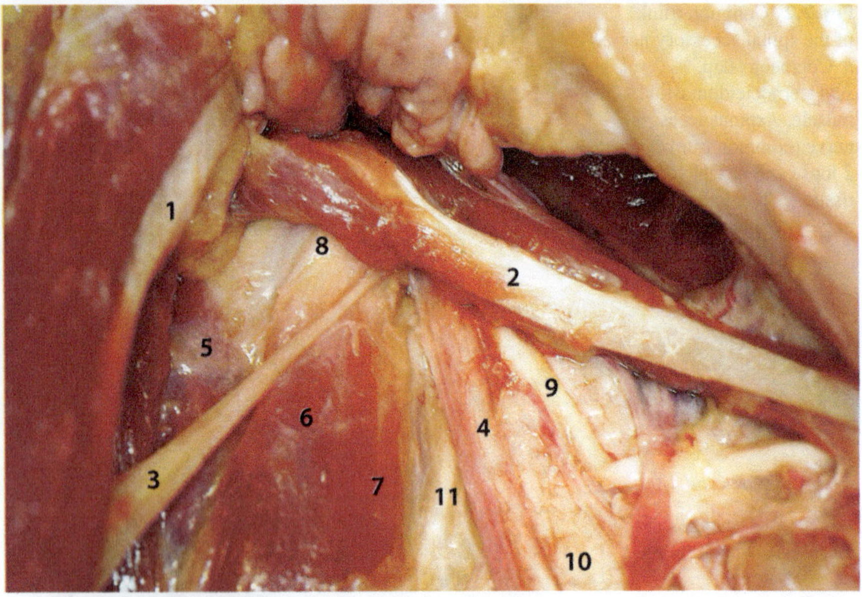

Fig. 7.3 Robbins level II. *1* Sternocleidomastoid muscle, *2* posterior belly of digastric muscle, *3* spinal accessory nerve (common trunk), *4* internal jugular vein, *5* splenius capitis muscle, *6* levator scapulae muscle, *7* anterior scalene muscle, *8* transverse process of atlas, *9* hypoglossal nerve, *10* carotid bifurcation, *11* branches of cervical plexus

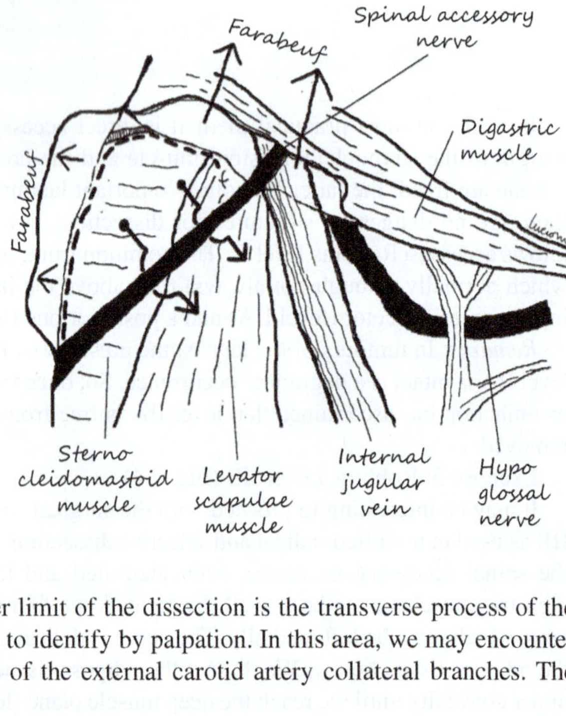

Fig. 7.4 Exercise 5: Robbins level IIB

capitis muscles). The upper limit of the dissection is the transverse process of the atlas, which is always easy to identify by palpation. In this area, we may encounter the occipital artery or one of the external carotid artery collateral branches. The adipose tissue is then elevated from the muscular plane from top to bottom and passed below the spinal nerve as a "bridge".

7.3 Sternocleidomastoid Muscle

The inferior insertions of the sternocleidomastoid muscle are now dissected, that is:

1. The head of the clavicle, corresponding to the deepest muscular portion, more wide and thin, which inserts on the medial quarter of the clavicle
2. The sternal head, corresponding to the most superficial and most consistent portion, which inserts on the anterior face of the sternal manubrium with a conoid tendon

The two components of the sternocleidomastoid muscle have different functions. The contraction of the head of the clavicle induces the flexion of the head onto the clavicle; the contraction of the sternal head induces the rotation and the contralateral extension of the head. The two muscle heads mark off a small triangular space with the base at the bottom (fossa supraclavicularis minor), which corresponds in depth to a length of the common carotid artery.

The sternocleidomastoid muscle is dissected and everted up to the exit point of the spinal accessory nerve, and the omohyoid muscle is entirely isolated (Fig. 7.5).

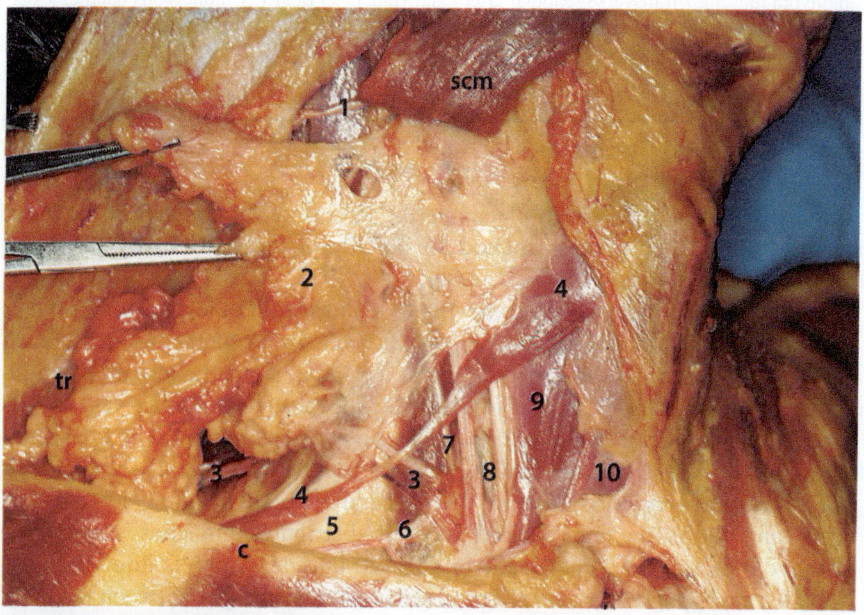

Fig. 7.5 Lateral deep plane. *scm* Sternocleidomastoid muscle, *tr* trapezius muscle, *c* clavicle, *1* lesser occipital nerve, *2* cervical plexus nerves (dissected), *3* transverse cervical artery and vein, *4* omohyoid muscle, *5* brachial plexus, *6* transverse scapular artery and vein, *7* phrenic nerve, *8* cervical vasculonervous bundle, *9* sternothyroid muscle, *10* sternohyoid muscle

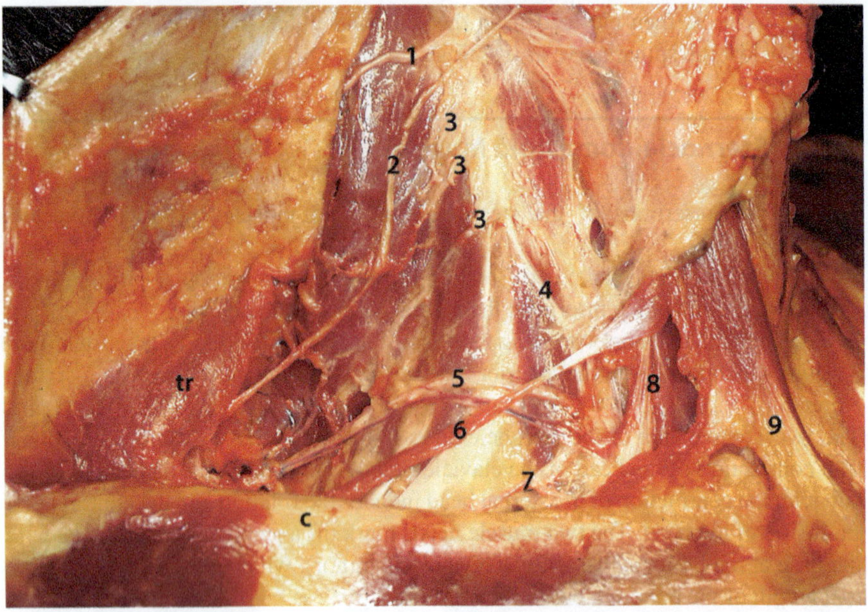

Fig. 7.6 Cervical plexus. *tr* Trapezius muscle, *c* clavicle, *1* lesser occipital nerve, *2* spinal accessory nerve, *3* cervical plexus nerve, *4* phrenic nerve, *5* transverse cervical artery and vein, *6* omohyoid muscle, *7* transverse scapular artery and vein, *8* cervical vasculonervous bundle, *9* sternal head of sternocleidomastoid muscle

7.4 Cervical Plexus

The dissection from the deep plane and the latero-medial shifting of the loose and fascial cellular tissues isolated so far allows the cervical plexus to be revealed.

It is composed of three anastomosing loops formed by the anterior branches of the upper four cervical nerves (C1–C4). It gives rise to cutaneous, superficial, and sensory nerves (lesser occipital, great auricular, supraclavicular, and cutaneous nerves of the neck) and to muscular, deep, and motor nerves (nerves serving the sternocleidomastoid and trapezius muscles, phrenic, and descending cervical nerves). The descending cervical nerve, lying laterally to the cervical vasculonervous bundle, joins the descending branch of the hypoglossal nerve and forms the hypoglossal loop. It distributes branches serving the infrahyoid muscles, except for the thyrohyoid muscle, which is directly innervated by the hypoglossal nerve. The cervical plexus emerges between the anterior and medial scalene muscles; the cutaneous branches are dissected while the phrenic nerve and descending cervical nerve are identified and preserved (Fig. 7.6).

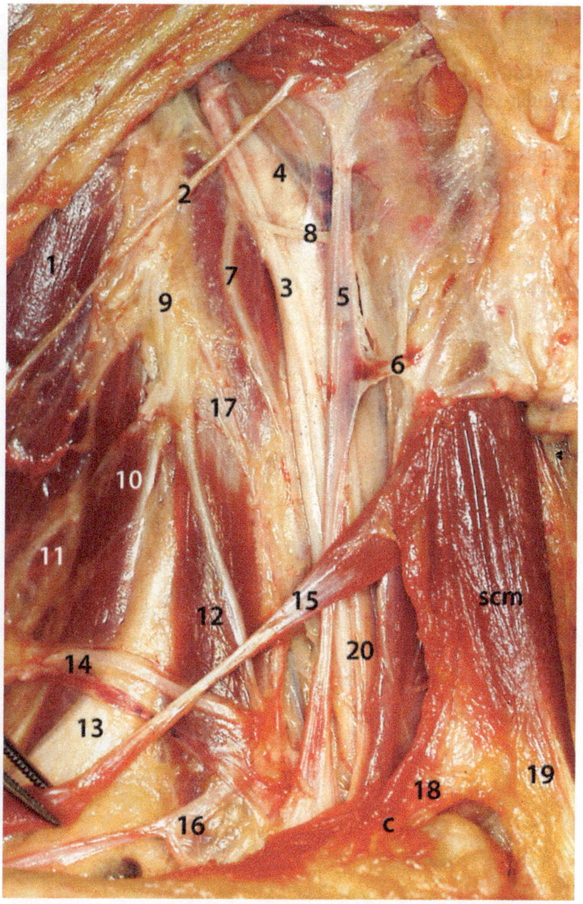

Fig. 7.7 Prescalene plane. *scm* Sternocleidomastoid muscle, *c* clavicle, *1* levator scapulae muscle, *2* spinal accessory nerve, *3* vagus nerve, *4* internal carotid artery, *5* internal jugular vein, *6* thyrolinguofacial trunk, *7* descending branch of hypoglossal nerve, *8* hypoglossal nerve, *9* cervical plexus, *10* medial scalene muscle, *11* posterior scalene muscle, *12* anterior scalene muscle, *13* brachial plexus, *14* transverse cervical artery and vein, *15* omohyoid muscle, *16* transverse scapular artery and vein, *17* descending cervical nerve, *18* clavicular head of sternocleidomastoid muscle, *19* sternal head of sternocleidomastoid muscle, *20* common carotid artery

7.5 Cervical Vasculonervous Bundle (Upper Part)

Dissection in a latero-medial direction takes us to the cervical vasculonervous bundle, formed laterally by the internal jugular vein, medially by the common carotid artery, and posteriorly, in the dihedral angle formed by the two vessels, by the vagus nerve. It should be remembered that the omohyoid muscle is a good guide in this phase, as it is invariably situated above the internal jugular vein. The only significant item on the surface of the omohyoid muscle is the external jugular vein.

The vessels are freed, as in routine surgical practice, from the tunica externa and consequently separated from the celluloadipose tissue and perijugular lymph nodes immerged therein. The specimen is pulled gently upwards, and the perivasal fasciae are cut with scissors. When working on the right side of the neck, right-handed operators proceed more easily from right to left (Fig. 7.7).

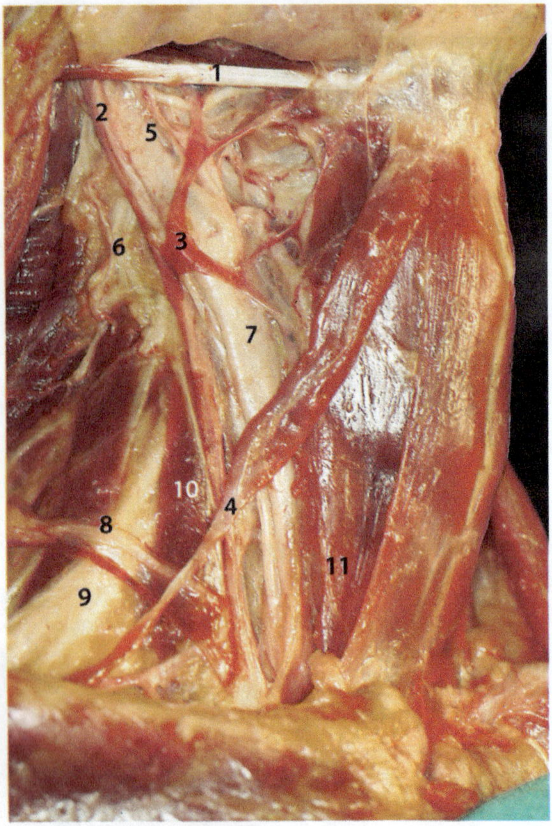

Fig. 7.8 Robbins levels II, III, and IV (I). *1* Posterior belly of digastric muscle, *2* internal jugular vein, *3* thyrolinguofacial trunk, *4* superior belly of omohyoid muscle, *5* internal carotid artery, *6* cervical plexus, *7* common carotid artery, *8* transverse cervical artery, *9* brachial plexus, *10* anterior scalene muscle, *11* sternothyroid muscle

As a rule, the internal jugular vein does not present lateral tributaries. Medially, we seek and isolate the main tributary, the thyrolinguofacial trunk, which enters it with a superior acute angle. Our exercise will isolate each of the major trunk components, that is, the superior thyroid, lingual, and facial veins.

The specimen, composed of adipose tissue and of the cervical fascia, generally contains numerous lymph nodes. The perijugular lymph nodes are the most important group in the cervical lymphatic system. Most superior respiratory–digestive tract lymphatics empty into them and lie subfascially between the sternocleidomastoid muscle and lateral surface of the internal jugular vein. They extend from the stylohyoid process to the point where the internal jugular empties into the brachiocephalic vein and are habitually divided by anatomists into three groups: (1) superior jugular (or jugular–digastric) lymph nodes, lying between the posterior belly of the digastric muscle and the thyrolinguofacial trunk; (2) middle jugular lymph nodes, lying between the thyrolinguofacial trunk and the superior belly of the omohyoid muscle; and (3) inferior jugular lymph nodes, situated below the omohyoid muscle. In the clinical practice, they are much better defined by the Robbins level system (Fig. 7.8).

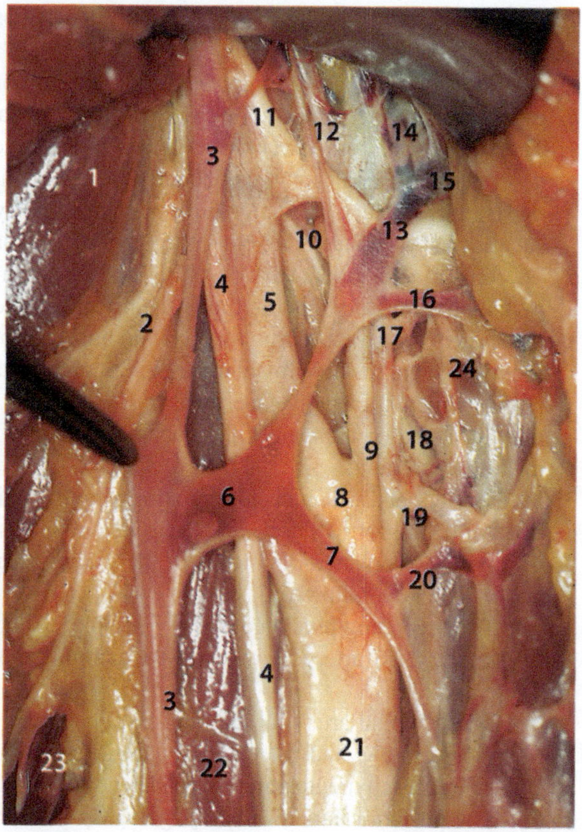

Fig. 7.9 Carotid bifurcation (I). *1* Levator scapulae muscle, *2* cervical plexus nerves, *3* internal jugular vein, *4* vagus nerve, *5* internal carotid artery, *6* thyrolinguofacial trunk, *7* superior thyroid vein, *8* carotid bifurcation, *9* external carotid artery, *10* superior laryngeal nerve, *11* hypoglossal nerve, *12* occipital artery, *13* facial vein (common trunk), *14* retromandibular vein, *15* facial vein, *16* lingual vein, *17* lingual artery, *18* cervical sympathetic chain, *19* superior thyroid artery, *20* superior laryngeal vein, *21* common carotid artery, *22* anterior scalene muscle, *23* medial scalene muscle, *24* greater cornu of hyoid bone

Dissection of the cervical vasculonervous bundle continues in a superior direction. From the sternoclavicular joint to the ipsilateral parotid cavity, it has variable relations with the sternocleidomastoid muscle, according to the position of the head. In the anatomic position, or with limited rotation, the carotid axis tends to protrude from the anterior margin of the sternocleidomastoid muscle and occupies the so-called carotid triangle.

At the superior margin of the thyroid cartilage, the common carotid artery divides into the internal and external carotid. The former runs laterally and is recognisable by the absence of collateral branches, while the latter runs medially. Higher up, the two vessels rotate and invert their position.

We may try to show the carotid glomus at the posterior wall of the carotid bifurcation. Its appearance is that of a reddish corpuscle as large as a grain of wheat, invested in a perivascular fibrous sheath, lying adjacent to the posterior surface of the bifurcation. It is innervated by the glossopharyngeal, cervical sympathetic, and vagus nerves. It functions as a chemoreceptor because it transmits variations in blood oxygen and carbon dioxide content to the nerve centres, thanks to extensive vascularisation and innervation (Fig. 7.9).

7.6 External Carotid Artery and Hypoglossal Nerve

Dissection proceeds by isolating, again from top to bottom and back to front, the external carotid artery and, one by one, its anterior collateral branches:

1. The superior thyroid artery, which arises immediately after the carotid artery, describes a descending curve of about 100° and medially branches into the superior laryngeal artery. The latter artery accompanied by the homonymous vein and superior laryngeal nerve together form the superior laryngeal pedicle.
2. The lingual artery, situated above the superior laryngeal nerve, is accompanied by the homonymous vein and runs superomedially, embedding itself just above the apex of the greater cornu of the hyoid bone and passing inferiorly to the posterior border of the hyoglossal muscle. Although parallel to the lingual artery, the lingual vein takes a more superficial course, running anteriorly to the hyoglossal muscle. It thus accompanies the hypoglossal nerve in the direction of the submandibular region.
3. The facial artery, which in close proximity to its origin, runs behind the posterior belly of the digastric muscle and heads towards the submandibular region.
4. The posterior collateral branches of the external carotid artery are also sought and their course followed:
   (a) The ascending pharyngeal artery, arising just above the origin of the superior thyroid artery; it rests on the middle constrictor of the pharynx and reascends.
   (b) The occipital artery, originating just below the junction between the hypoglossal nerve and the external carotid artery; it ascends laterally, passing behind the hypoglossal nerve and heading towards the mastoid.
   (c) The posterior auricular artery, going upwards beneath the external meatus acusticus (Fig. 7.10).

We now look for the hypoglossal nerve, which arches anteriorly, passing between the internal jugular vein and the two carotid arteries. At this point, it gives rise to a descending branch, which accompanies the anterior scalene muscle. This branch is anastomosed inferiorly with the descending cervical branch of the cervical plexus, forming the hypoglossal loop. On medially reaching the extreme posterior of the greater cornu of the hyoid bone, the hypoglossal nerve gives rise to a branch serving the thyrohyoid muscle, which passes anteriorly to the greater cornu of the hyoid bone.

Remarks: The hypoglossal nerve is the most important structure in this region from the functional point of view, of course excluding the internal carotid artery. Any surgical manoeuvre in this site should identify the hypoglossal nerve beforehand so as not to damage it.

Complications: The hypoglossal nerve conducts motor nerves to the tongue. Its paralysis deviates the protruded tongue towards the side of the lesion and the resting tongue towards the unimpaired side. Iatrogenic injury to the hypoglossal nerve is a very serious event in view of the importance of tongue movement in speech,

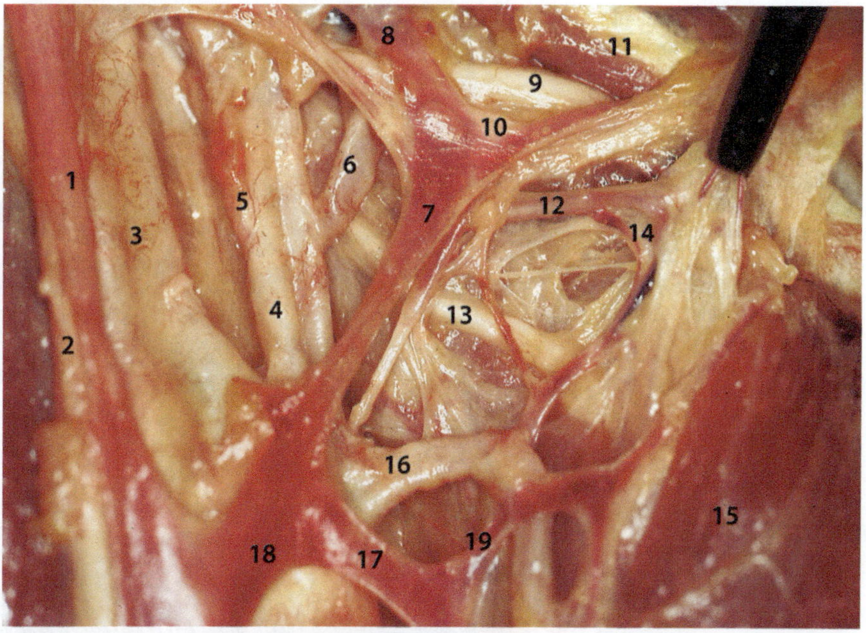

Fig. 7.10 Carotid bifurcation (II). *1* Internal jugular vein, *2* vagus nerve, *3* internal carotid artery, *4* external carotid artery, *5* ascending pharyngeal artery, *6* lingual artery, *7* common facial trunk, *8* retromandibular vein, *9* hypoglossal nerve, *10* facial vein, *11* hypoglossal muscle, *12* lingual vein, *13* superior laryngeal nerve, *14* greater cornu of hyoid bone, *15* sternohyoid muscle, *16* superior thyroid artery, *17* superior thyroid vein, *18* thyrolinguofacial trunk, *19* superior laryngeal vein

mastication, and deglutition; the lack of backward movement by the base of the tongue to project ingested food towards the hypopharynx causes dysphagia, which is more severe after partial supraglottic surgery. There is a risk of injuring the hypoglossal nerve particularly (1) in submandibular surgery, when periglandular scarring processes or excessive bleeding make it difficult to identify the nerve; (2) in functional laryngeal surgery, during isolation of the greater cornu of the hyoid bone, where dissection is not performed in very close contact to the bone; and (3) in functional supraglottic surgery, during suturing of the supra-infrahyoid muscles when the hyoid bone has been resected, as the nerve runs in the immediate vicinity.

The greater cornu of the hyoid bone, treated separately in view of its importance as a major landmark within this region, is above all a landmark in identifying the lingual artery and hypoglossal nerve. It is also an important landmark in seeking and ligating the superior laryngeal pedicle and, in surgical practice, in accessing the piriform recess. Finally, it is a precious indicator of the starting point for external carotid artery ligature (Fig. 7.11).

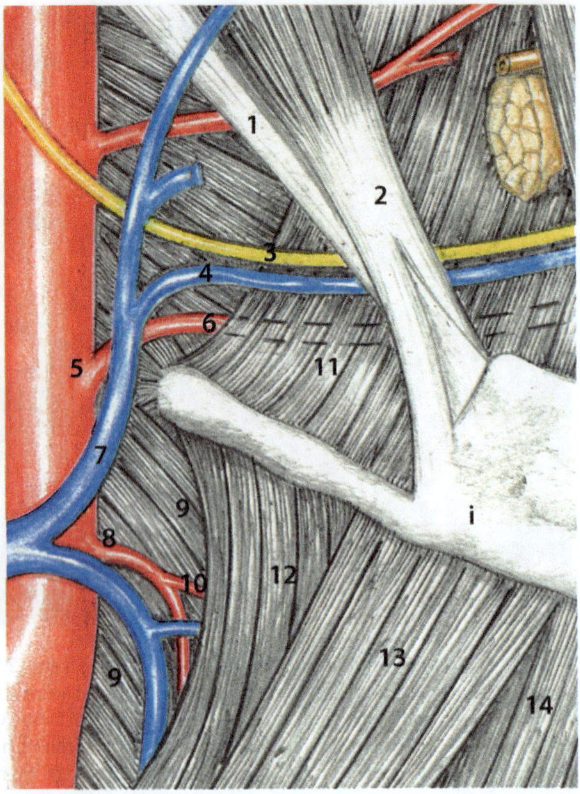

Fig. 7.11 Greater cornu of the hyoid bone. *i* Hyoid bone, *1* posterior belly of digastric muscle, *2* stylohyoid muscle, *3* hypoglossal nerve, *4* lingual vein, *5* external carotid artery, *6* lingual artery, *7* linguofacial trunk, *8* superior thyroid artery, *9* inferior constrictor muscle of pharynx, *10* superior laryngeal artery, *11* hyoglossal muscle, *12* thyrohyoid muscle, *13* omohyoid muscle, *14* sternohyoid muscle

7.7 Farabeuf's Triangle

The external carotid artery or its branches may be ligated to arrest otherwise uncontrollable cervicofacial haemorrhaging or for preventive purposes in ablative oncological surgery of the upper respiratory and digestive tracts. The entire external carotid system may be sacrificed when englobed by neoplasms or tumorous metastases with rupture of the lymph node capsule. Ligation or excision of one side of the external carotid does not lead to functional impairment, given the extensive anastomotic network with the contralateral arterial system.

Normally, the external carotid artery is bound between the origin of the superior thyroid artery and the origin of the lingual artery. One very precious landmark is the lateral extremity of the greater cornu of the hyoid bone, lying immediately medially. In order to identify the external carotid artery at its origin, anatomists describe a "triangular window" bounded by the medial wall of the internal jugular vein, the lateral wall of the thyrolinguofacial trunk, and, at the

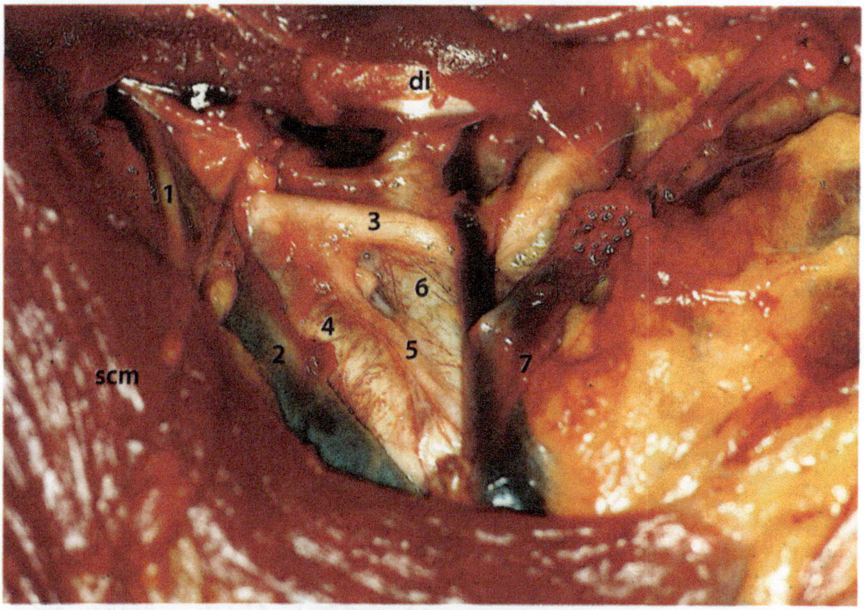

Fig. 7.12 Farabeuf's triangle. *scm* Sternocleidomastoid muscle, *di* digastric muscle, *1* spinal accessory nerve, *2* internal jugular vein, *3* hypoglossal nerve, *4* internal carotid artery, *5* descending branch of hypoglossal nerve, *6* external carotid artery, *7* thyrolinguofacial trunk

top, the hypoglossal nerve; this space is referred to as Farabeuf's triangle (Fig. 7.12).

It is certainly a useful exercise, to be carried out when the sternocleidomastoid muscle is still intact, to find and ligate the external carotid in Farabeuf's triangle. It is a typical emergency surgical procedure carried out in cases of haemorrhage of the pharynx or of the oral cavity that cannot be controlled in any other way.

The first thing is to find the anterior margin of the sternocleidomastoid muscle which, in its medial portion, will be isolated at top and bottom for about 10 cm. Following the muscle medially, we arrive at the internal jugular vein. This too must be well exposed. On the anterior margin of the vein, we look for the thyrolinguofacial venous trunk. At the top, transversely and under the vein, we look for and isolate the hypoglossal nerve. In this triangle, formed by the anterior margin of the internal jugular vein, the lateral margin of the thyrolinguofacial venous trunk, and the hypoglossal nerve, we look in depth for the carotid arteries. The external carotid artery is the one with collateral branches.

Exercise 6: Farabeuf's Triangle (Fig. 7.13)

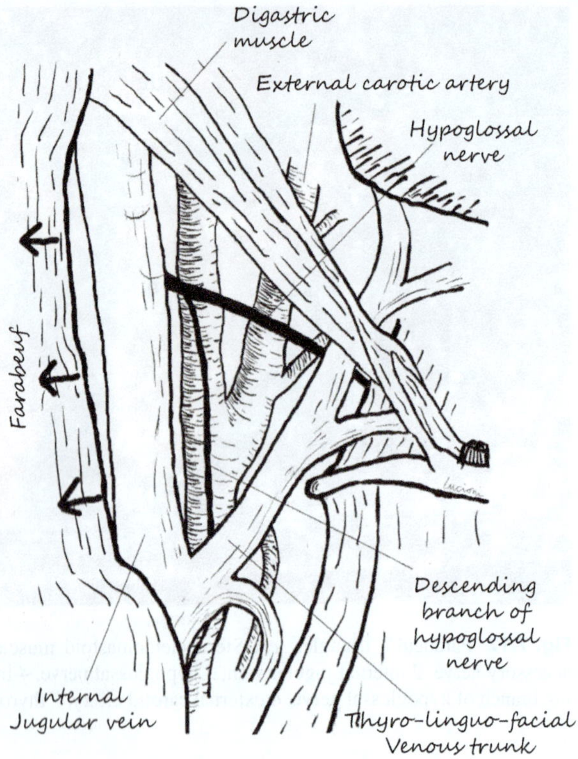

Fig. 7.13 Exercise 6: Farabeuf's triangle

7.8 Cervical Vasculonervous Bundle (Lower Part)

At this point, most of the sternocleidomastoid region is revealed. Medially, dissection reaches the subhyoid muscles and stops here because this is the medial limit of the region. Dissection continues in a downward direction, following the vasculonervous axis of the neck. We distinguish and skeletalise its three components. The fragility of the wall of the internal jugular vein, which is exposed with a subadventitious approach, can be clearly seen.

Complications: The limited laceration of the wall of the internal jugular vein may be recomposed with a vascular suture, followed by the application of gauzes soaked in very hot water, which induces the formation of a vascular thrombus. The complete interruption of an internal jugular vein normally occurs in radical neck dissection, and this does not involve severe functional consequences. The problem may arise when both internal jugular veins are infiltrated by metastases and therefore have to be sacrificed; in this case, the neck dissection operations are carried out at an interval of at least 1 month to allow time for the establishment of sufficient lymphatic drainage.

The dissection of the common carotid artery is extended down as far as the confluence with the subclavian artery. Its wall is robust and does not tear easily. It lies very close to the prevertebral plane, particularly the transverse processes of the cervical vertebrae. It may be easily compressed against these for temporary hemostatic purposes in the event of haemorrhage. In particular, the artery lies close to the transverse process of the sixth cervical vertebra, which is more prominent on palpation. It is located two transverse fingers above the junction between the carotid and inferior thyroid artery. This bony prominence takes the name of carotid tubercle and is the landmark for ligating the common carotid artery.

Complications: Internal carotid artery ligation is an exceptional clinical requirement. It is only required in the event of cataclysmic haemorrhage secondary to irreparable, spontaneous, or provoked wall rupture. Rupture of the common or of the internal carotid generally occurs in ablative neck operations in patients irradiated at maximum dosages or in neck cancer-related wall consumption. In these cases, ligation of the carotid axis is performed up- and downstream from the lesion. In the majority of cases, carotid englobement by neoplasia or tumorous metastases does not involve the arterial wall, and the vessel may be isolated subadventitiously. In cases where the tumour has infiltrated the carotid wall, preventing isolation, the arterial segment concerned must be excised and replaced by a vascular prosthesis or the reversed saphenous vein, autotransplanted after temporarily clamping the carotid axis. This surgical procedure, not devoid of danger from potential cerebral ischemia-related functional injury, does not produce long-term satisfactory oncological results.

Temporary or permanent interruption of blood flow in the internal carotid artery, in the absence of sufficient contralateral compensatory circulation through the arterial circle of the cerebrum, causes homolateral cerebral ischemia, with alteration in the state of consciousness and contralateral hemiparesis. Sometimes, common carotid artery ligation does not produce neurological deficits, because it is already appreciably excluded by arteriosclerotic plaques or by neoplastic thrombi (Fig. 7.14).

As has already been seen in the vasculonervous bundle, the internal jugular vein is laterally positioned and the common carotid artery medially. The vagus nerve lies in the dihedral angle behind the two vessels. It can be easily isolated from the posterior foramen lacerum (where it exits from the cranium together with the glossopharyngeal and accessory nerves) to its entrance to the thorax.

The vagus is a mixed nerve. It contains motor fibres (muscles of the velum palatinum, middle and inferior constrictors of the pharynx, muscles of the larynx, and cervical oesophagus), parasympathetic fibres (extensive splanchnic innervation: heart, respiratory, and digestive tracts, involuntary muscles, and glandular secretion), and sensory fibres (general sensitivity of part of the external auditory meatus, velum palatinum, pharynx, larynx, and trachea; chemopressor reflex arcs). Its most important cervical branch is the superior laryngeal nerve, which separates posteriorly very high up, accompanies the pharyngeal muscles, and running posteriorly to the carotid arteries, converges towards the larynx to form the superior laryngeal pedicle.

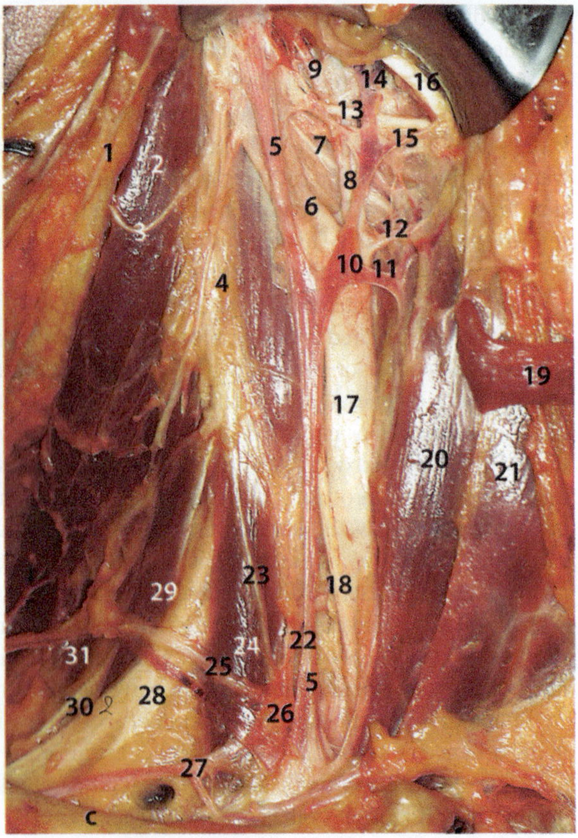

Fig. 7.14 Robbins levels II, III, and IV (II). *c* Clavicle, *1* spinal accessory nerve, *2* levator scapulae muscle, *3* lesser occipital nerve, *4* cervical plexus, *5* internal jugular vein, *6* internal carotid artery, *7* superior laryngeal nerve, *8* external carotid artery, *9* occipital artery, *10* thyrolinguofacial trunk, *11* superior thyroid vein, *12* superior thyroid artery, *13* hypoglossal nerve, *14* facial vein, *15* lingual vein, *16* intermediate tendon of digastric muscle, *17* common carotid artery, *18* vagus nerve, *19* omohyoid muscle, *20* sternothyroid muscle, *21* sternohyoid muscle, *22* inferior thyroid artery, *23* phrenic nerve, *24* anterior scalene muscle, *25* transverse cervical artery, *26* thyrocervical trunk, *27* transverse scapular artery, *28* brachial plexus, *29* anterior scalene muscle, *30* dorsal scapulae nerve, *31* posterior scalene muscle

The nerve filaments for the striated pharyngeal muscles are hard to isolate; together with the terminal branches of the glossopharyngeal nerve, they govern the deglutition mechanism and receive pharyngolaryngeal sensitivity.

Complications: Sectioning the vagus is fully compatible with life since numerous anastomoses between the two vagal hemi-systems permit any necessary compensatory action, thus avoiding the appearance of clinical symptoms, except obviously for paralysis of the hemilarynx and corresponding hemivelum palatinum or hemipharyngolaryngeal anaesthesia. Conversely, dissecting both vagus nerves is not compatible with life (Fig. 7.15).

7.9 Subclavian Artery Region

We now come to the lower portion of the sternocleidomastoid region where some important anatomic structures are identified and followed.

In the left laterocervical region, the thoracic duct is located in the laterally open dihedral angle formed by the internal jugular and subclavian veins. This is much

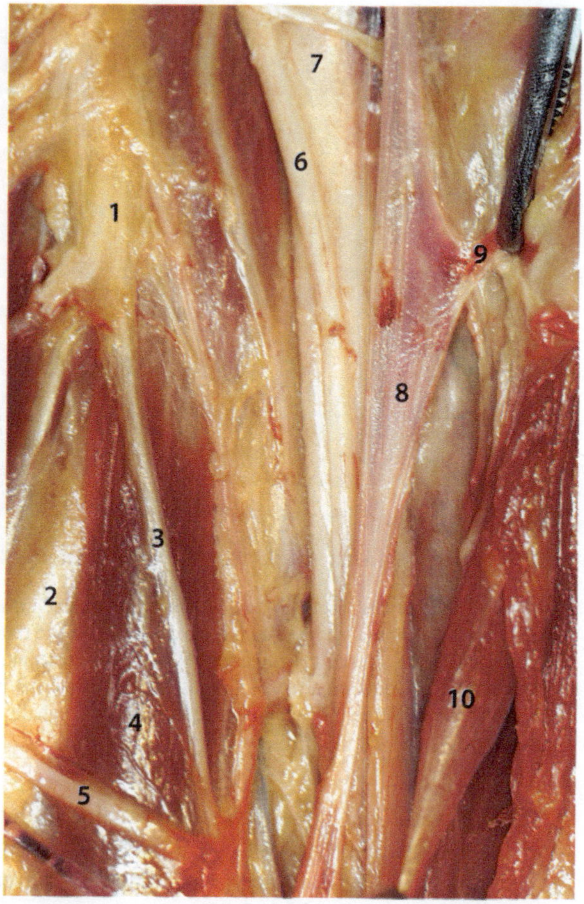

Fig. 7.15 Cervical vasculonervous bundle. *1* Cervical plexus, *2* brachial plexus, *3* phrenic nerve, *4* anterior scalene muscle, *5* transverse cervical artery, *6* vagus nerve, *7* common carotid artery, *8* internal jugular vein, *9* thyrolinguofacial trunk, *10* superior belly of omohyoid muscle

larger than the right great lymphatic vein since it collects lymph from the entire subdiaphragmatic area and from the left half of the supradiaphragmatic region. The duct posteriorly surrounds the subclavian vein and, making a 180° reverse turn in direction, empties into it.

Complications: Lymphorrhage may be favoured by anatomic anomalies (high outlet of the thoracic duct, up to 5 cm from the clavicle) or by surgical manoeuvres on metastases at level IV.

Usually, it is autolimited with compressive medications and gravity drainage. If it exceeds 600 ml per day and persists for more than a week, surgical revision is indicated to avoid general complications and granulations and scars in the surgical bed of neck dissection. The latter occurrence would pose problems for subsequent re-exploration [2].

In relation to the medial margin of the anterior scalene muscle, it is easy to find the thyrocervical trunk, which arises in the subclavian artery and branches out at this point into secondary arteries, namely:

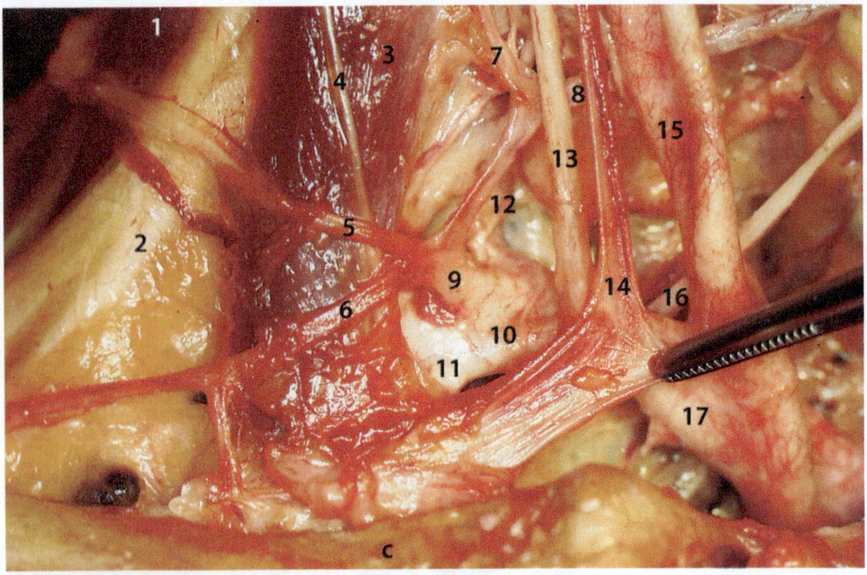

Fig. 7.16 Thyrocervical trunk. *c* Clavicle, *1* medial scalene muscle, *2* brachial plexus, *3* anterior scalene muscle, *4* phrenic nerve, *5* transverse cervical artery, *6* transverse scapular artery, *7* ascending cervical artery, *8* inferior thyroid artery, *9* thyrocervical trunk, *10* subclavian artery, *11* internal thoracic artery, *12* vertebral artery, *13* vagus nerve, *14* internal jugular vein, *15* common carotid artery, *16* recurrent nerve, *17* innominate artery (brachiocephalic trunk)

1. The transverse scapular artery, which becomes intrathoracic at the junction with the brachial plexus.
2. The transverse cervical artery, which laterally traverses the phrenic nerve, scalene muscles, and brachial plexus
3. The ascending cervical artery
4. The inferior thyroid artery, which arches medially, passing the common carotid artery posteriorly, and heads towards the recurrent region

Often, as appears in the anatomic specimen in the figure, the ascending cervical and inferior thyroid arteries have a common origin (Fig. 7.16).

We also consider that at this level, the largest lower branch of the subclavian artery is the internal thoracic artery (or internal mammary artery), which gives rise to the perforating branches that feed the deltopectoral reconstructive flap. The myocutaneous flap of the major pectoral is instead fed by the thoracoacromial artery, a branch of the axillary artery.

In the triangular space bounded by the clavicular and sternal head tendons of the sternocleidomastoid muscle, which anatomists refer to as the fossa supraclavicularis minor, the common carotid artery is separated from the skin solely by interposition of subcutaneous tissue, superficial cervical fascia, and middle cervical fascia.

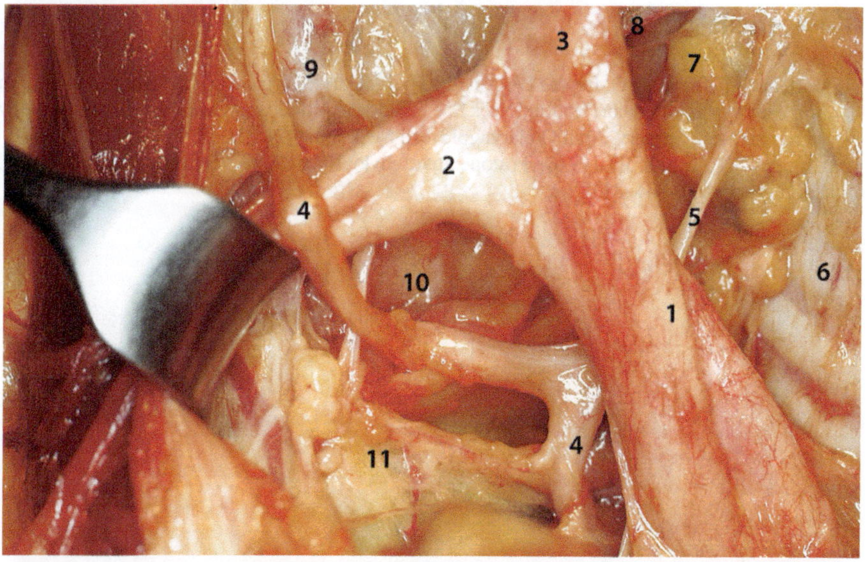

Fig. 7.17 Vagus nerve and recurrent nerve. *1* Innominate artery (brachiocephalic trunk), *2* subclavian artery, *3* common carotid artery, *4* vagus nerve, *5* recurrent nerve, *6* trachea, *7* recurrent region, *8* inferior thyroid artery, *9* middle cervical ganglion (cervical sympathetic chain), *10* stellate ganglion (sympathetic chain), *11* apex of the lung

We conclude the dissection of this region by assessing below the origins of the common carotid artery and of the subclavian artery from the anonymous artery. We observe the course of the vagus nerve, which passes the subclavian artery anteriorly (on the right and the aortic arch on the left). Last, we seek the origin of the inferior laryngeal nerve or recurrent nerve, which, passing behind the artery, reascends towards the larynx (Fig. 7.17).

Take-Home Messages

- In the dissection of the carotid axis above the bifurcation, the vessel encountered laterally is the internal carotid artery. One must always consider the possibility of anomalies of the arteries, known as "kinking", especially in elderly patients. Though they are rare, the failure to recognise them promptly in this site may be very dangerous.
- The ligation of the internal jugular vein must be tightened only after having ensured that the vagus nerve is outside the tie.
- The sternocleidomastoid muscle and the trapezius have a double innervation (C3 and C4 of the cervical plexus and spinal accessory nerve). This explains how shoulders without functional deficits may be observed after ascertained resections of the spinal accessory nerve.

References

1. Robbins KT, Clayman G, Levine PA et al (2002) Neck dissection classification update: revision proposed by the American Head and Neck Society and the American Academy of Otolaryngology Head and Neck Surgery. Arch Otolaryngol Head Neck Surg 128:751–758
2. Crumley RL, Smith JD (1976) Postoperative chylous fistula prevention and management. Laryngoscope 86:804–813

Anterior Region (Robbins Level VI: Inferior Part)

8

Core Messages

- In this chapter, we shall discuss above all the surgical anatomy of the thyroid. The essence of the exercise consists of removing the gland after having identified and followed the inferior laryngeal nerve (or recurrent nerve) with the intention of preserving it. The correct preparation of the area of operation and the precise knowledge of the landmarks must ensure that the finding of the nerve is absolutely not arrived at by chance.
- The cervical trachea will then be examined, and we shall make a few considerations on tracheotomies. The dissection of this region will conclude with the exploration of the large vessels at the base of the neck and of the cervical oesophagus.

8.1 Anatomic Layout

The anterior region which we shall explore in this chapter and in the following one corresponds to what anatomists call the anterior infrahyoid region, since the suprahyoid region, which we called submandibular and submental, has already been dealt with in a previous chapter.

It coincides approximately with Robbins level VI and has as its upper limit the hyoid bone and lower limit the medial end of the clavicles, the acromioclavicular articulation, and the suprasternal notch. Laterally it extends from the anterior margin of one sternocleidomastoid muscle to that of the contralateral muscle. Robbins classification specifies superficial limits, which are the lateral margins of the sternohyoid muscles, and the deep limits which are the common carotid arteries.

Robbins level VI is also called the central compartment of the neck, and its precise definition with reference to its lymph node content is not constant in the literature. Robbins original definition talks about pretracheal and paratracheal nodes, precricoid (Delphian) node, and the perithyroidal nodes including the lymph nodes

M. Lucioni, *Practical Guide to Neck Dissection*,
DOI 10.1007/978-3-642-33977-6_8, © Springer-Verlag Berlin Heidelberg 2013

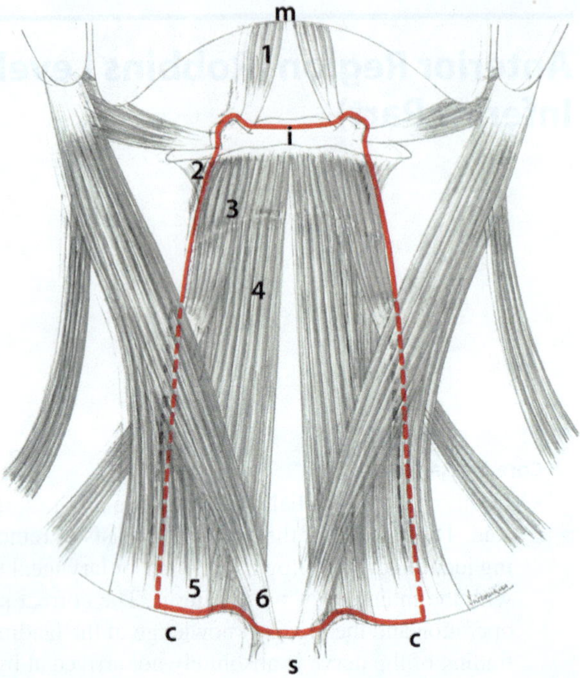

Fig. 8.1 Boundaries of the anterior region. *m* Mandible, *i* hyoid bone, *c* clavicle, *s* sternum, *1* anterior belly of digastric muscle, *2* thyrohyoid muscle, *3* omohyoid muscle, *4* sternohyoid muscle, *5* sternocleidomastoid muscle (clavicular head), *6* sternocleidomastoid muscle (sternal head)

along the recurrent laryngeal nerves [1]. In other articles, he only defines the paratracheal lymph nodes (PTLNs) as those contained in an anatomical space defined as follows: the lateral border is the medial edge of the common carotid artery, the medial border the trachea, and the cranial border the cricoid. The least well-defined border is the suprasternal notch, which is considered to be the caudal border, but the PTLN and the superior mediastinal lymph nodes are not well separated by an anatomic landmark [2]. To clarify the nomenclature and boundaries of the nodes in the "central compartment" of the neck, an important group of authors says that it is best to refer to these nodes by their anatomic location, that is, prelaryngeal, pretracheal, or paratracheal lymph node. It is also advisable to describe dissection (SND) with an annotation about the specific lymph node groups removed [3]. Finally, we have to consider the last Robbins upgrade (2008) where the Robbins level VII indicates the extension of the chain of paratracheal nodes below the suprasternal notch (the dividing line between levels VI and VII) to the level of the innominate artery [4].

In order to balance out the topic more evenly for teaching purposes, in our dissection, we have divided the median region into an *inferior part*, corresponding to the trachea, oesophagus, and thyroid gland, and a *superior part*, corresponding to the larynx and hypopharynx (Fig. 8.1).

Significant Anatomical Structures: central compartment, pretracheal and paratracheal nodes, perithyroidal nodes, paratracheal lymph nodes, Robbins level VII, infrahyoid linea alba, anterior jugular vein, Gruber's recess, middle cervical fascia, infrahyoid muscles, hypoglossal loop, thyroid gland, superior thyroid artery, cricothyroid artery, Lalouette's lobe, thyroglossal duct, thyropericardial lamina, ima thyroid artery, recurrent nerve, inferior thyroid artery, cricothyroid joint, parathyroid glands, Berry–Gruber's ligament, subclavian artery, vagus nerve, vertebral artery, innominate artery, tracheostomy, Bjork's flap, cervical oesophagus, trachea, pars membranacea of the trachea, tracheal muscle

Landmarks: Lorè's triangle, Killian's mouth, carotid tubercle

8.2 Fascial Plane and Infrahyoid Muscles

First of all, we identify the main landmarks of this region, that is, the body of the hyoid bone and its greater cornu, the laryngeal prominence, the cricoid ring and the cricothyroid space, and finally the jugulum (Fig. 8.2).

Dissection begins latero-medially by elevating the superficial and middle fasciae of the infrahyoid muscle plane (Fig. 8.3).

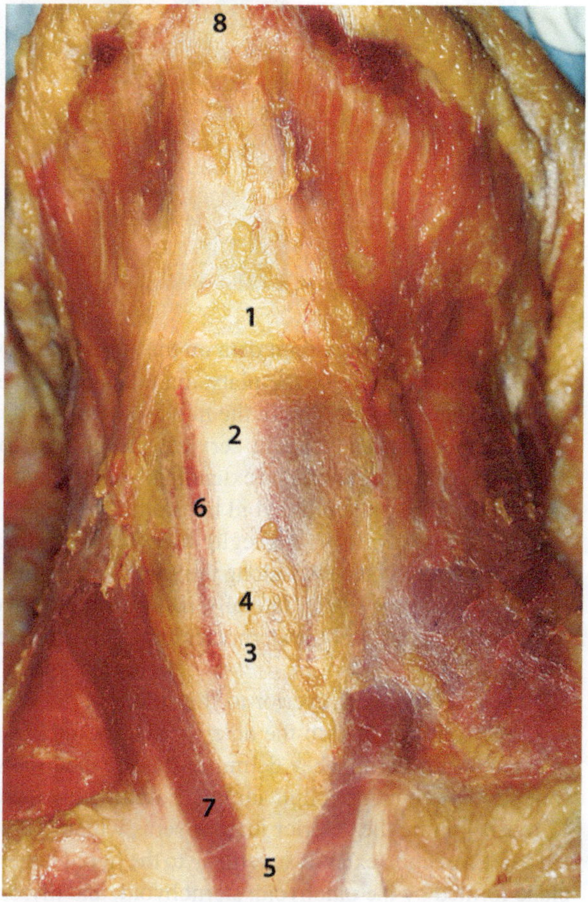

Fig. 8.2 Anterior region: orientation. *1* Body of hyoid bone, *2* laryngeal prominence, *3* cricoid ring, *4* intercricothyroid space, *5* jugular notch, *6* anterior jugular vein, *7* sternocleidomastoid muscle (sternal head), *8* mental prominence

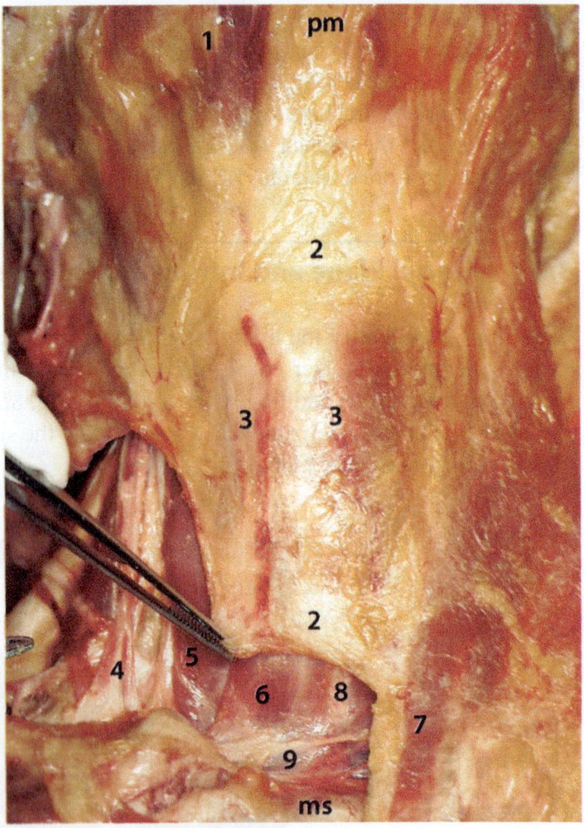

Fig. 8.3 Superficial fascial plane. *pm* Mental prominence, *ms* manubrium sterni, *1* platysma muscle, *2* superficial cervical fascia, *3* anterior jugular vein, *4* internal jugular vein, *5* sternothyroid muscle, *6* sternohyoid muscle, *7* sternocleidomastoid muscle (sternal head), *8* infrahyoid linea alba, *9* Gruber's recess

Below are some important data on the superficial fascial plane:

1. The medial margin of the platysma takes a divergent downward course and is consequently not present in the medio-inferior part of the region.
2. The superficial and middle cervical fasciae fuse on the midline into a single aponeurosis, a sort of raphe extending from the hyoid bone to the sternum, which is referred to as the infrahyoid linea alba.
3. The superficial vessels are negligible, except for the anterior jugular veins, which run vertically to the neck along the paramedian line. At approximately 2 cm from the sternum, they bend laterally and become embedded, passing posteriorly to the sternal tendon of the sternocleidomastoid muscle and empty into the brachiocephalic veins.
4. A few centimetres superior to the sternum, the cervical fascia divides into two sheets, one directed to the anterior and the other to the posterior border of the manubrium sterni. They delimit a space called the suprasternal space (Gruber's recess): it contains celluloadipose tissue with a few lymph nodes and an anastomosis joining the anterior jugular veins which cross it.

Fascia resection extends superiorly to the hyoid bone, thereby exposing the muscle plane formed by the omohyoid, sternohyoid, and thyrohyoid muscles (Fig. 8.4).

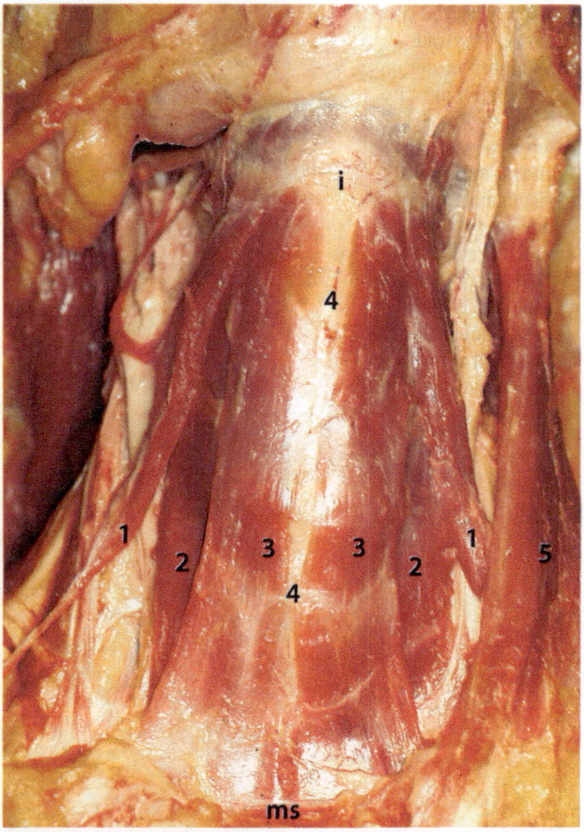

Fig. 8.4 Infrahyoid muscles plane. *i* Hyoid bone, *ms* manubrium sterni, *1* omohyoid muscle, *2* sternothyroid muscle, *3* sternohyoid muscle, *4* infrahyoid linea alba, *5* sternocleidomastoid muscle

We can see that the middle cervical fascia extends laterally from one omohyoid muscle to the other and that the sternothyroid muscle laterally overlaps more than the overlying sternohyoid muscle.

The infrahyoid muscles are then sectioned at sternoclavicular level and raised from the thyroid gland and cricoid and thyroid cartilages, applying cranial traction. The sternohyoid muscles are elevated up to the hyoid bone, and the sternothyroid muscles up to the line of attachment to the thyroid lamina. The innervation of these muscles derives from the hypoglossal loop, with the exception of the thyroid muscle which is directly innervated by a branch of the hypoglossal nerve. At the end of this manoeuvre, the thyroid gland is well revealed (Fig. 8.5).

8.3 Thyroid Gland

The next step is to examine and dissect the thyroid gland and parathyroid glands.

The thyroid is an endocrine gland lying medially to the base of the neck, whose front view has an open H shape and cross section a horseshoe shape, enclosing the cervical trachea in its concavity and the larynx and oesophagus laterally. It is

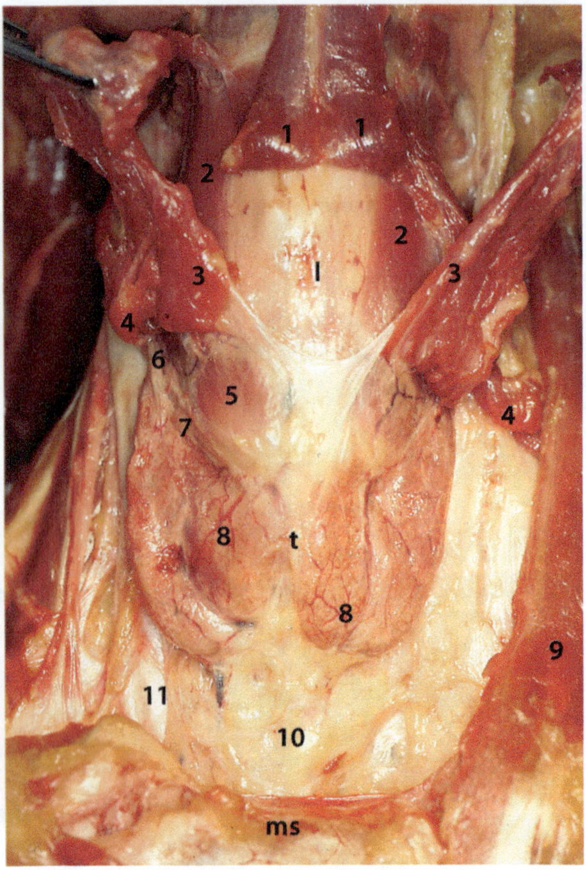

Fig. 8.5 Thyroid (I). *l* Larynx, *t* thyroid, *ms* manubrium sterni, *1* sternohyoid muscle, *2* thyrohyoid muscle, *3* sternothyroid muscle, *4* omohyoid muscle, *5* cricothyroid muscle, *6* superior thyroid artery, *7* medial branch of superior thyroid artery, *8* thyroid capsule vessel, *9* left sternocleidomastoid muscle, *10* pretracheal region, *11* common carotid artery

invested by a slender, fibrous perithyroid sheath which proceeds laterally along the pedicles and attaches to the cervical vasculonervous bundle. This covering is part of the vascular sheath and is independent of the superficial and middle cervical fascia [4]. Lying below the sheath is the thyroid capsule, which is an integral part of the parenchyma enclosing the gland's superficial vessels (Fig. 8.6).

As in clinical practice, the gland is dissected after identifying and ligating the superior vascular pedicles. The superior thyroid artery (and vein), an upper branch of the external carotid artery, initially runs horizontally, parallel to the greater cornu of the hyoid bone, then descends towards the homolateral thyroid lobe; medially it gives rise to the superior laryngeal artery and then divides into three branches: one medial, which is the largest and runs along the superior thyroid margin; one posterior; and one lateral, from which the cricothyroid artery arises and takes a medial course, perforating the homonymous membrane (Fig. 8.7).

Complications: In thyroid surgery, the superior thyroid pedicle must be ligated downstream from the laryngeal artery origin and, above all, should not involve the external branch of the superior laryngeal nerve. Once the upper pedicle has been ligated, we must avoid proceeding downwards with the elevation of the thyroid from the larynx because we would arrive immediately in the vicinity of the recurrent nerve just where it enters the larynx.

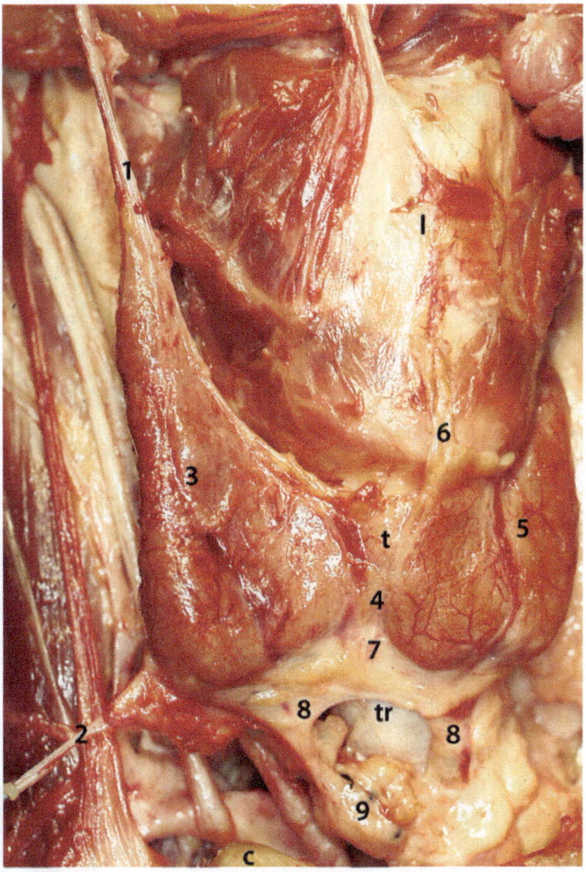

Fig. 8.6 Thyroid (II). *l* Larynx, *t* thyroid gland, *tr* trachea, *c* clavicle, *1* superior thyroid artery, *2* inferior thyroid artery, *3* right thyroid lobe, *4* isthmus of the thyroid gland, *5* left thyroid lobe, *6* pyramidal thyroid lobe (Lalouette's lobe), *7* ima thyroid artery, *8* inferior thyroid artery, *9* pretracheal lymph nodes

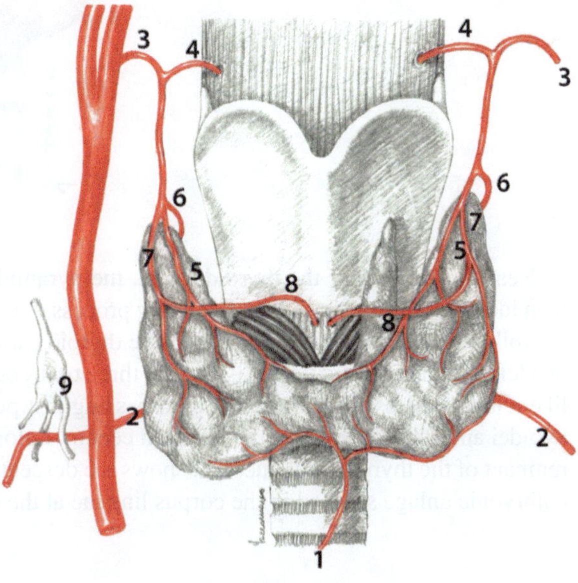

Fig. 8.7 Thyroid vascular pedicles. *1* Ima thyroid artery, *2* inferior thyroid artery, *3* superior thyroid artery, *4* superior laryngeal artery, *5* superior thyroid artery (medial branch), *6* superior thyroid artery (posterior branch), *7* superior thyroid artery (lateral branch), *8* cricothyroid artery, *9* middle cervical ganglion (sympathetic cervical chain)

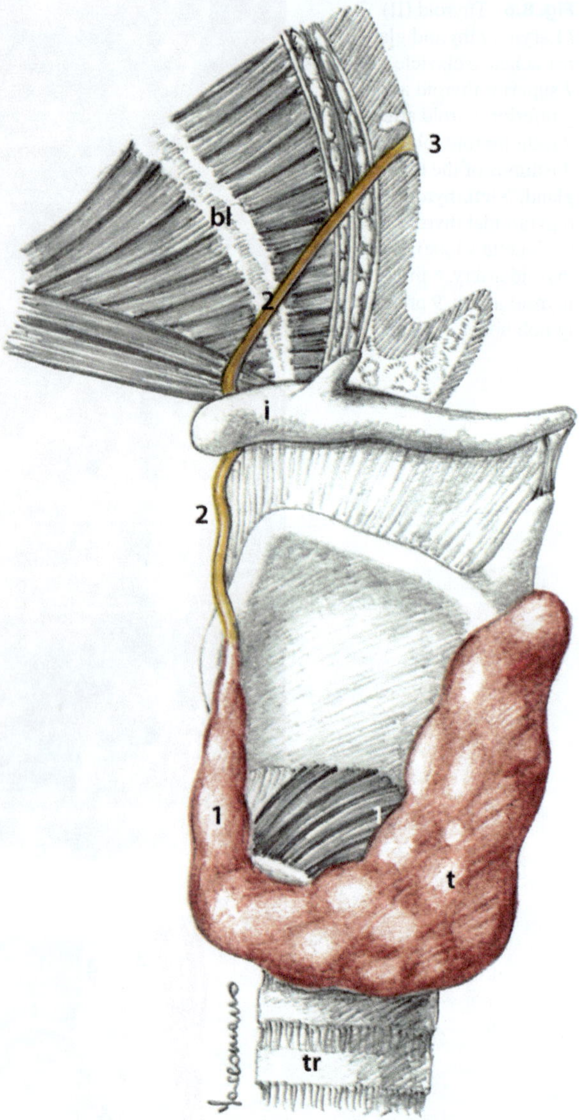

Fig. 8.8 Thyroglossal duct and Lalouette's lobe. *bl* Tongue base, *i* hyoid bone, *t* thyroid gland, *tr* trachea, *1* Lalouette's lobe, *2* thyroglossal duct, *3* foramen cecum

Near the isthmus of the thyroid gland, the pyramidal lobe (Lalouette's lobe) is then identified. It consists of an ascending process of the thyroid parenchyma. It has the following characteristics: it saddles the thyroid cartilage of the larynx, generally in a left paramedian position; it is present three times out of four; it extends upwards like a more or less evident fibrous cord passing just posteriorly to the corpus ossis hyoidei and ascends towards the foramen cecum. Lalouette's lobe is the embryonic remnant of the thyroglossal duct that shows the descent of the thyroid gland from its embryonic anlage situated in the corpus linguae at the base of the neck (Fig. 8.8).

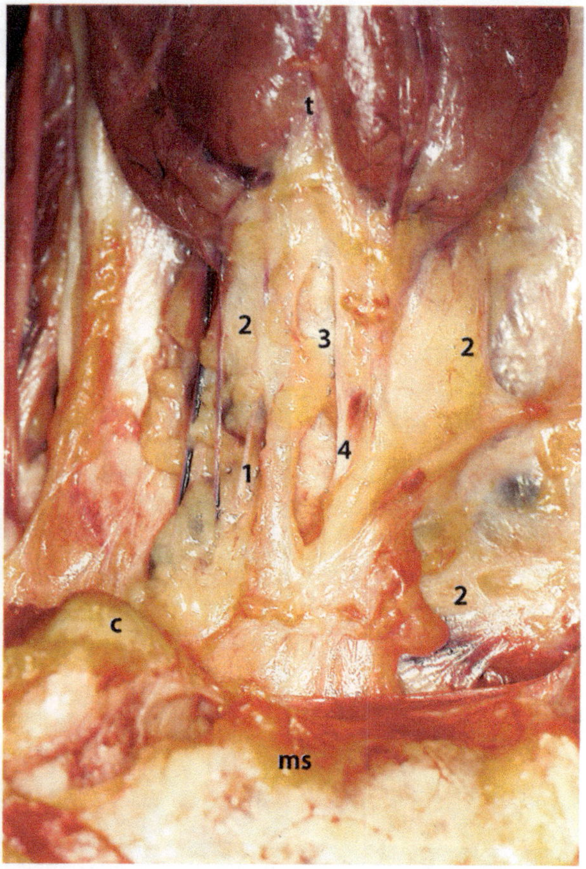

Fig. 8.9 Pretracheal area. *t* Thyroid gland, *c* clavicle, *ms* manubrium sterni, *1* inferior thyroid veins, *2* thyropericardial lamina, *3* trachea, *4* ima thyroid artery

Remarks: Cysts and median fistulae of the neck develop along the path of the thyroglossal duct, like "aberrant" thyroids or accessory thyroids. Their removal requires the complete exeresis of these structures and, to avoid recurrences, of the median portion of the hyoid bone with which the thyroglossal duct establishes close relations.

Before beginning to look for the recurrent nerves, we free the anterior surface of the trachea. The thyroid gland/cervical trachea complex needs to be stretched as far as possible cranially in order to expose an extensive tract of the trachea (Fig. 8.9).

The subthyroid pretracheal space is occupied by the so-called thyropericardial lamina, which is sectioned to expose the anterior trachea wall. We section the tissue which is on a more superficial plane than the anterior surface of the trachea, that is, we avoid going any deeper laterally because in doing so we would risk encountering the recurrent nerves (Fig. 8.10).

The middle cervical fascia is attached superiorly to the hyoid bone and laterally to the omohyoid muscles. Inferiorly, it adheres to the osteofibrous contour of the superior opening of the thoracic cavity (sternum, clavicle, and upper ribs). Inferiorly, the fascia continues downwards with more or less consistent thickness associated with the large vessels of the mediastinum and pericardial serosa. This median fascial

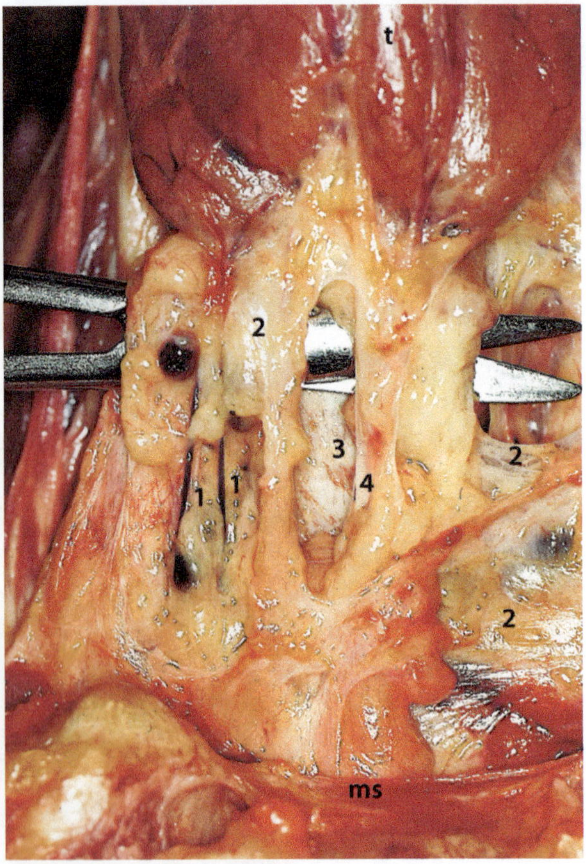

Fig. 8.10 Thyropericardial lamina. *t* Thyroid gland, *ms* manubrium sterni, *1* inferior thyroid veins, *2* thyropericardial lamina, *3* trachea, *4* ima thyroid artery

structure takes the name of thyropericardial lamina and encloses the following: (1) the inferior thyroid venous plexus; (2) the ima thyroid artery, which arises directly from the innominate artery or aortic arch, with inconstant presence and calibre; and (3) the pretracheal lymph nodes. On exposure, proceeding craniocaudally, the trachea can be seen increasingly embedding below the cutaneous plane.

Complications: Perfect familiarity with this anatomic site is essential to ensure a risk-free subthyroid tracheotomy. In some cases, the inferior thyroid nerves may be rather large and numerous. The accidental interruption and downward loss of a sectioned inferior thyroid vein, which naturally tends to retract into the mediastinic adipose tissue and to bleed, may become a serious problem.

8.4 Recurrent Region

At this point, we can turn our attention to the recurrent nerves. The inferior laryngeal nerve, or recurrent nerve, originates in the first intrathoracic tract of the vagus nerve: it arises more cranially to the right than to the left and immediately encloses the subclavian artery anteroposteriorly and inferosuperiorly. To the left, it takes a

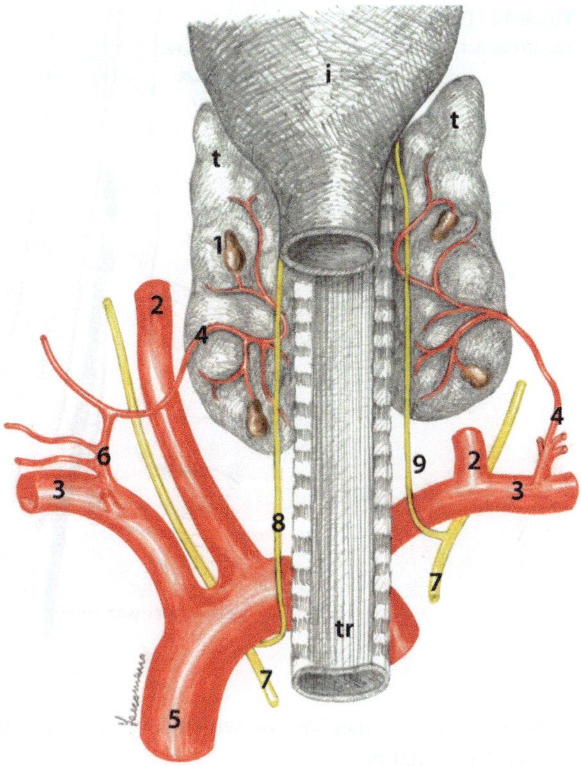

Fig. 8.11 Recurrent nerves. *i* Hypopharynx, *t* thyroid gland, *tr* trachea, *1* parathyroid gland, *2* common carotid artery, *3* subclavian artery, *4* inferior thyroid artery, *5* aortic arch, *6* thyrocervical trunk, *7* vagus nerve, *8* left recurrent nerve, *9* right recurrent nerve

similar course, enclosing the aortic arch. The recurrent nerves reascend, running through the dihedral angle between trachea and oesophagus, with slight asymmetry insofar as the oesophagus protrudes further to the left than the trachea. In this tract, it gives rise to numerous collateral branches (middle cardiac branches serving the cardiac plexus and pharyngeal branches serving the pharyngeal plexus, in addition to tracheal and oesophageal branches). It penetrates the larynx behind the articulation between the inferior cornu of the thyroid cartilage and the cricoid ring. The recurrent nerve is a mixed nerve: it innervates all intrinsic laryngeal muscles, except for the cricothyroid muscle, which is innervated by the superior laryngeal nerve; sensory fibres innervate the mucosa of the inferior aspect of the vocal folds, the subglottic region, and the upper tracheal rings (Fig. 8.11).

Complications: Thyroid and tracheal surgery present the surgeon with the risk of recurrent nerve injury. Such lesions are generally manifested by vocal fold fixity in a paramedian or intermediate position. If the lesion is not bilateral (in which case tracheotomy is often required with subsequent surgery to extend the glottis), the main symptom is dysphonia owing to incomplete glottal closure. When the lesion is incomplete, because, for example, the nerve has been excessively stretched, the paralysis may regress spontaneously.

Where, instead, paralysis persists, the voice may spontaneously improve through compensation by the healthy voice fold, which exceeds the midline during phonation.

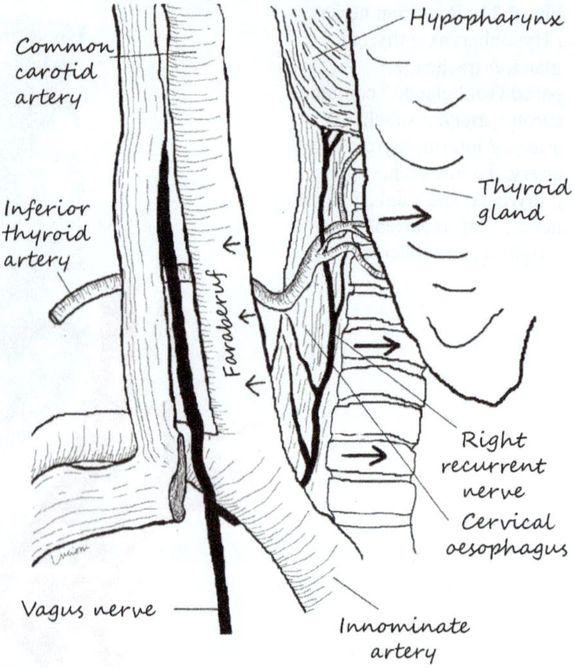

Fig. 8.12 Exercise 7: recurrent nerve

This compensatory mechanism, which develops over a period of months, is helped by speech rehabilitation.

First, we look for the inferior thyroid artery. It arises from the thyrocervical trunk and enters the recurrent region, passing posteriorly to the common carotid artery. Its relations with the recurrent nerve are important for the surgeon who, on ligating the inferior thyroid pedicle during thyroidectomy, should be careful not to impair the nerve. Unfortunately, relations between the two structures are variable: the artery is often already divided when it crosses the nerve, which may run between its branches. The right recurrent nerve more commonly runs anteriorly to the artery and the left one posteriorly. In routine surgical practice, ligation of the inferior thyroid pedicle should only be performed after definitely identifying and isolating the homolateral recurrent nerve.

Complications: In vivo, the inferior thyroid artery must be ligated with particular attention. It is a vessel of considerable calibre, and if its ligature comes undone, considerable difficulties may arise in recovering the interrupted and bleeding vessel. It is useful to remember that it enters our field of operation by passing posteriorly to the common carotid artery; this should avoid serious trouble in surgical movements which in these cases are often agitated.

Exercise 7: Recurrent Nerve (Fig. 8.12)

The search for and isolation of the inferior laryngeal nerve (recurrent nerve) is the focal point of this dissection exercise. To be successful, we must prepare the field of operation precisely.

First of all, we must apply traction medially on the thyroid lobe and identify, farther down, the hypopharynx and the cervical oesophagus. Laterally we seek the common carotid artery, which is lateralised with a Farabeuf. Deep down, by palpation, we can identify the prevertebral plane.

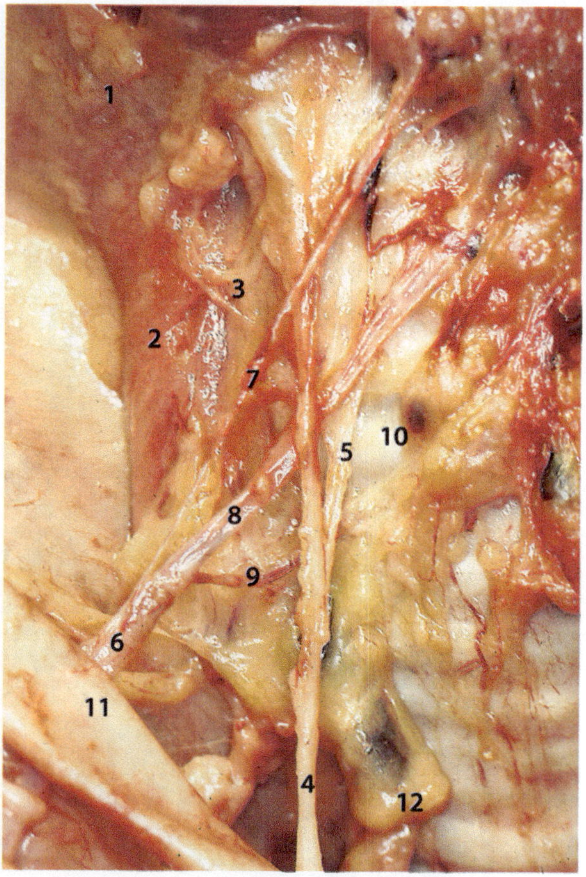

Fig. 8.13 Recurrent region. *1* Hypopharynx, *2* cervical oesophagus, *3* recurrent nerve (oesophagus branches), *4* recurrent nerve, *5* recurrent nerve (tracheal branches), *6* inferior thyroid artery, *7* inferior thyroid artery (superior branch), *8* inferior thyroid artery (inferior branch), *9* inferior thyroid artery (tracheal branches), *10* trachea, *11* common carotid artery, *12* recurrent lymph nodes

We then seek tangentially in the triangle between the trachea (medially), the common carotid artery (laterally) and the inferior thyroid artery (superiorly) whose name is Lorè's triangle. This is the recurrent region in which the recurrent nerve and inferior thyroid artery are found, embedded in the celluloadipose connective tissue, crossing each other at right angles. To seek the nerve, we divaricate the adipose tissue with scissors in the dihedral angle between the oesophagus and the trachea, proceeding craniocaudally. Once it has been found, it must be isolated and followed until it enters the larynx, posterior to the cricothyroid joint. In this region, we can also find some lymph nodes of the recurrent chain, which form the lymphatic drainage of the thyroid gland, of the hypoglossal region and of the cervical trachea. Lastly, we shall try to identify the parathyroid glands.

Complications: If it is difficult to identify the right recurrent nerve, we must also consider the possibility of a "nonrecurrent" recurrent nerve (0.5–1 % of cases) [5]. That means that, due to a congenital anomaly of the right subclavian artery, the right nerve starts directly from the vagus nerve next to the thyroid gland.

In the point where the recurrent nerve and the inferior thyroid artery cross, we can try to identify some of the parathyroid glands, of which there are generally four. The inferior ones are generally more voluminous; the dimensions are about the size of a lemon seed, and they are brown in colour (Fig. 8.13).

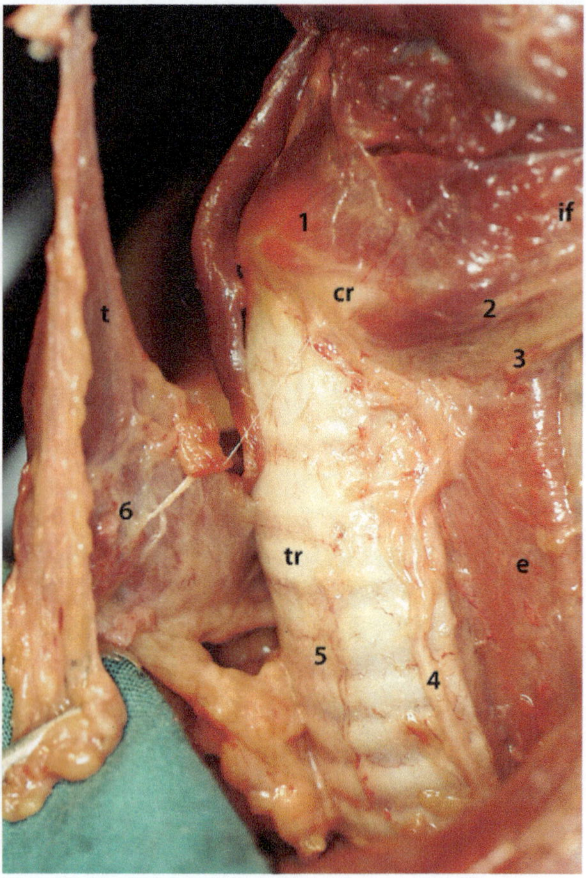

Fig. 8.14 Oesophagotracheal angle. *if* Hypopharynx, *cr* cricoid cartilage, *tr* trachea, *e* oesophagus, *t* thyroid, *1* cricothyroid muscle, *2* cricopharyngeal muscle, *3* Killian's mouth, *4* recurrent nerve, *5* tracheal vascular arches, *6* parathyroid gland

Complications: The removal of the thyroid gland must normally be performed preserving both the recurrent nerve and the parathyroid glands, which control the calcium and phosphorus metabolism through the parathyroid hormone. Their removal leads to tetany, and the replacement therapy must associate calcium and vitamin D. It is believed that their number can be halved without causing imbalance due to calcaemia. The correct procedure is to identify them and preserve them together with the actual vascular pedicle. If removed accidentally, they may be reimplanted in a niche in the sternocleidomastoid muscle, after having been finely chopped on a slide with a scalpel.

The completely isolated thyroid gland is now removed, after having sectioned the lateral Berry–Gruber's ligament, extending between the thyroid capsule and the cricoid perichondrium, and the residual pedicles. We follow the upward course of the recurrent nerve, checking in particular the point of embedment behind the crico-thyroid joint (Fig. 8.14).

To conclude the dissection of the thyroid gland, we recall that there are two methods of thyroidectomy, at least in the benign pathology, which differ in whether

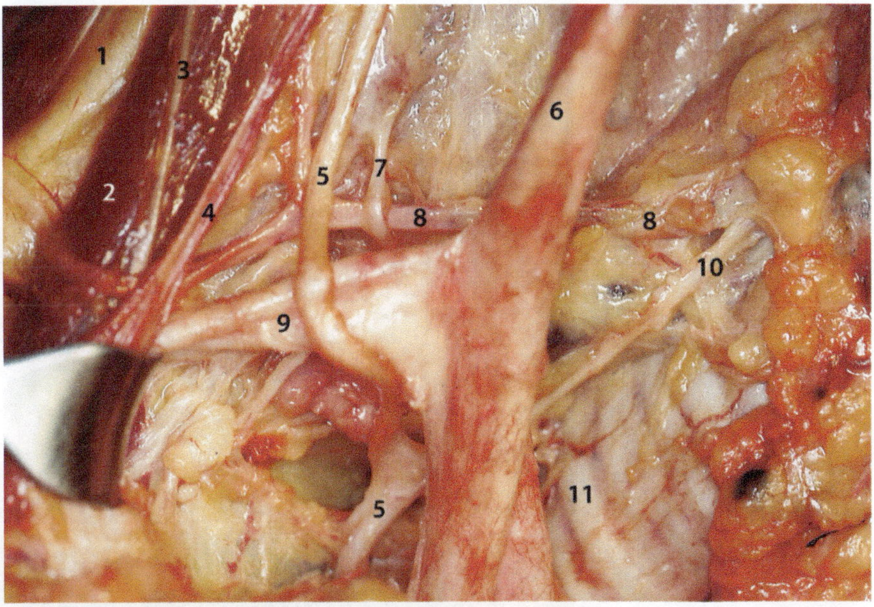

Fig. 8.15 Inferior perivisceral area. *1* Brachial plexus, *2* anterior scalene muscle, *3* phrenic nerve, *4* internal jugular vein, *5* vagus nerve, *6* common carotid artery, *7* sympathetic cervical chain, *8* inferior thyroid artery, *9* subclavian artery, *10* recurrent nerve, *11* trachea

or not the recurrent nerve is identified beforehand. The most recent case histories say that the percentage of paralysis is considerably lower where the recurrent nerve is sought, identified, and preserved.

8.5 Base of the Neck

Isolation of the major arteries of the base of the neck starts from the bottom, with exposure concentrating in particular on the common carotid artery in relation to the superior opening of the thorax. Following the common carotid caudal, the subclavian artery can be reached from the right and isolated. The first thing to control is the course of the vagus nerve, already observed in the dissection of the sternocleidomastoid region, which passes anteriorly to the artery and gives rise to the recurrent nerve in proximity to its inferior border; the recurrent nerve ascends posteriorly to the vessel towards the junction with the inferior thyroid artery, where it was isolated beforehand. The arterial branches of the subclavian artery, particularly the thyrocervical trunk, lying just medially to the anterior scalene muscle, are then isolated. The transverse cervical, transverse scapular, and ascending cervical and inferior thyroid arteries all arise from this main branch of the subclavian, and the latter two often have a common origin (Fig. 8.15).

The origin of the vertebral artery, which ascends medially (re-emerging in the prevertebral region), is sought at roughly the same level, on the posterosuperior

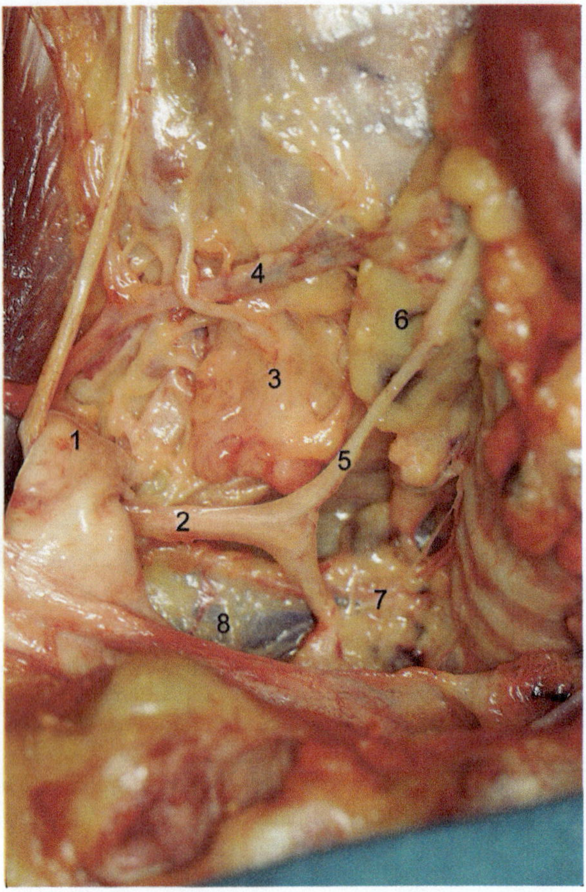

Fig. 8.16 Recurrent region and upper mediastinum. *1* Subclavian artery, *2* vagus nerve, *3* inferior cervical ganglion and first thoracic ganglion (stellate ganglion), *4* inferior thyroid artery, *5* recurrent nerve, *6* recurrent space, *7* upper mediastinum, *8* apex of the lung

border of the subclavian artery. Just after its origin, it accompanies the vertebral vein, which descends and passes anteriorly to the subclavian artery.

The internal thoracic artery, instead, arises from the inferior margin of the subclavian. The subclavian artery then embeds itself, passing posteriorly to the anterior scalene muscle and inferiorly to the brachial plexus, overstriding the first rib. The lateral portion has already been examined in supraclavicular dissection.

The subclavian artery and right common carotid artery arise from the brachiocephalic trunk or innominate artery; on the left, the origins of the subclavian and common carotid arteries are instead separate from the arch of the aorta. The adipose and fascial connective tissue enclosing the great paratracheal vessels continue with the mediastinal cellular tissue (upper mediastinum). It abounds in lymph nodes, some anthracotic, in continuity with the overlying recurrent lymph node chains (Fig. 8.16).

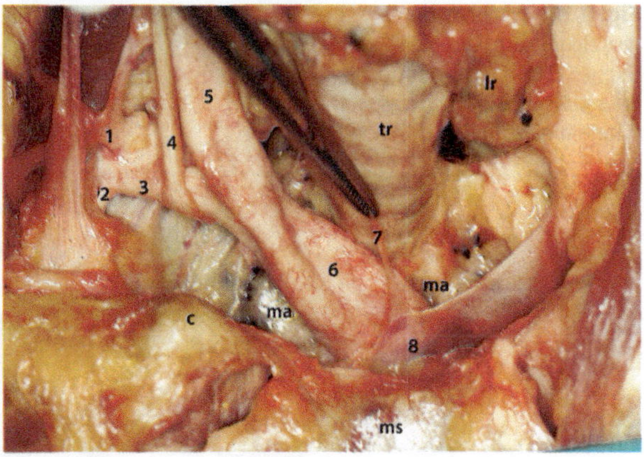

Fig. 8.17 Innominate artery. *tr* Trachea, *lr* recurrent lymph nodes, *c* clavicle, *ms* manubrium sterni, *ma* upper mediastinum, *1* thyrocervical trunk, *2* internal thoracic artery, *3* subclavian artery, *4* vagus nerve, *5* common carotid artery, *6* innominate artery, *7* anterior jugular vein, *8* brachiocephalic vein

8.6 Tracheostomy

The innominate artery, hidden below the manubrium sterni, is short and fat and may be a major source of danger during performance of low tracheotomy (Fig. 8.17).

Complications: Tracheostomy, as a preliminary stage in oncological surgery of the larynx and hypopharynx, is preferably performed below the isthmus of the thyroid gland. This mode of access is more exploratory than the trans- and supraisthmic routes since the trachea is deeper than the cutaneous plane, but there are two advantages: (1) it enables the surgeon to operate at a distance from the neoplasia, and, as has been demonstrated, this reduces the incidence of paratracheostomal recurrences; (2) the surgeon remains at a considerable distance from the subglottic cone, which is often the site of secondary cicatricial stenoses. This applies also in functional surgery of the larynx or for provisional tracheostomies, which are usually performed in the third/fourth tracheal ring.

Besides, the proximity of the tracheotomy to the innominate artery exposes the patient to the risk of this vessel rupturing, generally as a result of tracheotomy tube decubitus. This event invariably has a fatal prognosis both due to the extent of haemorrhaging and to the fact that there is no effective form of emergency compressive plugging. It has thus been decided to reduce this low-tracheotomy-related risk by protecting the upper mediastinum with an everted, lower-hinge tracheal flap sutured to the cutis (Bjork's flap) (Fig. 8.18).

Remarks: One disadvantage of systematic tracheostomy in preservative surgery of the larynx is the need for plastic surgery to close the tracheostoma, requiring a minor operation under local anaesthetic. One certain benefit is, instead, the ease with which the tracheotomy tube can be replaced in the postoperative period by nursing staff. The tracheal flap joined to the cutis provides a handy "slide" by which to access the tracheal lumen, and the risk of taking the wrong mediastinal route is practically non-existent.

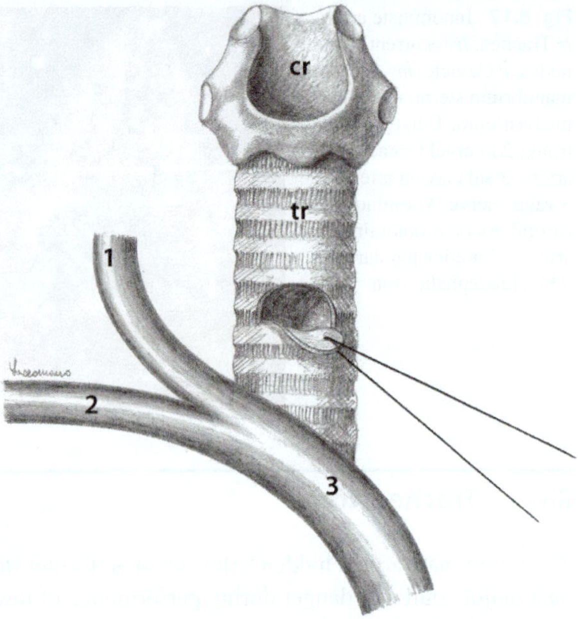

Fig. 8.18 Tracheostomy. *cr* Cricoid cartilage, *tr* trachea, *1* common carotid artery, *2* subclavian artery, *3* innominate artery

The current tendency regarding tracheotomies is the following:

1. Confirmation of the tracheostomic rather than the tracheotomic procedure, that is, the trachea is always joined to the cutis (for safety when changing the tube, even at home, and ease of managing the tracheostoma).
2. Tracheostomas are increasingly smaller and are closed much earlier.
3. The use of cuffed or fenestrated tubes is avoided, unless in exceptional cases.

The great veins of the base of the neck are sought in this area, anteriorly to the great arterial trunks. A useful guide for this purpose is the internal jugular vein, whose descending course leads to the subclavian vein and medial course to the bilaterally present brachiocephalic vein.

To conclude this dissection phase, the trachea is completely skeletised; the prevertebral plane is at once exposed laterally to the median organs (Fig. 8.19).

8.7 Oesophagus

After removing the thyroid gland, the complete cervical oesophagus can be examined. The oesophagus can immediately be seen protruding more to the left than the trachea. Its superior end can be clearly identified near the cricoid cartilage. Its calibre narrows at this point (more markedly than the constriction present on crossing the aortic arch and diaphragm) and takes the name of Killian's mouth (Fig. 8.20).

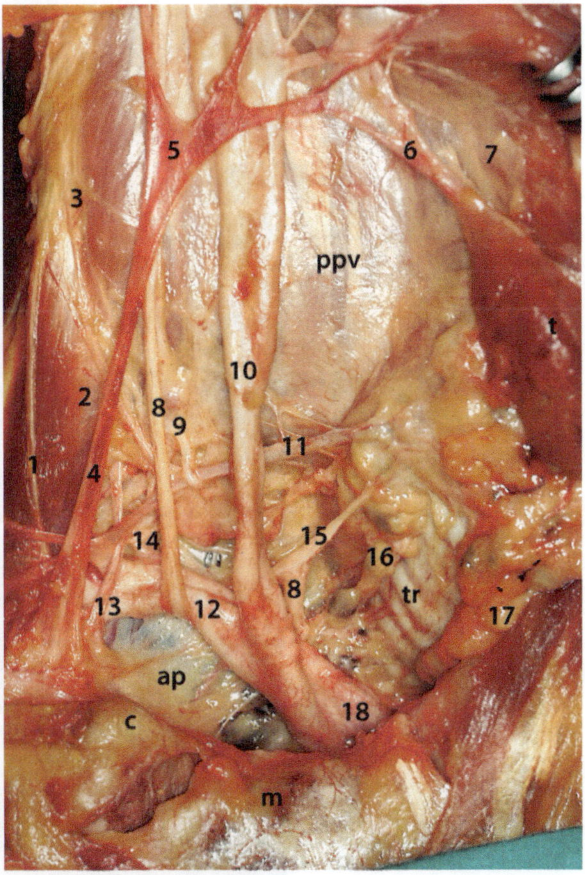

Fig. 8.19 Prevertebral perivisceral area. *ppv* Prevertebral plane, *t* thyroid gland, *tr* trachea, *ap* apex of the lung, *c* clavicle, *m* manubrium sterni, *1* phrenic nerve, *2* anterior scalene muscle, *3* branches of cervical plexus, *4* internal jugular vein, *5* thyrolinguofacial trunk, *6* superior thyroid pedicle, *7* lateral hypopharyngeal wall, *8* vagus nerve, *9* cervical sympathetic chain, *10* common carotid artery, *11* inferior thyroid artery, *12* subclavian artery, *13* vertebral vein, *14* vertebral artery, *15* recurrent nerve, *16* recurrent lymph nodes, *17* pretracheal lymph nodes, *18* innominate artery

Remarks: Killian's mouth has a sphincteric function provided by the action of the cricopharyngeal fibres of the inferior constrictor muscle of the pharynx. These muscle fibres play a part in establishment of the neoglottis, which produces the oesophageal voice in total laryngectomees. This is where selective myotomy is performed in the event of excessive oesophageal stricture after supraglottic, subtotal reconstructive, or total laryngectomy with a phonatory fistula.

The oesophagus is then isolated from both the prevertebral plane and the trachea. It is worth noting that the tracheal rings are interrupted along the contact surface with the oesophagus. The posterior wall (pars membranacea of the trachea) is devoid of cartilage: it does instead possess smooth muscle fibres (tracheal muscle) which, on contraction, approximate the ends of the cartilage arches, thereby decreasing the transverse diameter of the trachea during respiration. Lastly, we consider that the sixth cervical vertebra represents the level of passage between the hypopharynx and the cervical oesophagus, as also between the larynx and the trachea and, indirectly, between Robbins level III and IV. The transverse processes of the sixth cervical vertebra are known as the carotid tubercles.

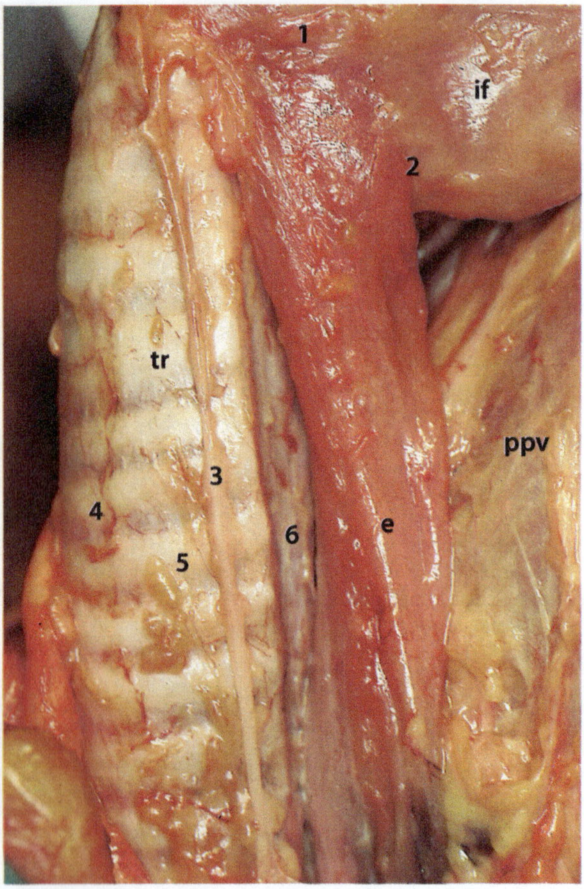

Fig. 8.20 Trachea and cervical oesophagus. *if* Hypopharynx, *tr* trachea, *e* oesophagus, *ppv* prevertebral plane, *1* cricopharyngeal muscle, *2* Killian's mouth, *3* recurrent nerve, *4* tracheal vascular arches, *5* tracheal ring, *6* pars membranacea of the trachea

Take-Home Messages

- The infrahyoid linea alba is an important landmark: (a) in thyroid surgery, as it allows access to the thyroid gland simply by divaricating the infrahyoid muscles on this line, without necessarily sectioning them, and (b) as the "tracheotomy rhombus", since in that spot only two planes, the cutis and the fascia, cover the laryngotracheal duct; the surgeon passes through these planes when he wants to open the trachea.
- The tracheotomy may be performed with a vertical or transverse incision of the cutis. We usually prefer the transverse incision for aesthetic reasons. The vertical incision tends to adhere to the scar of the infrahyoid linea alba, which is necessarily opened craniocaudally. In this way, an unattractive adherence is formed between the cutis and the muscular plane, which is clearly seen during deglutition.

- The trachea is nourished segmentarily, above all by branches of the inferior thyroid arteries: when a temporary tracheostomy is being performed, it is good practice to limit the skeletisation of the tracheal rings to what is strictly indispensable so as to avoid peritracheostomal chondronecrosis induced by ischemia, especially in patients treated with radiotherapy.
- The dissection of the recurrent region is very important and is often overlooked in laryngeal tumours with subglottic extension. This factor may explain cases of peritracheostomal recurrence after total laryngectomy. Moreover, it must not be forgotten that the recurrent region is in continuity with the upper mediastinum: hence the indication for postoperative radiotherapy even in this seat in subglottic tumours.

References

1. Robbins KT, Clayman G, Levine PA et al (2002) Neck dissection classification update: revision proposed by the American Head and Neck Society and the American Academy of Otolaryngology- Head and Neck Surgery. Arch Otolaryngol Head Neck Surg 128(7):751–758
2. de Bree R, Leemans CR, Silver CE et al (2011) Paratracheal lymph node dissection in cancer of the larynx, hypopharynx, and cervical esophagus: the need for guidelines. Head Neck 33(6): 912–916
3. Medina JE, Ferlito A, Robbins KT et al (2011) Central compartment dissection in laryngeal cancer. Head Neck 33(5):746–752
4. Testut L, Jacob O (1977) Trattato di Anatomia Topografica, 2nd Italian edn. UTET, Turin
5. Blondeau P, Neouze GL, Rene L (1977) The inferior non-recurrent laryngeal nerve; hazards of thyroid surgery (7 cases). Ann Chir 31(11):917–923, French

Anterior Region (Robbins Level VI: Superior Part)

9

Core Messages

- In this chapter, we shall explore especially the larynx and the hypopharynx. In all surgical specialities in recent years, there has been an increased preference for the conservative approach to malignant tumours. This is particularly true in oncological surgery of the larynx, thanks especially to endoscopic laser surgery and to the supracricoid (reconstructive) laryngectomies. The precise anatomical knowledge of the larynx, as regards both its surface appearance and its deep spaces, and of the preferential means of diffusion of neoplasias is of fundamental importance to allow the surgical indication to be as conservative as possible without decreasing the survival percentage.

9.1 Anatomic Layout

This area coincides with the superior part of Robbins level VI. We begin dissection from the top starting from the hyoid bone, which we identify by palpation to distinguish its parts: the body, the greater cornu, and the lesser cornu, the latter being close to the point of insertion of the stylohyoid muscles (Fig. 9.1).

Significant Anatomical Structures: hyoid bone, larynx, quadrangular membrane, conus elasticus, Galen's loop, superior laryngeal nerve, cricothyroid space, cricothyroid artery, preepiglottic space, inferior constrictor muscle of pharynx, cricopharyngeus muscle, Laimer's triangle, Zenker diverticula, piriform sinus, retrocricoid area, lingual "V", glossoepiglottic valleculae, Morgagni's ventricle, posterior commissure, paraglottic spaces, thyroepiglottic ligament, aryepiglottic fold, vocal process of arytenoid cartilage, Reinke's space, anterior commissure

Landmarks: mental prominence, carotid tubercle, greater cornu of hyoid bone, foramen cecum, threefold region, laryngeal corner

M. Lucioni, *Practical Guide to Neck Dissection*,
DOI 10.1007/978-3-642-33977-6_9, © Springer-Verlag Berlin Heidelberg 2013

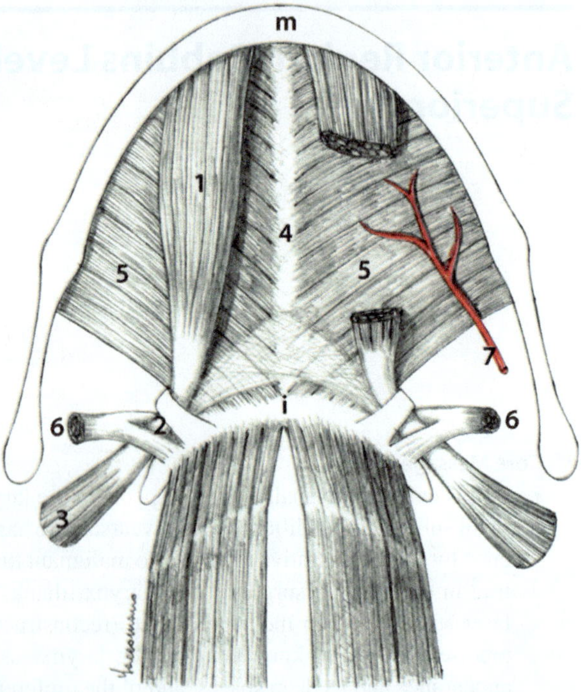

Fig. 9.1 Hyoid bone and interdigastric space. *m* Mandible, *i* hyoid bone, *1* anterior belly of the digastric muscle, *2* intermediate tendon of the digastric muscle, *3* posterior belly of the digastric muscle, *4* suprahyoid linea alba, *5* mylohyoid muscle, *6* stylohyoid muscle, *7* submental artery

9.2 Hyoid Area

We recognise the submental triangle, also referred to as the interdigastric space, which we already drained in a previous exercise and which corresponds to Robbins level Ia. The suprahyoid linea alba runs from the mental prominence to the hyoid bone and is formed by fusion of the mylohyoid muscles on the midline (Fig. 9.2).

The hyoid bone is the only bone that is not joined to the rest of the skeleton; instead, in most vertebrates it is joined through the ossification of that structure which, in exceptional cases, is ossified also in man, that is, the stylohyoid ligament. It is an important point for the support of the larynx during the ample craniocaudal excursions of deglutition (which may be as much as 2–3 cm) (Fig. 9.3).

The sternohyoid and sternothyroid muscles already interrupted at the bottom are removed. The thyrohyoid muscles are sectioned at the point of insertion in the thyroid cartilage and hyoid bone and then ablated, thereby exposing the thyrohyoid membrane (Fig. 9.4).

The larynx is thus completely exposed. Posteriorly, the hypopharynx is separated from the prevertebral plane (Fig. 9.5).

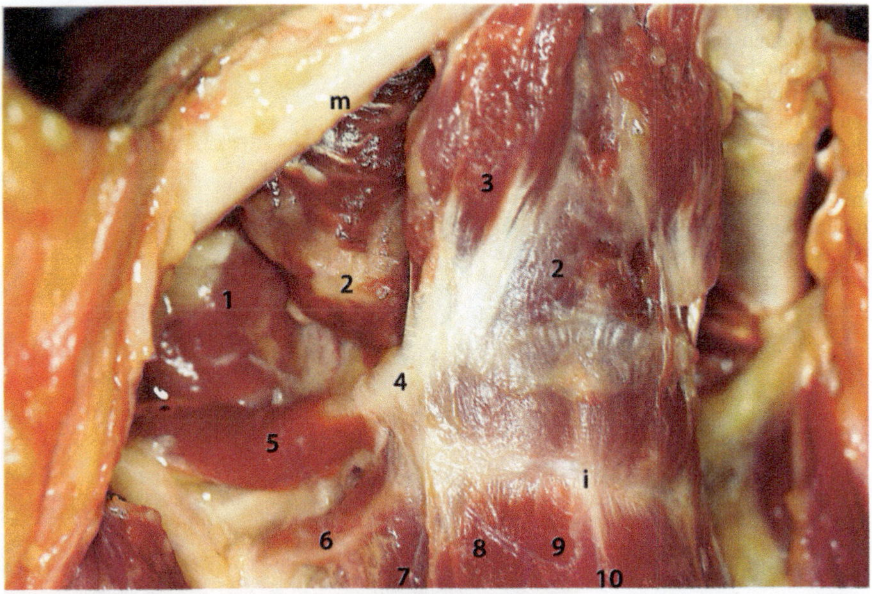

Fig. 9.2 Hyoid area (I). *m* Mandible, *i* hyoid, *l* hyoglossal muscle, *2* mylohyoid muscle, *3* anterior belly of the digastric muscle, *4* intermediate tendon of the digastric muscle, *5* posterior belly of the digastric muscle, *6* greater cornu of the hyoid bone, *7* thyrohyoid muscle, *8* omohyoid muscle, *9* sternohyoid muscle, *10* infrahyoid linea alba

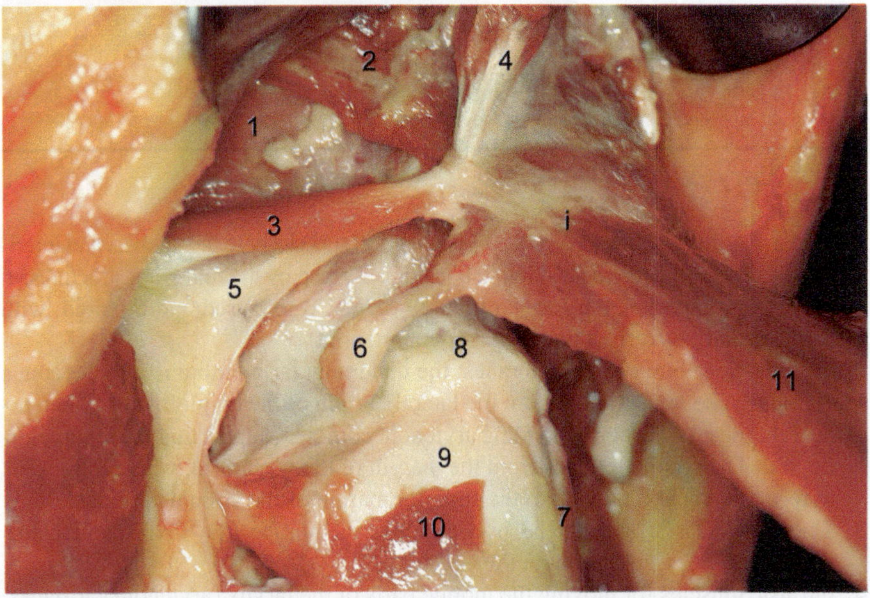

Fig. 9.3 Hyoid area (II). *i* Hyoid bone, *l* hyoglossal muscle, *2* mylohyoid muscle, *3* posterior belly of the digastric muscle, *4* anterior belly of the digastric muscle, *5* hypoglossal nerve, *6* greater cornu of the hyoid bone, *7* laryngeal prominence, *8* preepiglottic space, *9* thyroid lamina, *10* line of insertion of the thyrohyoid muscle, *11* sternohyoid muscle

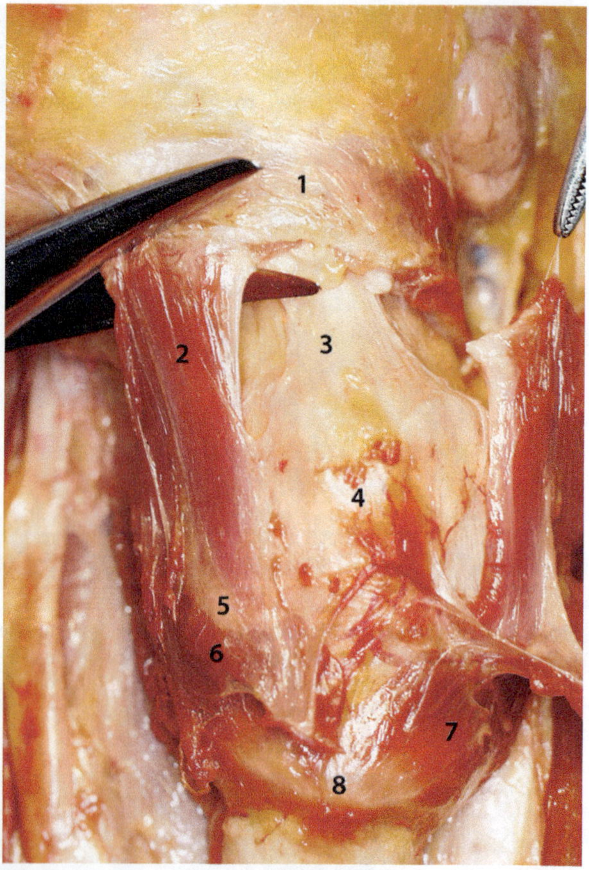

Fig. 9.4 Larynx and infrahyoid muscles. *1* Body of the hyoid bone, *2* thyrohyoid muscle, *3* thyrohyoid membrane, *4* laryngeal prominence, *5* line of insertion of the thyrohyoid muscle, *6* sternothyroid muscle (sectioned), *7* cricothyroid muscle, *8* cricoid ring

9.3 Larynx

The larynx is situated anteriorly to the hypopharynx, superiorly to the trachea, and inferiorly to the base of the tongue and hyoid bone. It is formed by the following structures:

1. A cartilaginous skeleton formed by 9 cartilages, 3 unpaired and 6 paired, with two articulations (cricothyroid and cricoarytenoid).
2. Two elastic submucous membranes: the quadrangular membrane and the conus elasticus. The first extends from the lateral margin of the epiglottis to the anterolateral aspect of the corresponding arytenoid cartilage, supporting the aryepiglottic fold. The second extends from the margin of the vocal fold, where it thickens to form the vocal ligament, to the superior margin of the cricoid cartilage.
3. Three fibroelastic sheets: the hyoepiglottic membrane, the thyrohyoid membrane, and the cricothyroid membrane.
4. Intrinsic muscles for moving the mobile parts of the larynx (arytenoid cartilages, vestibular folds, and vocal folds). Adduction movements are affected by the interarytenoid (transverse and oblique) and lateral cricoarytenoid muscles and abduction movements by the posterior cricoarytenoid muscles; vocal cord tension is provided by the thyroarytenoid and cricothyroid muscles (Fig. 9.6).

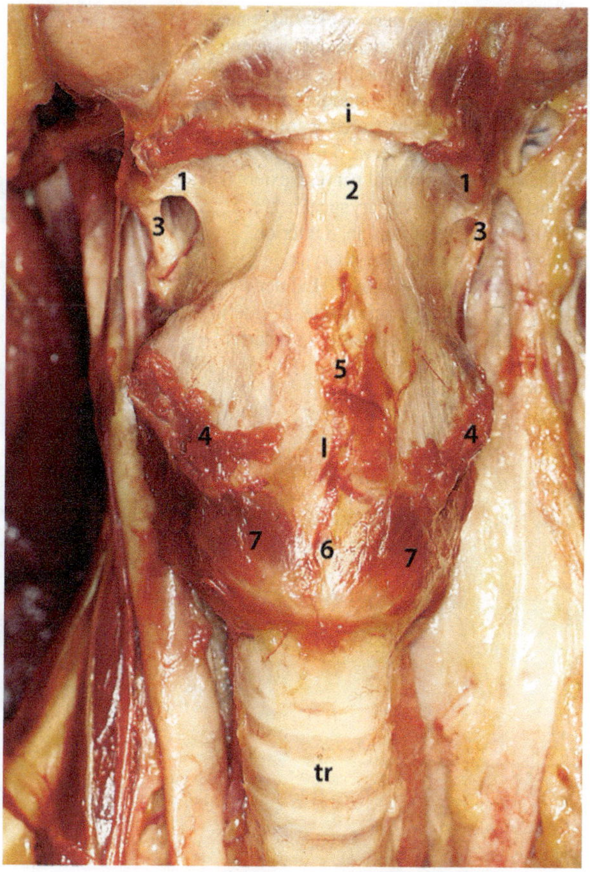

Fig. 9.5 Larynx: anterior view. *i* Hyoid bone, *l* larynx, *tr* trachea, *1* greater cornu of the hyoid bone, *2* thyrohyoid membrane, *3* thyrohyoid ligament, *4* line of insertion of the thyrohyoid muscle, *5* laryngeal protuberance, *6* cricothyroid membrane, *7* cricothyroid muscle

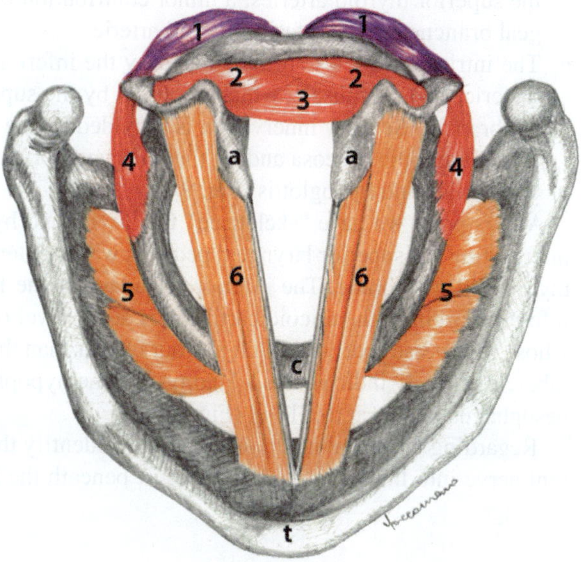

Fig. 9.6 Intrinsic laryngeal muscles. *c* Cricoid cartilage, *t* thyroid cartilage, *a* arytenoid cartilage, *1* posterior cricoarytenoid muscle, *2* interarytenoid muscle (oblique component), *3* interarytenoid muscle (transverse component), *4* lateral cricoarytenoid muscle, *5* cricothyroid muscles, *6* thyroarytenoid muscles

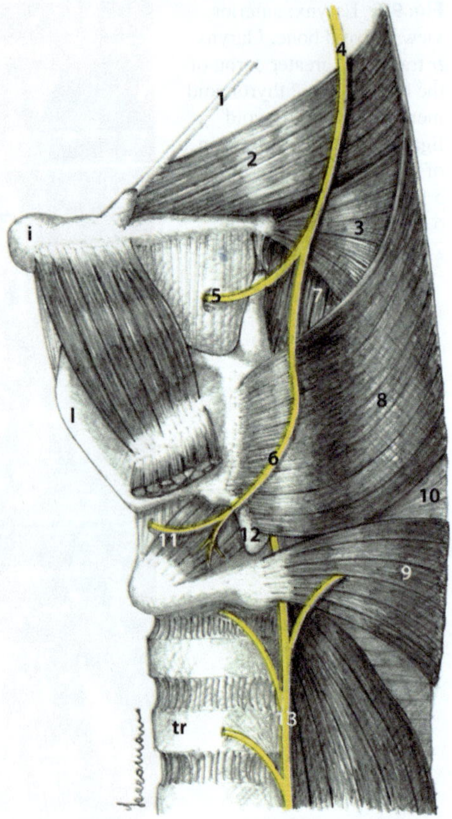

Fig. 9.7 Laryngeal nerves. *i* Hyoid bone, *l* larynx, *tr* trachea, *1* stylohyoid ligament, *2* middle constrictor muscle of the pharynx (superior component), *3* middle constrictor muscle of the pharynx (inferior component), *4* superior laryngeal nerve, *5* internal branch of the superior laryngeal nerve, *6* external branch of the superior laryngeal nerve, *7* palatopharyngeal muscle, *8* inferior constrictor muscle of the pharynx, *9* cricopharyngeal muscle, *10* Laimer's triangle, *11* cricothyroid muscle, *12* inferior cornu of the thyroid cartilage, *13* recurrent nerve

5. The larynx is vascularised by the superior laryngeal and cricothyroid branches of the superior thyroid arteries; a minor contribution is made by the inferior laryngeal branches of the inferior thyroid arteries.
6. The intrinsic muscles are innervated by the inferior laryngeal nerve, except for the cricothyroid muscle, which is served by the superior laryngeal nerve (external branch). Sensory innervation is provided by the superior laryngeal nerve for the supraglottic mucosa and by the recurrent nerve for the inferior aspect of the vocal folds and hypoglottis (Fig. 9.7).

At this point we have "skeletised" the larynx and hypopharynx, though without interrupting the superior laryngeal pedicles which enter the larynx at the level or the thyrohyoid membrane. The inferior boundary of the larynx, corresponding to the inferior margin of the cricoid cartilage, is at the level of the sixth cervical vertebra, whose transverse process, which juts out further than the others, is called the carotid tubercle (already indicated as a landmark). The hypopharynx ends and the cervical oesophagus begins at this level (Fig. 9.8).

Regarding laryngeal innervation, we now identify the point of entry of the recurrent nerve into the larynx, which lies just beneath the inferior cornu of the thyroid cartilage.

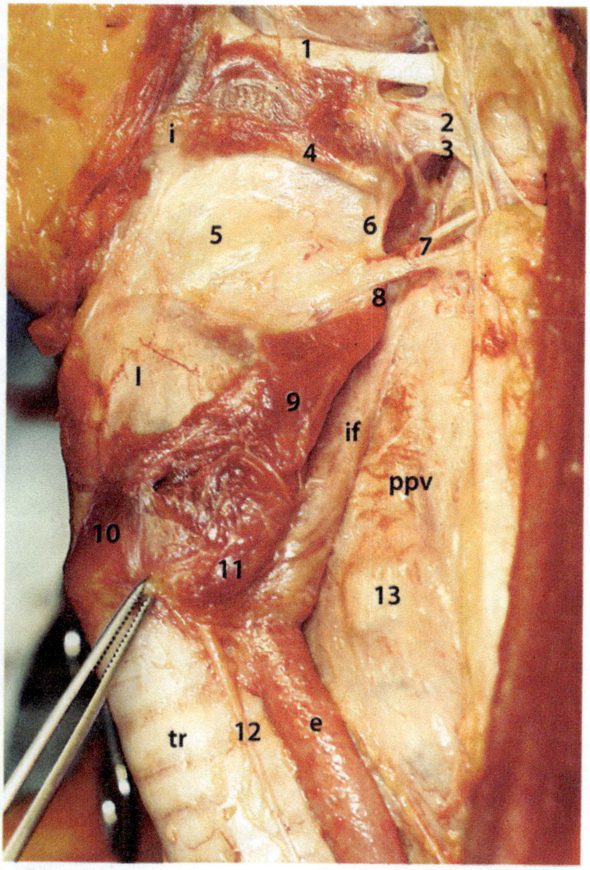

Fig. 9.8 Larynx: side view. *i* Hyoid bone, *l* larynx, *if* hypopharynx, *tr* trachea, *e* oesophagus, *ppv* prevertebral plane, *1* intermediate tendon of the digastric muscle, *2* hypoglossal nerve, *3* lingual artery, *4* greater cornu of the hyoid bone, *5* thyrohyoid membrane, *6* thyrohyoid ligament, *7* superior laryngeal pedicle, *8* superior cornu of the thyroid cartilage, *9* inferior constrictor muscle of the pharynx, *10* cricothyroid muscle, *11* cricopharyngeal muscle, *12* recurrent nerve, *13* VI cervical vertebra

Remarks: In supracricoid laryngectomies, the inferior cornu of the thyroid cartilage is generally sectioned, at least on the side where the arytenoid is preserved. In doing so the line of caudal resection is far from the recurrent nerve at its entrance into the larynx, and there is less risk of damaging it, which could cause fixity of the residual arytenoid.

In the larynx, the recurrent nerve divides into two terminal branches, a posterior one which serves the posterior cricoarytenoid muscle (sole abductor muscle of the vocal cords) and an anterior one which, with subsequent divisions, innervates all the other intrinsic muscles of the larynx.

With the branches of the superior laryngeal nerve, the sensory component of the recurrent nerve forms Galen's loop and, with afferents from the subglottic mucosa, runs into the extralaryngeal tract of the common recurrent trunk (Fig. 9.9).

The superior laryngeal nerve is then identified and followed. Laterally, at the apex of the greater cornu of the hyoid bone, it divides into an internal branch (component of the superior laryngeal pedicle), which penetrates the thyrohyoid membrane and conducts sensory fibres to the supraglottic portion of the larynx, and an

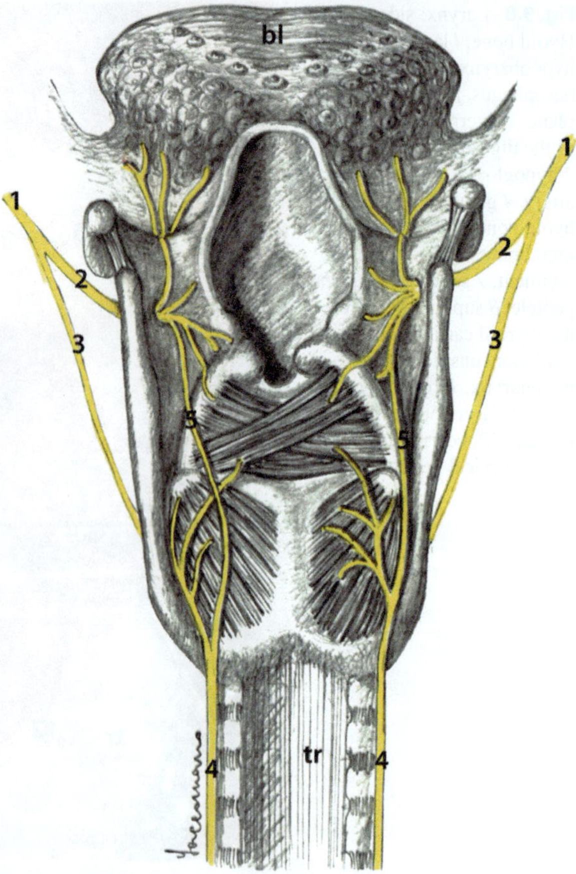

Fig. 9.9 Laryngeal nerves and Galen's loop. *bl* Tongue base, *tr* trachea, *1* superior laryngeal nerve, *2* internal branch of the superior laryngeal nerve, *3* external branch of the superior laryngeal nerve, *4* recurrent nerve, *5* Galen's loop

external branch, which runs obliquely in an anteroinferior direction to innervate the cricothyroid muscle; it then perforates the cricothyroid ligament and transmits sensory fibres to the glottis and the laryngeal ventricle (Fig. 9.10).

Remarks: The reason why the superior laryngeal nerve is preserved in functional laryngectomies is to preserve sensitivity as much as possible at glottic and subglottic level and in the piriform sinus. It is believed that this facilitates the functional recovery of deglutition in the postoperative period.

Complications: Accidental injury to the superior laryngeal nerve during ligation of the superior thyroid pedicle causes homolateral hypotonia of the vocal cord, which on laryngoscopy can be seen lying below the level of the contralateral one, leading to dysphonia of prevalently acute pitch. Another outcome is homolateral supraglottic hemilaryngeal anaesthesia.

The thyrohyoid membrane joins the superior margin of the thyroid cartilage and the hyoid bone. Its lateral margins thicken to form two fibrous ligaments (thyrohyoid ligaments), which join the superior cornu of the thyroid cartilage to the apex of the greater cornu of the hyoid bone.

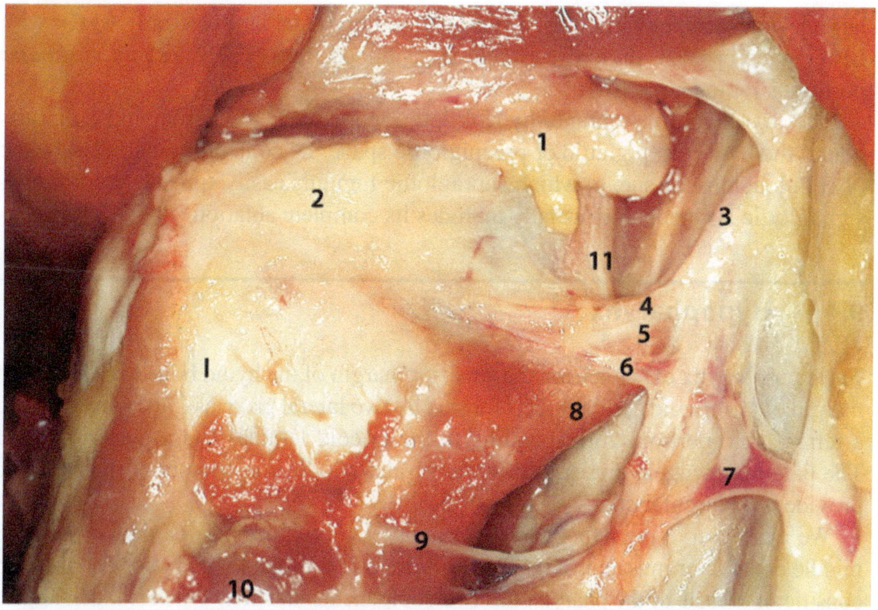

Fig. 9.10 Superior laryngeal pedicle. *I* Larynx, *1* greater cornu of the hyoid bone, *2* thyrohyoid membrane, *3* superior thyroid artery, *4* superior laryngeal artery, *5* internal branch of the superior laryngeal nerve, *6* superior laryngeal vein, *7* superior thyroid vein, *8* superior cornu of the thyroid cartilage, *9* external branch of the superior laryngeal nerve, *10* cricothyroid muscle, *11* thyrohyoid ligament

Remarks: The thyrohyoid ligaments and posterior margins of the thyroid cartilage correspond to the lateral hypopharyngeal walls and are a good landmark for lateral pharyngotomy, providing access to the piriform recesses and vestibule of the larynx in surgery in this region.

In the lateral portion of the thyrohyoid membrane, we identify the point where the laryngeal pedicle enters the larynx. It is formed by the superior laryngeal artery and by the corresponding vein and nerve. We see how the vessels come from the superior thyroid pedicle, while the superior laryngeal nerve, which is deeper down, comes from the vagus nerve. We ligate and section the pedicle.

The cricothyroid space is well identified and defined by palpation. This space is easily accessible in laryngotracheal lumen surgery, should severe respiratory stenosis require emergency incision (intercricothyroid laryngotomy). This is in fact the point where the respiratory lumen is closest to the skin of the neck. In performing the manoeuvre, the only significant vessel that may be encountered in this region is the cricothyroid artery, a ramification of the lateral branch of the superior thyroid artery. The same vessel may also be encountered in endoscopic laser surgery of the anterior glottis, when extending the subperichondrial exeresis downwards.

Remarks: A small depression internally corresponding to the anterior commissure is appreciable by palpation between the third superior and third median of the anterior profile of the thyroid cartilage. This landmark is important in functional laryngeal surgery, in particular in supraglottic horizontal laryngectomy when, with Lister's forceps, the thyroid cartilage is sectioned transversely and its vestibular portion is removed.

The thyrohyoid membrane, hyoepiglottic membrane, and infrahyoid epiglottis bound the hyothyroepiglottic or preepiglottic space. Its contents essentially consist of celluloadipose tissue, to which supraglottic tumours easily spread. Accordingly, supraglottic laryngectomy usually includes its complete ablation.

9.4 Hypopharynx

The hypopharynx extends from the superior margin of the hyoid bone to the inferior margin of the cricoid cartilage. It is formed by two lateral grooves (piriform sinuses), a retrocricoid portion (corresponding to the cricoid lamina) and a posterior wall. The palatopharyngeal muscles and inferior and middle constrictors of the pharynx provide the muscular coat.

We can observe the hypopharyngeal muscles morphology, in particular how the middle constrictor muscle inserts in the greater cornu of the hyoid bone. The inferior constrictor muscle of the pharynx has two components: the first, which is much larger, runs obliquely and inserts into the lateral margin of the thyroid cartilage, while the second, with horizontal fibres, inserts in the lateral margin of the cricoid cartilage. This portion of the inferior constrictor muscle of the pharynx is called the cricopharyngeus muscle. The middle and inferior constrictors are innervated by the vagus nerve.

Remarks: The inferior constrictor corresponds to the muscular plane in the reconstruction of the hypopharynx after total laryngectomy. Longitudinal myotomy of the inferior constrictor muscle is an operation frequently associated with the creation of a phonatory fistula after total laryngectomy. A spasm or a fibrosis of the muscular sheath may cause aphonia. In this case the operation tends to obtain a greater elastic compliance of the hypopharynx and therefore optimum functioning of the phonatory prosthesis. It should also be remembered that, posteriorly, the oblique fibres of the superior portion of the inferior constrictor muscle of the pharynx and the horizontal fibres of its inferior portion (cricopharyngeus muscle) describe a triangle with an inferior base, lacking in muscular fibres and with lower resistance. Said space, referred to as Laimer's triangle, is the place where Zenker's diverticula are formed (Fig. 9.11).

We rotate the larynx just enough to reveal the profile of the lateral margin of the thyroid cartilage under the inferior constrictor: the sectioning of this muscle along this margin is the preliminary surgical step to obtain access to the piriform sinus.

9.5 Laryngectomy

At this point, it is preferable to remove the larynx and hypopharynx en bloc in order to explore internal laryngeal morphology. A horizontal incision is made at the level of the second tracheal ring, where the trachea and cervical oesophagus are at full thickness.

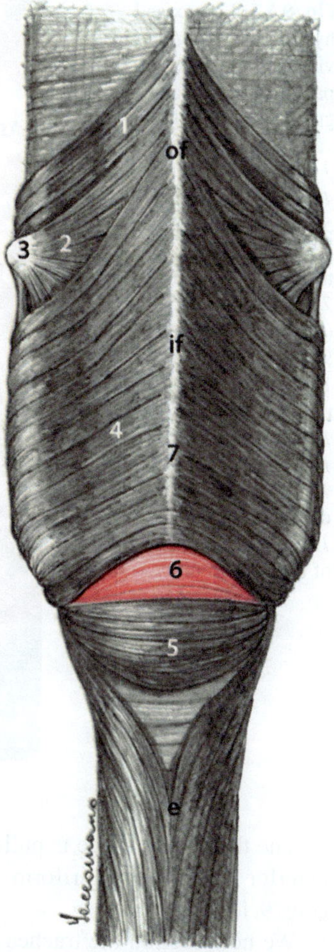

Fig. 9.11 Constrictor muscles of the pharynx. *of* Oropharynx, *if* hypopharynx, *e* oesophagus, *1* middle constrictor muscle of the pharynx (superior component), *2* middle constrictor muscle of the pharynx (inferior component), *3* apex of the greater cornu of the hyoid bone, *4* inferior constrictor muscle of the pharynx, *5* cricopharyngeal muscle, *6* Laimer's triangle, *7* posterior pharyngeal raphe

The laryngo-hypopharyngeal block is separated from the vertebral plane up to the hyoid bone. After opening the lateral hypopharyngeal wall, the tongue base is sectioned horizontally 4–5 cm. superior to the epiglottis and the posterior pharyngeal wall a few centimetres caudal, thereby separating the orohypopharynx and larynx en bloc. On the removed piece, a vertical incision is made in the posterior wall of the hypopharynx, thereby exposing the vestibule of the larynx and the retrocricoid area (Fig. 9.12).

Exercise 8: laryngectomy (Fig. 9.13)

Alternatively, for those who are interested, we propose the ablation of the larynx as is performed in the classic procedure of total laryngectomy. The manoeuvres described are performed on both sides.

After having gripped and laterally rotated the larynx, the inferior constrictor and the perichondrium are sectioned along the lateral margin of the thyroid cartilage (Fig. 9.13a).

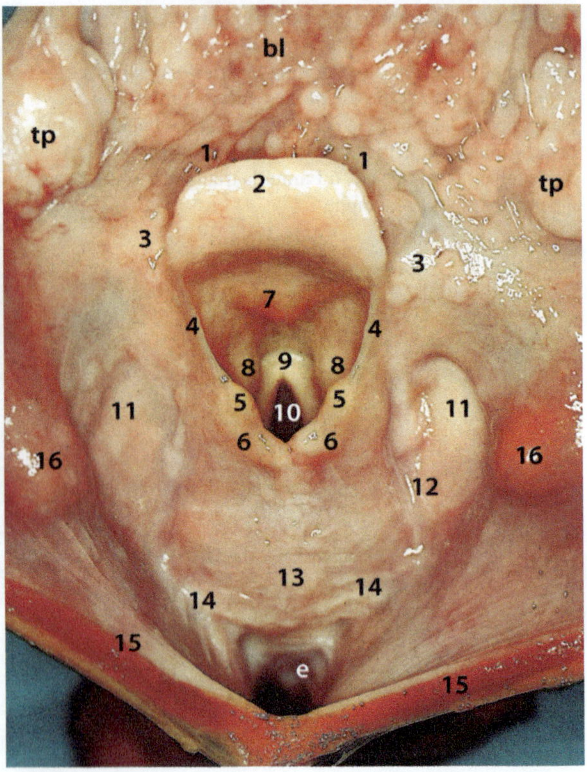

Fig. 9.12 Larynx and hypopharynx: intraluminal view (I). *bl* Tongue base, *tp* palatine tonsil, *e* oesophagus, *1* glossoepiglottic vallecula, *2* epiglottis, *3* pharyngoepiglottic fold, *4* aryepiglottic fold, *5* cuneiform tubercle (Wrisberg), *6* corniculate tubercle (Santorini), *7* epiglottic tubercle (petiole), *8* ventricular fold, *9* anterior commissure, *10* glottis, *11* piriform recess, *12* Galen's loop, *13* retrocricoid area, *14* Killian's mouth, *15* inferior constrictor muscle of the pharynx, *16* apex of the greater cornu of the hyoid bone

The thyroid cartilage is pulled up with a hook. We then proceed to separate the anterior wall of the piriform sinus with an internal subperichondrial approach (Fig. 9.13b).

We now section the trachea between the cricoid and the first tracheal ring. The hook pulls the cricoid ring upwards. The pars membranacea of the trachea is then dissected, without going to deep because this would take us into the oesophagus.

We go up posteriorly as far as the arytenoid cartilages, where we cut right through the mucosa and enter the hypopharynx (Fig. 9.13c).

Still pulling the larynx upwards, we continue to section the hypopharyngeal mucosa, keeping close to the larynx. The laryngectomy is concluded by transversely sectioning the mucosa of the glossoepiglottic valleculae (Fig. 9.13d).

Completion of the dissection caudal enables the three anatomic subareas of the hypopharynx to be extensively explored: that is, the retrocricoid area, piriform sinus, and posterior wall. A thread-like relief can be discerned, traversing the antero-superior part of each piriform sinus in a craniocaudal direction. It is the Galen's loop, the anastomosis between the internal branch of the superior laryngeal nerve and the recurrent nerve.

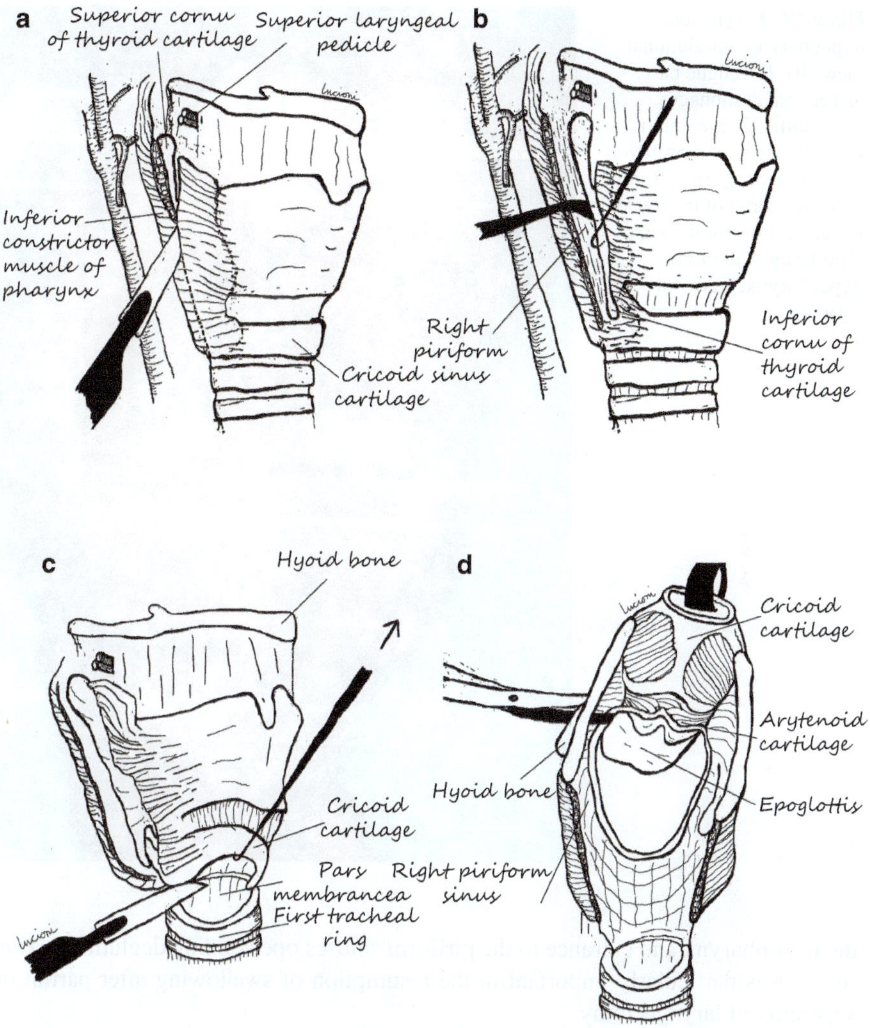

Fig. 9.13 Exercise no. 8: laryngectomy. (**a**) Dissection of inferior constrictor muscle. (**b**) Isolation of piriform sinus. (**c**) Dissection of trachea. (**d**) Ablation of larynx

Remarks: Tumours of the piriform sinus generally cause reflex otalgia: algogenic stimuli run along the superior laryngeal nerve and vagus nerve and reverberate in the external auditory canal. Stimulation of the external auditory canal cutis causes coughing via the same reflex arc (Fig. 9.14).

The lateral end of the greater cornu of the hyoid bone can be found by palpation laterally and superiorly at the entrance to the piriform sinus. The hyoid arch keeps

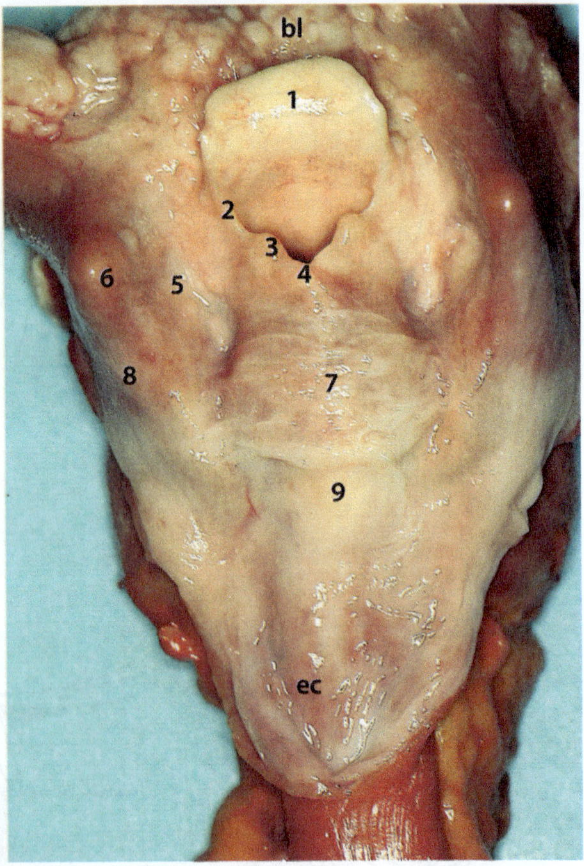

Fig. 9.14 Larynx and hypopharynx: intraluminal view (II). *bl* Tongue base, *ec* cervical oesophagus, *1* epiglottis, *2* aryepiglottic fold, *3* arytenoid, *4* posterior commissure, *5* piriform sinus, *6* greater cornu of the hyoid bone, *7* retrocricoid area, *8* posterior wall of the hypopharynx, *9* cricoid cartilage

the hypopharynx and entrance to the piriform sinuses open, aiding deglutition. This function is particularly important in the resumption of swallowing after partial or supracricoid laryngectomy.

The lingual "V" can be seen on observation of the anterior oropharynx. It is formed by the circumvallate papillae and separates the body from the base of the tongue and, at its apex, the foramen cecum. The lingual tonsil, formed by numerous more or less developed lymphatic follicles, can be seen just posteriorly. The foramen cecum may be the site of an ectopic thyroid and the point of onset of thyroglossal duct remnants (fistulas and congenital median cysts).

Remarks: In laryngeal surgery extending to the tongue base, the foramen cecum is considered the maximum limit of lingual exeresis to avoid severe dysphagia.

The pharyngoepiglottic fold is also clearly identifiable and represents the boundary between the oropharynx and hypopharynx and therefore also the superior limit of the piriform sinus (Fig. 9.15).

Between the base of the tongue and the epiglottis, the median and lateral glossoepiglottic folds delimit two depressions: the glossoepiglottic valleculae.

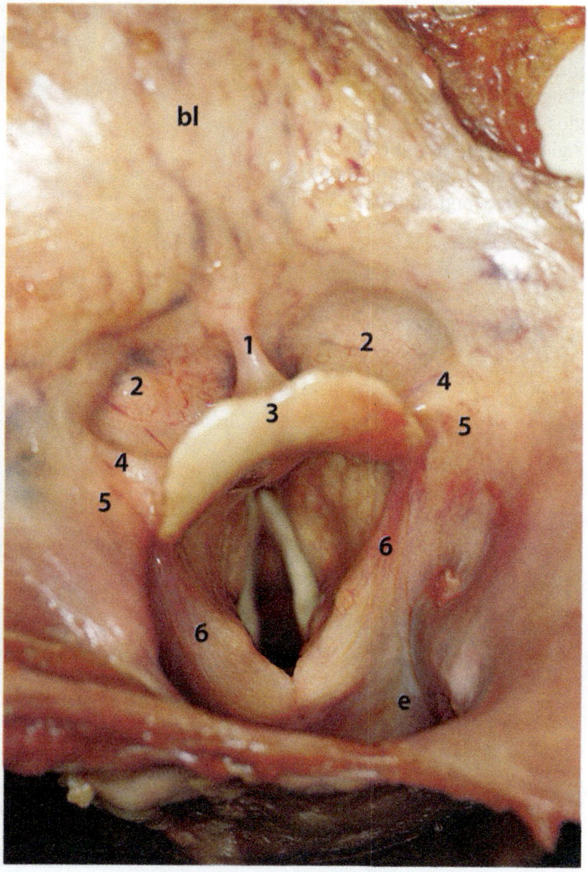

Fig. 9.15 Larynx and hypopharynx: intraluminal view (III). *bl* Tongue base, *e* oesophagus, *1* median glossoepiglottic fold, *2* glossoepiglottic vallecula, *3* suprahyoid epiglottis, *4* lateral glossoepiglottic fold, *5* pharyngoepiglottic fold, *6* aryepiglottic fold

Remarks: The glossoepiglottic valleculae mark the roof of the preepiglottic cavity, often invaded by tumours of the laryngeal lamina of the epiglottis; the neoplasia occasionally perforates the epiglottis and emerges anteriorly in the form of a "swelling" in the glossoepiglottic valleculae (Fig. 9.16). A potential site of pharyngolaryngeal tumours is the so-called "threefold region" (pharyngoepiglottic, aryepiglottic, and lateral glossoepiglottic folds) (Fig. 9.17).

We also examine the laryngeal aditus, as it will be observed during direct suspension laryngoscopy. It is bounded by the epiglottic margin, the aryepiglottic folds, the cuneiform and corniculate tubercles, and the posterior commissure between the two arytenoid cartilages. The cricoid lamina, situated inferiorly to the arytenoid cartilages, is between the two piriform sinuses. The vocal process of the arytenoids, where the vocal ligament is inserted, can be felt by palpation. From this position we identify the entrance to Morgagni's ventricle and assess how much the view of the ventricle floor is prevented by the ventricular fold. If we shift it to the side with forceps, we can have a better view (Fig. 9.18).

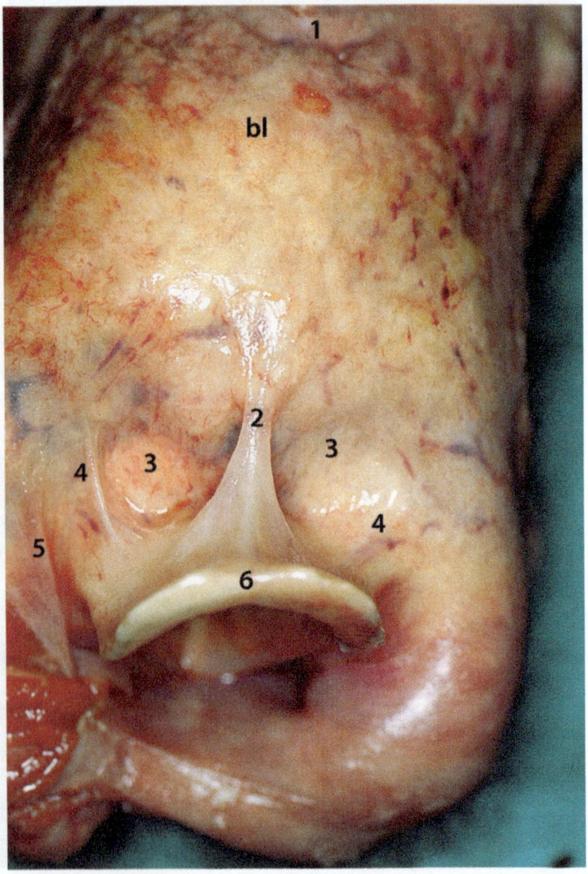

Fig. 9.16 Larynx and tongue base. *bl* Tongue base, *1* foramen cecum (apex of the "lingual V"), *2* median glossoepiglottic fold, *3* glossoepiglottic vallecula, *4* lateral glossoepiglottic fold, *5* pharyngoepiglottic fold, *6* epiglottis

Remarks: The preliminary stage of endoscopic laser cordectomy is nearly always vestibulectomy, that is, the ablation of the ventricular band. It is done simply in order to have a better exposure of the ventricle floor, the mucosa of which is very often completely removed. This also facilitates the cordectomy stage. The endoscopic follow-up is also more sure.

We now shift our angle of vision towards the anterior glottis, as though we were lifting the epiglottis with the laryngoscope. In this way we can examine the anterior commissure, the petiole, and the region of the laryngeal corner (Fig. 9.19).

The posterior laryngeal wall is then sectioned vertically along a line passing through the posterior commissure and involving the centre of the cricoid lamina. The vestibule of the larynx, the glottic plane, and the hypoglottis are exposed by divaricating the dissection margins with a self-retaining retractor (Fig. 9.20).

The anterior commissure region is also clearly evident (Fig. 9.21).

Fig. 9.17 Region of the three folds. *ep* Epiglottis, *bl* tongue base, *1* median glossoepiglottic fold, *2* lateral glossoepiglottic fold, *3* pharyngoepiglottic fold, *4* aryepiglottic fold

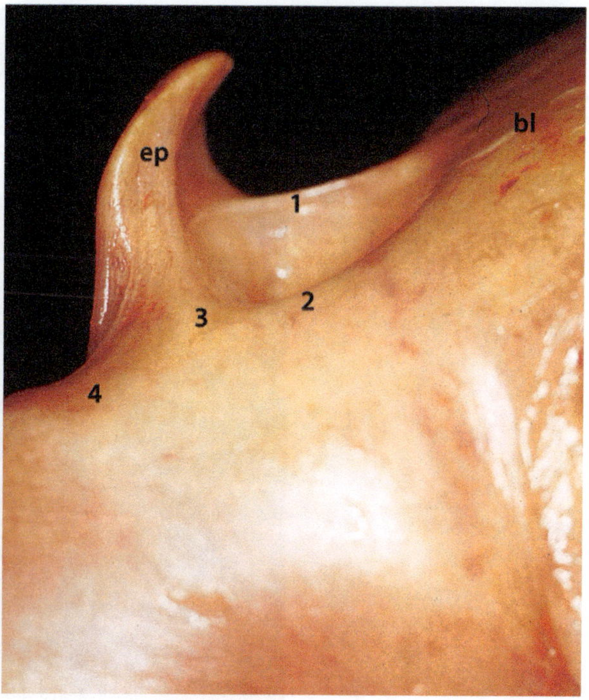

The exposure of the anterior commissure also depends on the size of the angle between the two thyroid laminas: it is usually obtuse in females and in children, approximately a right angle in adult males.

Morgagni's ventricles can be explored with a dissecting forceps. These lie between the ventricular fold and the vocal cords that separate in depth the superior and inferior infraglottic spaces. By palpation we identify the arytenoid cartilages and the cuneiform and corniculate accessory cartilages (Fig. 9.22).

Remarks: In TNM staging, the arytenoid cartilages are a subsite of the supraglottis. However, it appears clear that the arytenoid cartilage is a structure that belongs both anatomically and functionally to the glottic region [1].

Up until now, we have examined the external conformation of the larynx. We shall now try to consider the paraglottic spaces and the structures that bound them. To do this, we remove the portion of the base of the tongue which is in front of the hyoid bone and also the piriform sinuses.

9.6 Inside the Larynx

First Step: At this point the perichondrium of the thyroid cartilage is incised along its superior margin and separated from the cartilage on both sides, arriving anteriorly at the anterior commissure and laterally at the bottom of the laryngeal ventricles (Fig. 9.23a).

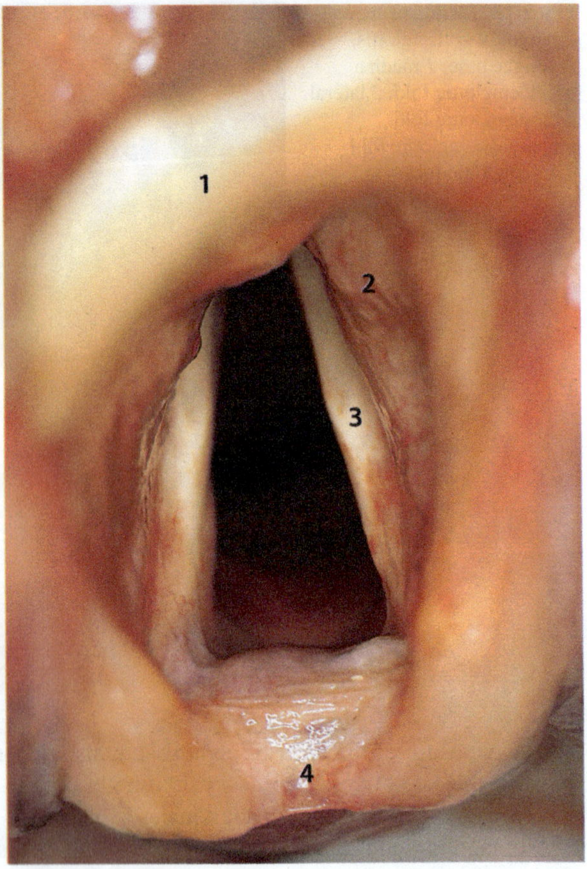

Fig. 9.18 Larynx: glottic plane. *1* Epiglottis, *2* ventricular fold, *3* vocal cord, *4* posterior commissure

Now we make a median sagittal incision which, starting from the tongue base, sections the hyoid bone in the centre and the epiglottis, dividing it in two halves, as far as the anterior commissure. We have thus exposed the adipose tissue of the preepiglottic space; we evaluate the conformation of the epiglottis cartilage and the consistency of the thyroepiglottic ligament (Fig. 9.23b).

We then section the aryepiglottic fold with forceps just in front of the arytenoid cartilage, proceeding in a craniocaudal direction, from its apex towards the vocal process. Then, straightening the forceps, we resect the mucosa of the bottom of the ventricle, until we arrive just up the anterior commissure. At this point we shall have removed the supraglottic larynx (mucosa, quadrangular membrane, submucosa, internal perichondrium) at one side (Fig. 9.23c).

Remarks: The laryngeal ventricle (Morgagni's ventricle) is no longer considered a subsite of TNM staging since it is considered to be formed by the inferior surface of the ventricular band and the superior surface of the vocal cord.

Second Step: We now consider the glottic plane. We grip the epithelium of the vocal cord with forceps near the anterior commissure and, pulling it medially and dissecting it with the scalpel, we expose the vocal ligament which appears as a thin

Fig. 9.19 Larynx: anterior glottis. *1* Right ventricular fold, *2* petiole of the epiglottis, *3* anterior commissure, *4* subglottis, *5* left vocal cord, *6* infrahyoid epiglottis, *7* laryngeal corner

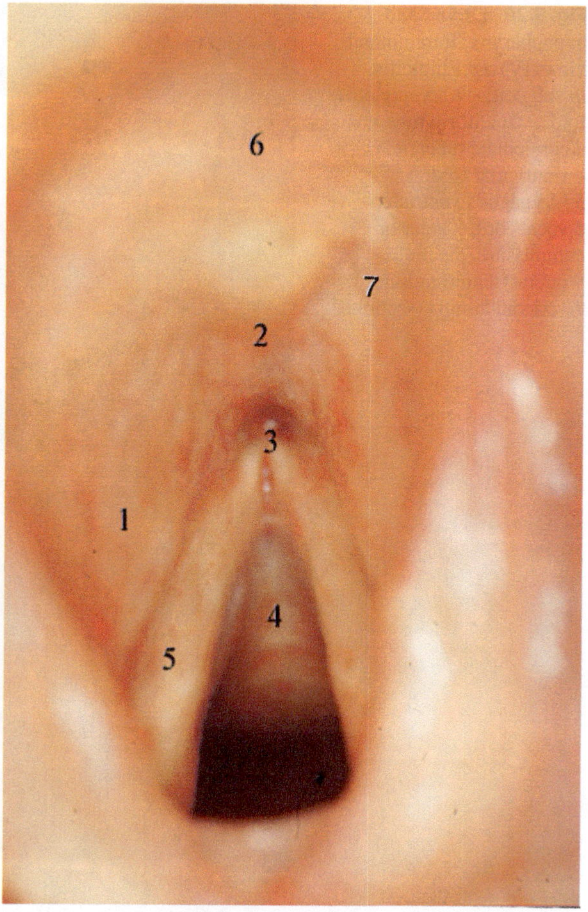

fibrous tendon extending as far as the vocal process of the arytenoid cartilage. Laterally, the arytenoid cartilage presents instead a muscular process into which the vocal and cricoarytenoid muscles insert (Fig. 9.23d (1)).

Remarks: In so doing, we have reproduced what is normally called "peeling" or "decortication" or "stripping" of the vocal cord, that is, the removal of the epithelium and of the tunica propria (Reinke space), leaving the vocal ligament intact.

At the level of the anterior commissure, by palpation we can check that the mucosa is very close to the thyroid cartilage. In fact the submucosa is not represented in this site.

Remarks: This fact introduces various considerations on the endoscopic laser treatment of the neoplasias affecting the anterior commissure, since the distance that the tumour can travel to infiltrate the cartilage is really minimum.

For the sake of precision, we point out that the anterior commissure is conventionally defined as an area of mucosa interposed between the vocal cords, bounded superiorly by an imaginary line joining the corner of the ventricles and extending inferiorly for 3 mm [2].

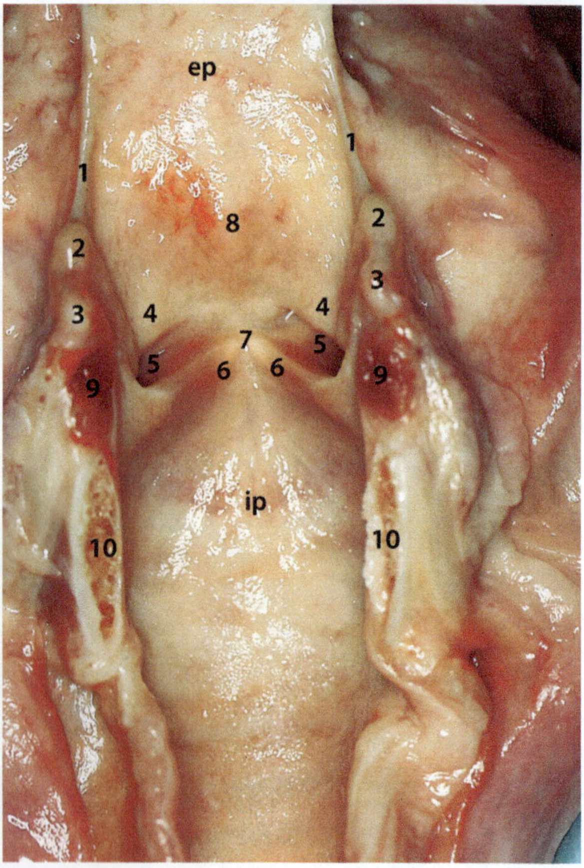

Fig. 9.20 Larynx and hypopharynx: intraluminal view (IV). *ep* Epiglottis, *ip* subglottis, *1* aryepiglottic fold, *2* cuneiform tubercle, *3* corniculate tubercle, *4* ventricular fold, *5* Morgagni's ventricle, *6* vocal cord, *7* anterior commissure, *8* petiole, *9* interarytenoid muscle, *10* cricoid lamina (sectioned)

Third Step: We section the vocal cord midway between the anterior commissure and the vocal process of the arytenoid cartilage, cutting down until we reach and interrupt the internal perichondrium. We observe and evaluate the five planes that we encounter in this section and remember that the stratification of the vocal cord is composed as follows: epithelium and tunica propria (or Reinke's space; together defined as the mucosa), vocal ligament, vocal muscle, submucosa (or inferior preepiglottic space), internal perichondrium, and finally thyroid cartilage (Fig. 9.23d (2)).

Remarks: The inferior limit of the glottis is conventionally established at 1 cm from the free edge of the vocal cord. This limit corresponds approximately to the point in which the elastic cone divides inferiorly into two components, one following the mucosa and the other enclosing the cartilage [1]. The attempt to have the glottic mucosa coincide with the pavement epithelium cannot be sustained because this covers the vocal cord with a maximum extent of 5 mm in its middle third and it is reduced, and often disappears, at the level of the anterior commissure.

Fourth Step: To conclude the dissection of this region, we transversely incise the cricothyroid membrane as in emergency tracheotomy (intercricothyroid laryngectomy).

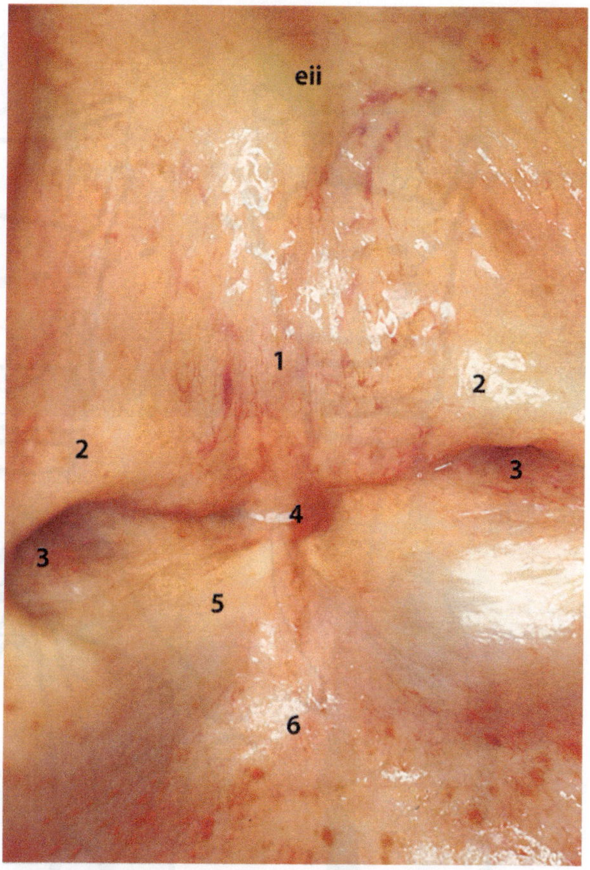

Fig. 9.21 Anterior commissure. *eii* Infrahyoid epiglottis, *1* petiole, *2* ventricular fold, *3* Morgagni's ventricle, *4* anterior commissure, *5* vocal cord, *6* subglottis

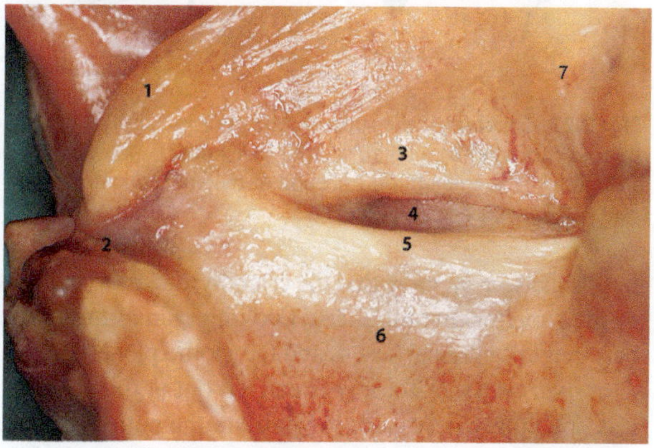

Fig. 9.22 Morgagni's ventricle. *1* Arytenoid cartilage, *2* posterior commissure, *3* ventricular fold, *4* Morgagni's ventricle, *5* vocal cord, *6* subglottis, *7* laryngeal corner

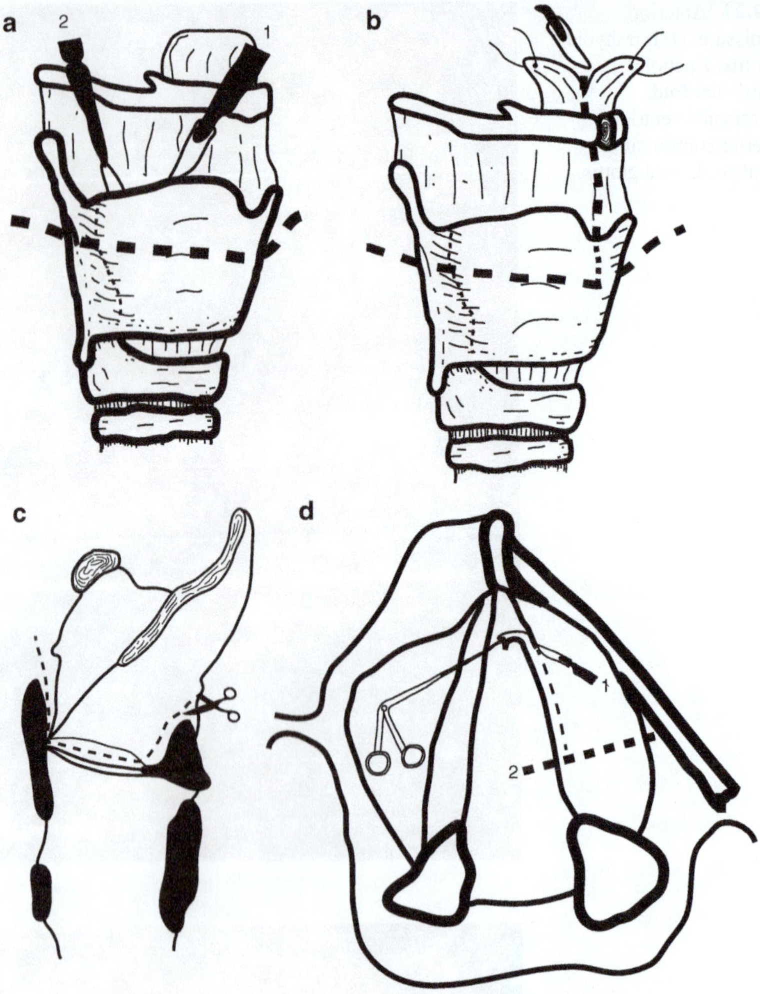

Fig. 9.23 Exeresis of the supraglottis, (**a**) *1* incision of the perichondrium, *2* its separation from thyroid cartilage (**b**) median sagittal incision (**c**) supraglottic horizontal hemilaryngectomy (**d**) *1* stripping of the epithelium, *2* section of the vocal cord

In this incision we may find the cricothyroid artery which comes from the lateral branch of the superior thyroid artery. Looking cranial into the lumen, we assess the distance between this surgical access and the glottic plane.

Remarks: Tracheotomy is carried out at this level in emergency situations where it is absolutely necessary to ventilate the patient in the shortest time possible. Once it is certain that the airways are controlled, the patient must be anaesthetised, and the tracheotomy must be moved more caudal, preferably under the thyroid isthmus. Tracheostomas that are too cranial invariably cause permanent stenosis of the subglottis.

Take-Home Messages

- The anatomo-pathological observation of the stratification of the vocal cord and clinical and surgical evaluations have led to new protocols for the treatment of tumours of this larynx subsite.
- The concept of functional surgery of the vocal cord was officially introduced in 2002 [3]. Considering that most glottic tumours do not go beyond the depth of the vocal ligament, it was deemed that the subperichondrial cordectomy systematically carried out for all T1a tumours was overtreatment in most cases. Endoscopic laser surgery takes this consideration into account and classifies cordectomies according to the depth of resection programmed for the various degrees of tumour infiltration. The result is a lower morbidity rate and often much less accentuated dysphonia.

References

1. Carlon G (1990) Il carcinoma della laringe. Piccin ed, Padova
2. Olofsson J (1974) Growth and spread of laryngeal carcinoma. Can J Otolaryngol 3:446–459
3. Remacle M, Eckel HE, Antonelli A et al (2000) Endoscopic cordectomy: a proposal for a classification by the Working Committee, European Laryngological Society. Eur Arch Otorhinolaryngol 257:227–231

10 Prevertebral Region

Core Messages

- The prevertebral plane is the deep limit of our dissection. It is usually exposed in demolitive surgery of the pharynx or in the drainage of retropharyngeal lymph node stations.

10.1 Anatomic Layout

The prevertebral plane is exposed on exeresis of the median-region viscera. Said plane is bounded laterally by the transverse processes of the cervical vertebrae, superiorly by the occipital bone, and inferiorly by the first thoracic vertebra.

The region consists of a slender musculoaponeurotic layer covering the cervical column. The most important structures are the cervical sympathetic chain and the vertebral artery, which cross the region from top to bottom (Fig. 10.1).

Significant Anatomical Structures: deep cervical fascia, cervical sympathetic chain, stellate ganglion, Claude Bernard–Horner's syndrome, vertebral artery, prevertebral plane

Landmarks: stellate ganglion, carotid tubercle, transverse process of the atlas

10.2 Cervical Sympathetic Chain

The dissection exercise begins by considering the prevertebral muscular plane and the deep cervical fascia that covers it. The pharynx, the oesophagus, and the vascular nerve bundle of the neck can be easily separated from this plane. The complex of these structures is lifted with one hand while the other dissects the thin layer of loose cellular tissue that connects it to the deep plane.

The deep muscle plane is invested by the deep cervical fascia that continues laterally over the scalene muscles. This fascia divides into two to hold the cervical sympathetic chain, located just medially to the anterior tubercles of the transverse vertebral processes.

M. Lucioni, *Practical Guide to Neck Dissection*,
DOI 10.1007/978-3-642-33977-6_10, © Springer-Verlag Berlin Heidelberg 2013

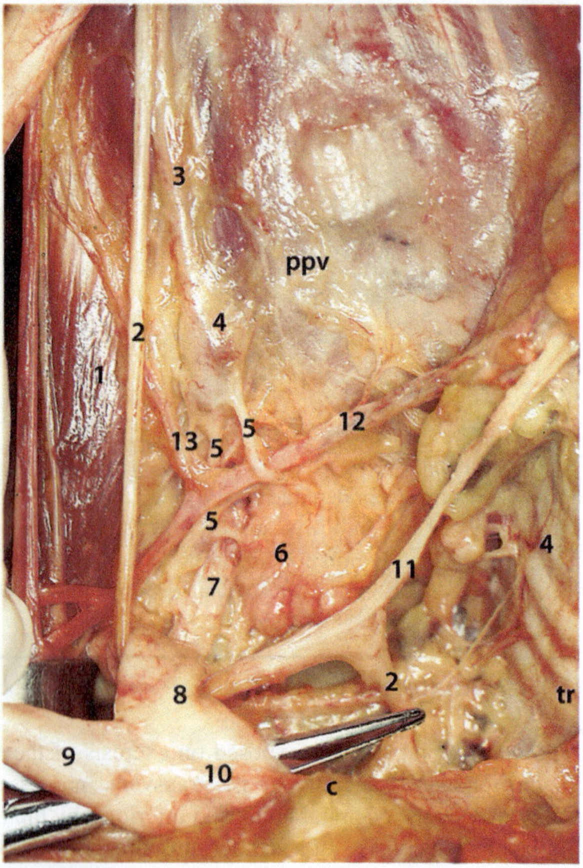

Fig. 10.1 Prevertebral plane. *ppv* Prevertebral plane, *tr* trachea, *1* anterior scalene muscle, *2* vagus nerve, *3* cervical sympathetic chain (superior portion), *4* middle cervical ganglion, *5* cervical sympathetic chain (inferior portion), *6* inferior cervical ganglion and first thoracic ganglion (stellate ganglion), *7* vertebral artery, *8* subclavian artery, *9* common carotid artery, *10* innominate artery, *11* recurrent nerve, *12* inferior thyroid artery, *13* ascending cervical artery

This nerve cord adheres to the deep muscular plane, thereby distinguishing it from the vagus that, albeit adjacent, is an integral part of the cervical vasculonervous bundle, invested by a vascular sheath shared with the carotid artery and internal jugular vein.

Remarks: The cervical sympathetic chain extends from just beneath the external orifice of the carotid canal to the level of the first rib, where it continues with the thoracic tract. It presents three ganglia: the superior ganglion is 3–4 cm long, fusiform, and located just beneath the base of the skull; the middle, inconstant ganglion lies where the inferior thyroid artery crosses the sympathetic trunk; the inferior ganglion is the most voluminous, being fused with the first thoracic ganglion to form the stellate ganglion, and lies just posteriorly to the origin of the vertebral artery. Afferent distribution to the cervical sympathetic ganglia arises from the thoracic sympathetic ganglia, which receive white (myelinated) rami communicantes from the spinal cord through spinal nerves (preganglionic fibres). Efferent impulses, through grey (unmyelinated) rami communicantes, are conveyed by the spinal nerves to the periphery and distributed to the various organs (postganglionic fibres), innervating their involuntary muscles and regulating secretory activity. The cervical sympathetic chain has a powerful vasomotor action, in the sense that its stimulation produces vasoconstriction and its interruption produces vasodilatation [1].

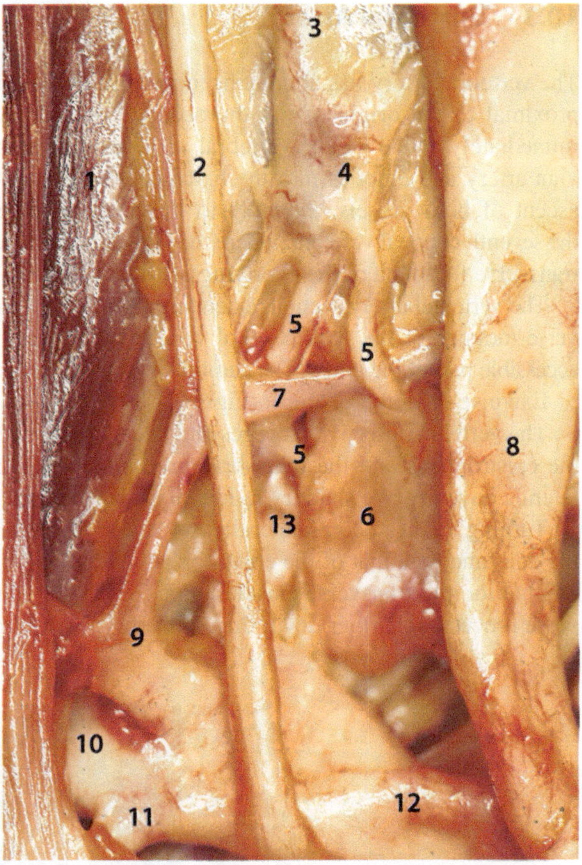

Fig. 10.2 Cervical sympathetic chain. *1* Anterior scalene muscle, *2* vagus nerve, *3* cervical sympathetic chain (superior portion), *4* middle cervical ganglion, *5* cervical sympathetic chain (inferior portion), *6* inferior cervical ganglion and first thoracic ganglion (stellate ganglion), *7* inferior thyroid artery, *8* common carotid artery, *9* thyrocervical trunk, *10* subclavian artery, *11* internal thoracic artery, *12* innominate artery, *13* vertebral artery

Complications: Injury to the iatrogenic cervical sympathetic chain is a very rare occurrence, less than 1 % [2]. Instead, the neoplastic infiltration of the deep plane following metastatic adenopathies or tumours of the apex of the lung is more frequent. We must also consider the section of the cervical sympathetic chain during radical neck dissection when the adenopathy involves the structure. In all these cases, a clinical syndrome is found (Claude Bernard–Horner's syndrome), characterised by ptosis of the eyelid, miosis, and enophthalmos and rarely associated with an increase in saliva viscosity, alterations of the cerebral flow, and pressor instability [3]. The enophthalmos is caused by paralysis of the eye bulb detrusor and ptosis of the eyelid by paralysis of the tarsal muscle. Miosis is caused by paralysis of the dilator pupillae. The innervating fibres run a long course: they exit from the spinal cord with the first thoracic nerve (brachial plexus) and, through a communicating branch, reach the stellate ganglion, from which they ascend to the eye along the cervical sympathetic trunk. This course explains how pupillary alterations can also result from lesions to the brachial plexus, involving the first thoracic nerve at its origin (apex of the lung, upper mediastinum).

In the dissection, the three sympathetic ganglia and some communicating branches are identified and isolated; in particular there is the constant presence of a communicating branch between the middle ganglion and the stellate ganglion, which forms an eyelet around the inferior thyroid artery (Fig. 10.2).

10.3 Vertebral Artery

The vertebral artery has already been identified at its origin, which is immediately proximal to the origin of the thyrocervical trunk. The inferior thyroid artery is immediately above it. The vertebral vein, instead, passes anteriorly to the subclavian artery and empties into the brachiocephalic vein. Our dissection follows the ascent of both vessels, medially to the anterior scalene muscle, to the level of the seventh cervical vertebra, where they bend medially and embed by penetrating the transverse foramina of the overlying cervical vertebrae.

The vertebral artery section, extending from the origin to the entrance to the transverse foramen of the sixth cervical vertebra, is the surgical portion and most easily accessible part of the artery. The carotid tubercle is an excellent landmark (Fig. 10.3).

Exercise 9: vertebral artery (Fig. 10.4)

The vertebral artery re-emerges and lateralises between the transverse process of the epistropheus and the transverse process of the atlas, describing a curve with lateral convexity. We shall try to identify it between these two structures.

Turning the head contralaterally, we shall first identify the transverse process of the atlas and then that of the epistropheus. We look for the artery below, dissecting the interior intertransverse muscles, along a line that joins the apex of the two transverse processes. Farther down than the artery, with an oblique downward path, we

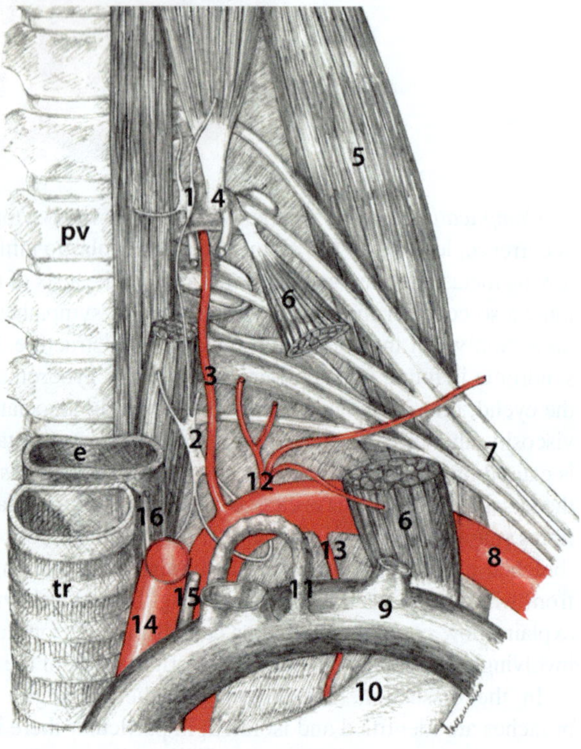

Fig. 10.3 Vertebral artery and carotid tubercle. *pv* Vertebral plane, *e* oesophagus, *tr* trachea, *1* middle cervical ganglion, *2* inferior cervical ganglion and first thoracic ganglion (stellate ganglion), *3* vertebral artery, *4* carotid tubercle, *5* medial scalene muscle, *6* anterior scalene muscle, *7* brachial plexus, *8* subclavian artery, *9* subclavian vein, *10* first rib, *11* thoracic duct, *12* thyrocervical trunk, *13* internal thoracic artery, *14* common carotid artery, *15* vagus nerve, *16* recurrent nerve

Fig. 10.4 Exercise 9: vertebral artery

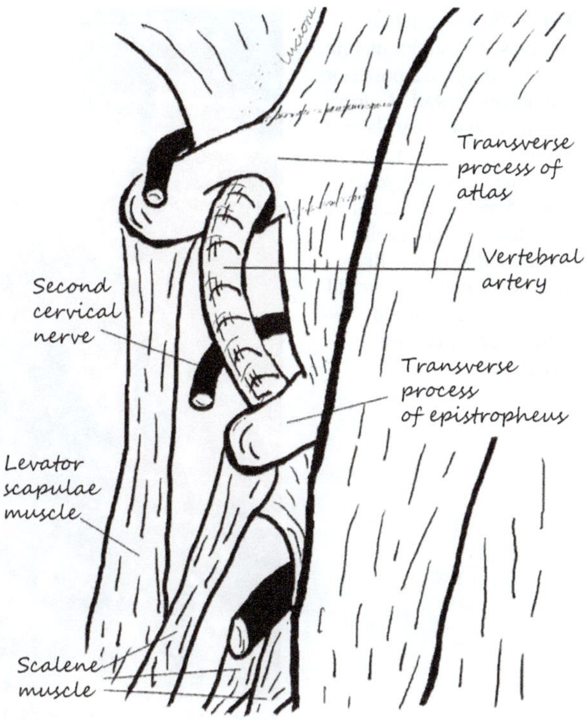

can identify the anterior branch of the second cervical nerve, which will form the cervical plexus lower down.

This procedure may also be carried out between the transverse processes of the underlying vertebrae, but it is easier to reach the artery between the atlas and the epistropheus.

10.4 Prevertebral Plane

At the end of dissection, the composition of the prevertebral plane should be examined. Inferiorly to the deep cervical fascia, it comprises four muscle groups:

1. The rectus capitis anterior muscles, extending from the basal surface of the occipital bone to the transverse processes of the atlas
2. The longus capitis muscles, extending from the basal surface of the occipital bone to the anterior tubercles of the third through sixth cervical vertebrae
3. The longus colli muscles, which are composite and extend from the transverse processes of the atlas to those of the fourth through sixth cervical vertebrae and second and third thoracic vertebrae
4. The intertransverse muscles, extending from one transverse vertebral process to the next (Fig. 10.5)

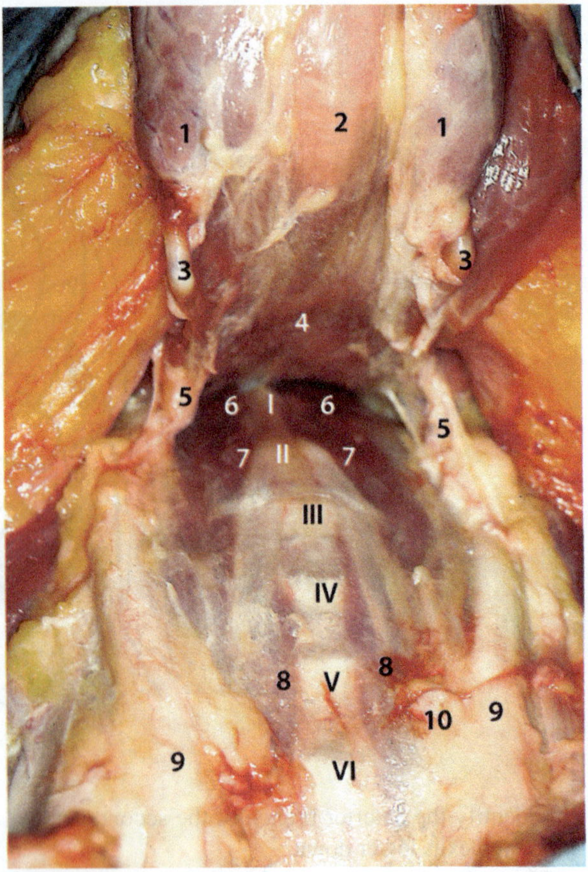

Fig. 10.5 Prevertebral muscles. *1* Posterior margin of thyroid lobes, *2* posterior hypopharynx wall, *3* superior cornu of thyroid cartilage, *4* posterior oropharyngeal wall, *5* greater cornu of hyoid bone, *6* rectus capitis anterior muscles, *7* longus capitis muscles, *8* longus colli muscles, *9* common carotid artery, *10* right carotid tubercle, *I–VI* cervical vertebrae

10.5 Epilogue

On the operating field there remain the prevertebral muscular plane, the internal carotid artery and the internal jugular vein, the cervical oesophagus and the open hypopharynx.

It is time to turn out the lights on our dissection exercise. All that is left for us to do is reconstitute the cadaver and wash the surgical instruments. Then sit down and think things over in front of a steaming hot cup of coffee.

Take-Home Messages

- The cervical sympathetic chain does not come from the skull but originates in the thorax and ends at the top just below the base of the skull.
- The deep cervical fascia that covers the prevertebral muscles may be used in demolitive surgery of the neck as an aid for the reconstruction of the hypopharynx.

References

1. Testut L, Jacob O (1977) Trattato di Anatomia Topografica. UTET, Turin
2. Calearo C, Teatini G (1983) Functional neck dissection: anatomical grounds, surgical techniques, clinical observations. Ann Otol Rhinol Laryngol 92:215–222
3. Stern SJ (1992) Cervical sympathetic trunk at the root of the neck. Head Neck 12:506–509

Microscopic Anatomy of the Larynx

11

Duilio Della Libera

Core Messages

- The structural organisation of the larynx is quite similar to that of the trachea. Schematically, the following may be distinguished: (a) a mucous membrane (epithelium and tunica propria), (b) an elastic lamina, (c) a submucous membrane (loose fibroadipose tissue, with the seromucous glands), and (d) a cartilaginous skeleton.
- The manner in which the laryngeal carcinoma grows and spreads depends on the anatomical site in which it originates. Knowing the compartmental organisation of the organ is useful for staging and for planning therapy.

Significant Anatomical Structures: quadrangular membrane, conus elasticus, cricoid cartilage, thyroid cartilage, arytenoid cartilage, Santorini's corniculate tubercle, Wrisberg's cuneiform tubercle, Reinke's space, thyrohyoid membrane, thyroepiglottic ligament, preepiglottic space, paraglottic space, Broyle's tendon, crista arquata, oblong fovea, subglottic region, arytenoid cartilage, anterior commissure.

11.1 Anatomical Layout

The larynx is the first highly differentiated segment of the lower airways with sphincteral, respiratory, and phonatory functions (voice box). It has a complex structural organisation [1], similar to that of the trachea, with a framework characterised by an external fibrocartilaginous skeleton and an internal elastic membrane, defining a submucosal membrane variously developed in the different regions of the organ [2, 3]. This framework, together with muscles, ligaments, and cartilages, is able, at least initially, to condition the means of growth and spread of neoplasias of the larynx [4, 5].

M. Lucioni, *Practical Guide to Neck Dissection*,
DOI 10.1007/978-3-642-33977-6_11, © Springer-Verlag Berlin Heidelberg 2013

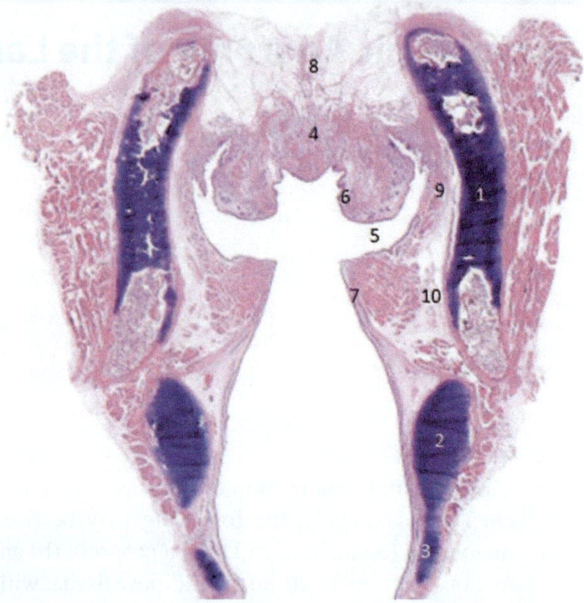

Fig. 11.1 Compartmental organisation of the larynx (coronal macrosection). *1* Thyroid cartilage, *2* cricoid cartilage, *3* trachea, *4* epiglottis, *5* Morgagni's ventricle, *6* quadrangular membrane, *7* conus elasticus, *8* preepiglottic space, *9* superior paraglottic space, *10* inferior paraglottic space

11.1.1 Elastic Skeleton

The most superficial support structure consists of an elastic membrane known as the quadrangular membrane in the supraglottic larynx and as conus elasticus in the glottosubglottic larynx (Fig. 11.1). The quadrangular membrane forms the skeleton of the aryepiglottic folds and extends down into the thickness of the ventricular bands right against the mucosal surface. It is attached at the bottom to the re-entrant angle of the thyroid cartilage, at the bottom to the corniculate cartilage, and the anterior margin of the arytenoid. It thickens at the bottom, forming the ventricular ligaments. In the anterior wall of the supraglottic larynx, the epiglottis, which is an elastic cartilage, represents the median focal chondrification of the quadrangular membrane.

At the glottosubglottic level [6], the elastic membrane forms the conus elasticus, a robust lamina with an external concavity which is reinforced at the top forming the vocal ligament, extending from the re-entrant angle of the thyroid cartilage to the vocal process of the arytenoid. Inferiorly, at the inferior margin of the vocal cord, the conus elasticus doubles forming two bands: the side one anchored to the superior margin of the cricoid cartilage, the median one in continuity with the elastic membrane of the trachea.

11.1.2 Fibrocartilaginous Skeleton

More externally, the framework of the larynx consists of the fibrocartilaginous skeleton which is formed of the thyroid, cricoid, and arytenoid cartilages with a

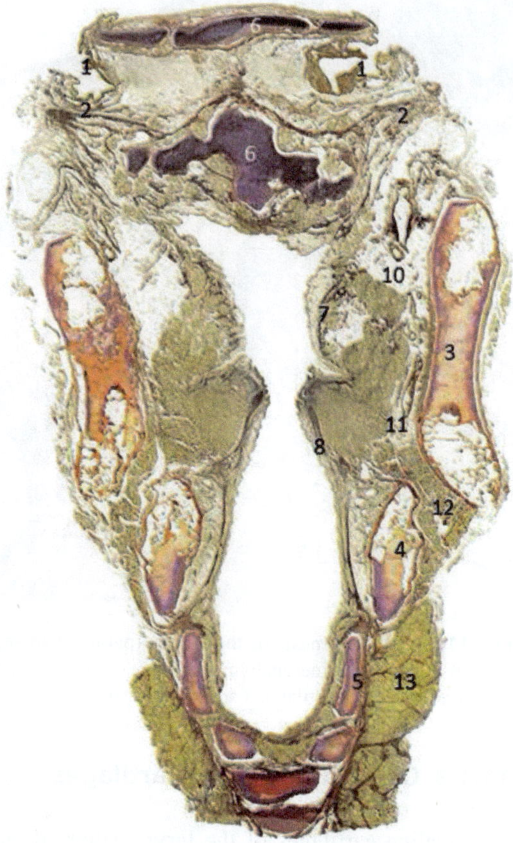

Fig. 11.2 Cartilages of larynx (coronal macrosection). *1* Glossoepiglottic valleculae, *2* hyoepiglottic membrane, *3* thyroid cartilage, *4* cricoid cartilage, *5* trachea, *6* epiglottis, *7* quadrangular membrane, *8* conus elasticus, *9* preepiglottic space, *10* superior paraglottic space, *11* inferior paraglottic space, *12* cricothyroid space, *13* thyroid gland

hyaline structure, the base in which the intrinsic and extrinsic musculature of the larynx are inserted and joined by ligaments and membranes (Fig. 11.2). The cricoid cartilage is the fundamental part of the organ which supports the entire framework of the larynx. It is articulated cranially with the inferior cornu of the thyroid cartilage with which it connects through the cricothyroid membrane with the arytenoids bilaterally. Caudally, it connects to the first tracheal ring through the cricotracheal ligament [7]. The thyroid cartilage forms the anterosuperior wall of the larynx by means of two laminae that join anteriorly at an angle and have at the posterior margin a superior cornu in which the lateral thyrohyoid ligament is inserted and an inferior cornu that is in contact with the cricoid. The arytenoid cartilages, skeletal datum points of the vocal cords, are pyramid-shaped and articulate at the bottom with the cricoid, continuing at the top with two elastic fibrocartilaginous nodule known as the Santorini's corniculate tubercle and Wrisberg's cuneiform tubercle.

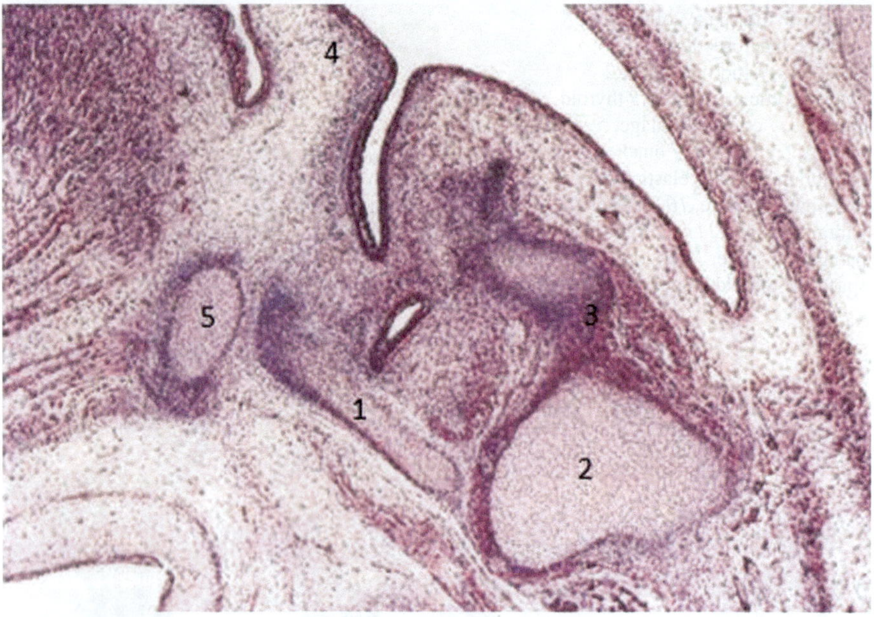

Fig. 11.3 Embryogenesis of the larynx (paramedian sagittal section of human foetus, 30 mm; condensation nuclei, mesenchyme cartilage of the fourth, fifth, and sixth branchial arch). *1* Thyroid cartilage, *2* cricoid cartilage, *3* arytenoid cartilage, *4* epiglottis, *5* hyoid bone

11.1.3 Ossification of the Cartilages

The hyaline cartilages of the larynx (thyroid and cricoid cartilages) [8] originate from a chondrification process of circumscribed condensation areas of the fourth, fifth, and sixth branchial arch (Fig. 11.3) which begins early in the first phases of life of the embryo. Within the third month of gestation, the hyaline cartilages have already reached a conformation similar to that of an adult (Fig. 11.4). The epiglottis and the corniculate and cuneiform cartilages are elastic cartilages and do not ossify. The arytenoids have a composite structure.

The hyaline cartilages tend to ossify in adulthood as from the third decade of life with an endochondral ossification mechanism as a response whereby the fibrocartilaginous skeleton of the larynx adapts to tensile stress at the point of insertion of the muscles, ligaments, and membranes. The ossification process [9] of the thyroid cartilage generally starts in the lower half of the dorsal margin of the lamina, near the origin of the inferior cornu. From there, it extends upwards along the posterior margin towards the superior cornu and along the inferior margin towards the median line, where a further centre of ossification generally appears that tends to merge prematurely with the previous one. The superior edge is then involved, and, in the advanced stages of the process, the entire lamina may be ossified with the exception of its central part. In the cricoid cartilage [10], the ossification process begins at the superior edge of the arch, anteriorly with respect to the cricothyroid articulation, and spreads to the arch and to the bezel in an anteroposterior direction.

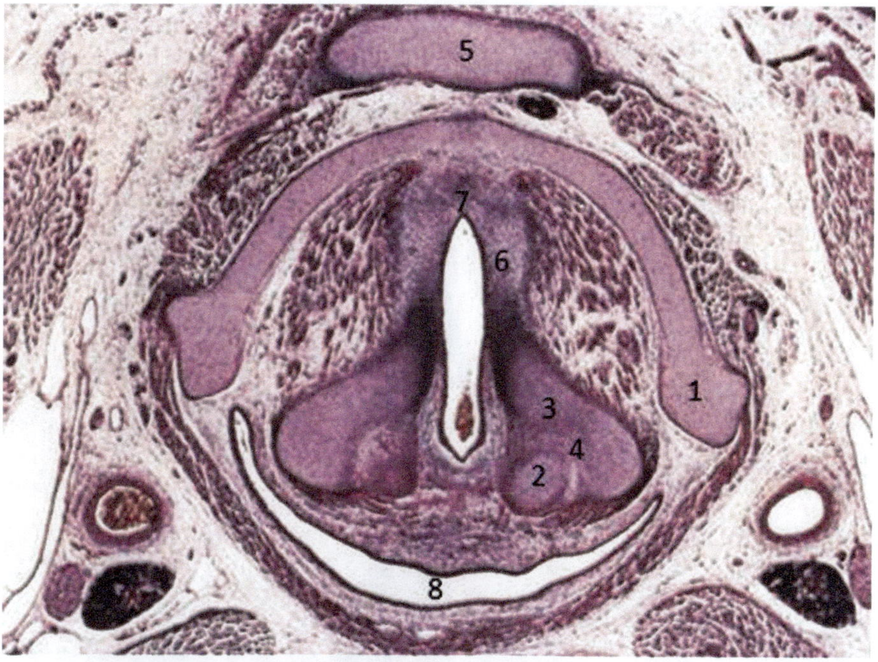

Fig. 11.4 Embryogenesis of the larynx (transverse section of human foetus, 53 mm; the laryngeal cartilages have reached the typical conformation of an adult). *1* Thyroid cartilage, *2* cricoid cartilage, *3* arytenoid cartilage, *4* cricoarytenoid joint, *5* hyoid bone, *6* vocal cord, *7* anterior commissure, *8* oesophagus

The ossification of the arytenoid begins at its muscular process and spreads to the entire cartilage, sparing the elastic apex and vocal process.

11.2 Larynx Mucosa

From the embryological point of view, the organogenesis of the larynx derives from the union of two buds, a superior supraglottic bud of branchial origin and an inferior glottosubglottic bud of non-branchial origin. The supraglottic larynx, a sphincteral structure of the aerodigestive crossroads, presents a cylindrical epithelial coating of a respiratory type with variously developed areas of squamous metaplasia, a rich supply of seromucous glands, and a developed lymphatic network. The glottic larynx is characterised by a mucous coating consisting of a composite pavement epithelium and by the negative or poor presence of seromucous glands. Lymphatic vascularisation is poor at cord level, more evident at the level of the anterior commissure, and very rich at the level of the posterior commissure. Lastly, the subglottic larynx presents a structure similar to the trachea as regards the cylindrical coating of a respiratory type, the distribution of the glands, and the lymphatic network. The epithelium lies on the tunica propria, which at cord level is known as Reinke's space. The elastic lamina, which thickens at cord level to form the vocal ligament, separates the mucosa from the submucosa.

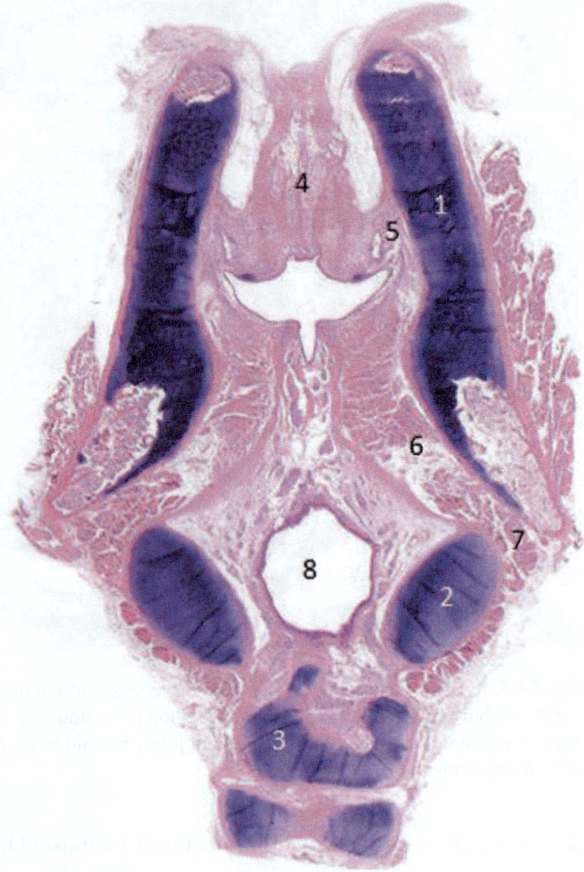

Fig. 11.5 Submucosal spaces of the larynx (coronal macrosection). *1* Thyroid cartilage, *2* cricoid cartilage, *3* trachea, *4* preepiglottic space, *5* superior paraglottic space, *6* inferior paraglottic space, *7* cricothyroid space, *8* subglottis

11.3 Larynx Submucosa

The submucosa of the larynx (Fig. 11.5) consists of loose and adipose connective tissues containing vascular and nerve structures [11]. It forms a single submucosal compartment, subdivided by ligaments and membranes into spaces or cavities, variously developed in the different regions of the larynx [12, 13]. It reaches its greatest development in the supraglottis, being formed anteriorly by the preepiglottic space and laterally by the superior paraglottic or paralaryngeal spaces.

11.3.1 Preepiglottic Space

The preepiglottic space appears bounded cranially by collagen fibres lying in an anteroposterior direction, stretching from the tongue base to the periosteum of the superior edge of the hyoid bone [14]. From the tongue base, along the mucous profile of the glossoepiglottic valleculae, some fibres lead posteriorly towards the anterior perichondrium of the free part

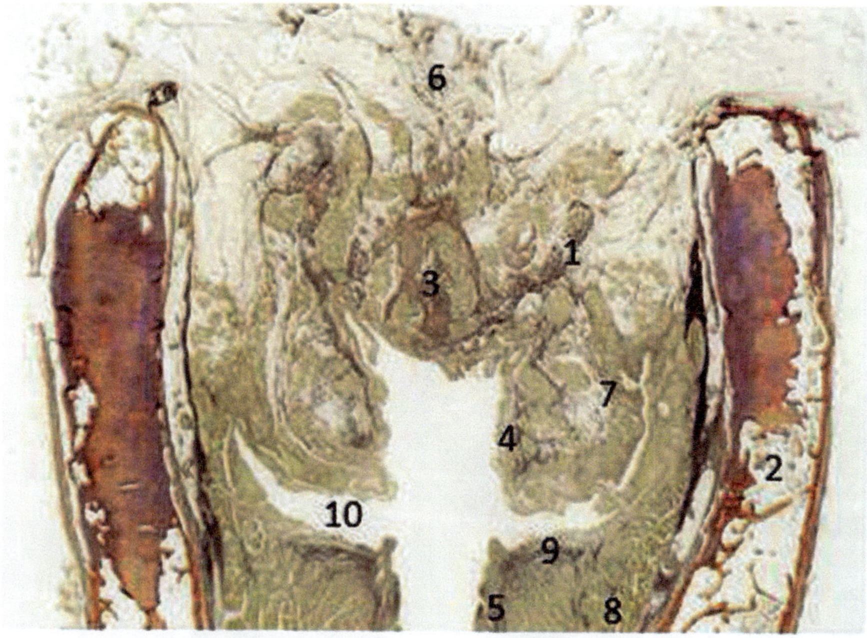

Fig. 11.6 Preepiglottic space (coronal macrosection). *1* Thyroepiglottic ligament, *2* thyroid cartilage, *3* epiglottis, *4* quadrangular membrane, *5* conus elasticus, *6* preepiglottic space, *7* superior paraglottic space, *8* inferior paraglottic space, *9* thyroglottic ligament, *10* Morgagni's ventricle

of the epiglottis, forming the hyoepiglottic membrane. The central part of the membrane is reinforced forming the median hyoepiglottic ligament which extends caudally to the lingual surface of the epiglottis and contains numerous groups of glands. Medially, some fibres of the ligament reflect cranially and are radially anchored to the perichondrium of the free part of the epiglottis [15]. Bilaterally, collagen fibres lead from the free margin of the epiglottis to the greater cornua of the hyoid bone.

Anteriorly, the preepiglottic space is bounded by the thyrohyoid membrane extending between the hyoid bone and the superior margin of the thyroid cartilage [16]. At this level, the median thyrohyoid ligament constitutes a median thickening of the membrane. The lateral parts of the thyrohyoid membrane insert in the inferior edge of the greater cornua of the hyoid bone. The thyrohyoid ligament is anchored to the superior margin of the body of the hyoid bone, merging with the fibres of the median hyoepiglottic ligament. Inferiorly, the thyrohyoid membrane is anchored to the perichondrium of the anterior surface of the superior margin of the thyroid cartilage [17]. Caudally, the preepiglottic space is bounded by the fibres of the thyroepiglottic ligament which medially connect the foot of the epiglottis to the superior edge of the incisure of the thyroid cartilage and continue laterally with the fibres of the quadrangular membrane. Dorsally, at the level of the hyoid bone, the preepiglottic space continues from the lateral margins of the epiglottis towards the aryepiglottic folds, without anatomical boundaries.

The preepiglottic space (Fig. 11.6) has roughly the shape of a truncated cone, decreasing in volume in a craniocaudal direction. Furthermore, its superior portion, the one corresponding to the greater cornu of the hyoid bone, is composed prevalently

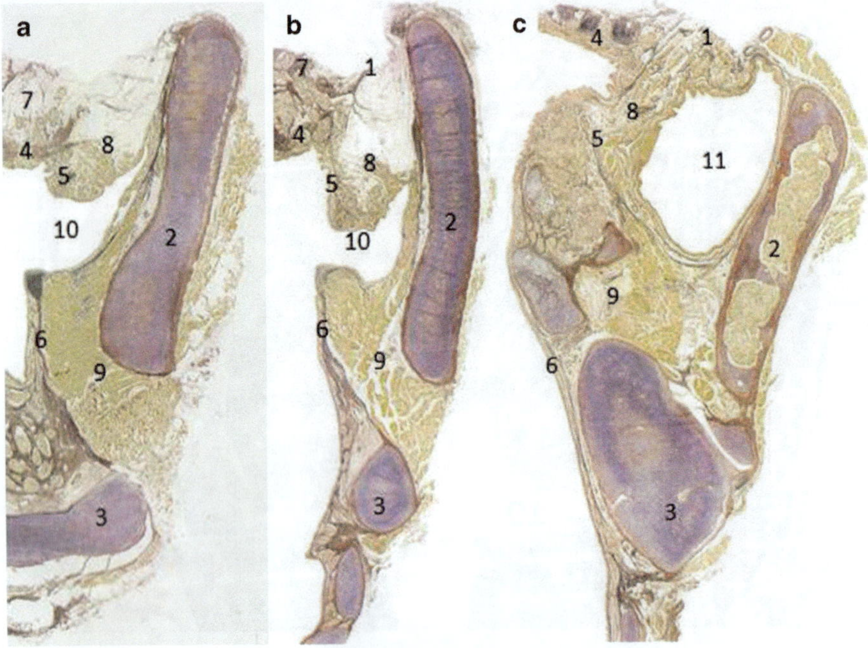

Fig. 11.7 Paraglottic spaces (coronal macrosection of the larynx along anterior third (**a**), middle third (**b**), and posterior third (**c**) of glottic plane). *1* Thyroepiglottic ligament, *2* thyroid cartilage, *3* cricoid cartilage, *4* epiglottis, *5* quadrangular membrane, *6* conus elasticus, *7* preepiglottic space, *8* superior paraglottic space, *9* inferior paraglottic space, *10* Morgagni's ventricle, *11* piriform sinus

of adipose tissue subdivided into lobules and rich in blood vessels, especially in the median portion [18]. Caudally, the space is subdivided into a central portion and two lateral ones by two discontinuous sagittal fibroconnective septums that originate from the thickening of the connective fibres that surround the seromucous glands present along the lateral margin of the epiglottis cartilage. Dorsally, these septums continue bilaterally along the quadrangular membranes. Ventrally, they are anchored to the incisure of the thyroid cartilage, continuing with the collagen fibres of the thyroepiglottic ligament. The lateral portions of the space contain adipose tissue in immediate contact with the perichondrium of the thyroid cartilage which reduces in a craniocaudal direction down to the level of the thyroepiglottic ligament.

11.3.2 Paraglottic Spaces

The paraglottic space (Fig. 11.7), a submucosal compartment characterised by a prevalent vertical craniocaudal orientation, is bounded laterally by the thyroid cartilage, dorsally by the piriform recess, and medially by the quadrangular membrane and the elastic cone [19, 20]. Cranially, it continues with the preepiglottic space. It contains loose connective tissue, adipose tissue with vessels, and includes Morgagni's

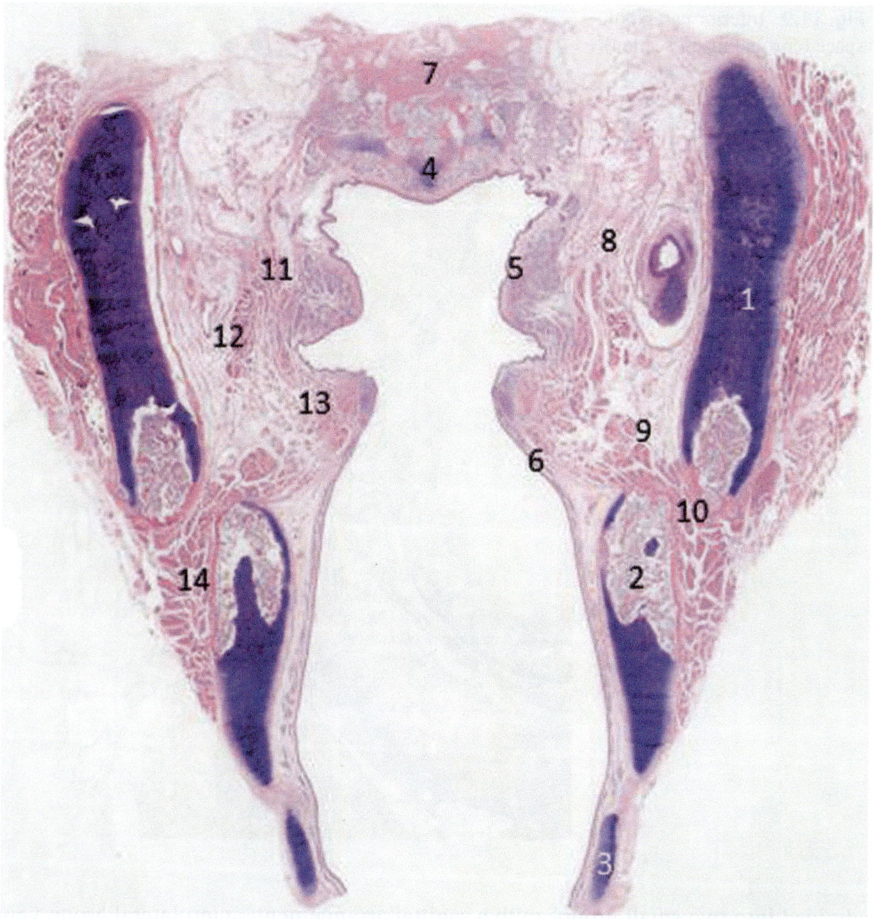

Fig. 11.8 Intrinsic muscles of the larynx (coronal macrosection). *1* Thyroid cartilage, *2* cricoid cartilage, *3* trachea, *4* epiglottis, *5* quadrangular membrane, *6* conus elasticus, *7* preepiglottic space, *8* superior paraglottic space, *9* inferior paraglottic space, *10* cricothyroid space, *11* anterolateral muscular system of ventricular fold, *12* anteromedial muscular system of ventricular fold, *13* thyroarytenoid muscle, *14* cricothyroid muscle

ventricle and its saccule and develops entirely in the superior vestibular portion of the ventricular bands. Laterally, the paraglottic space adjoins the lateral lamina of the thyroid cartilage. Inferomedially, it is bounded by the conus elasticus. Inferoanteriorly, it contacts the median cricothyroid ligament and the cricothyroid membrane. Its superior portion, principally in contact with the ventricular bands, continues without interruption superomedially with the preepiglottic space [21]. Sometimes a fibrous thickening is present discontinuously in the passage between the two compartments at the level of the angular region of the vestibule.

Supporting the intrinsic laryngeal musculature (Fig. 11.8), a minor muscular system with adductor functions is located at the level of the ventricular bands. It is

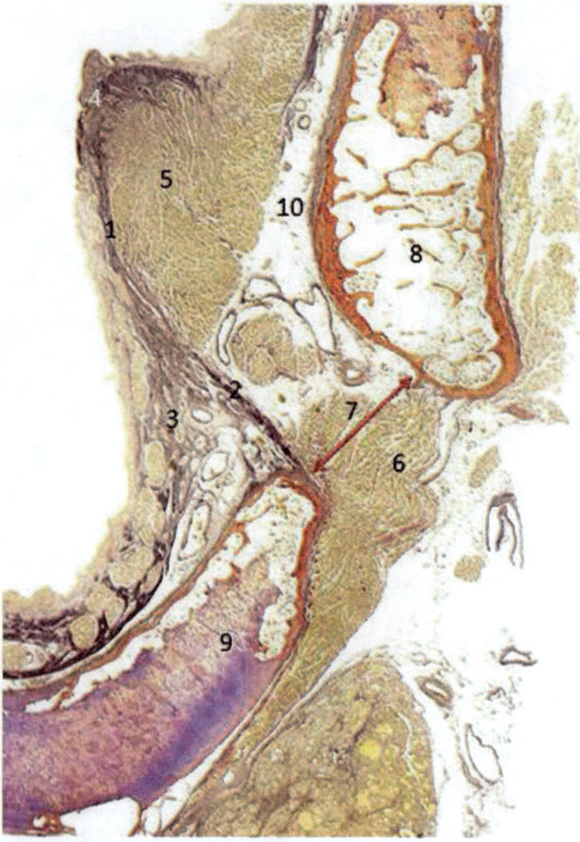

Fig. 11.9 Inferior paraglottic space (coronal macrosection). *1* Conus elasticus, *2* lateral portion of conus elasticus, *3* medial thin portion of conus elasticus, *4* vocal ligament, *5* vocal muscle, *6* cricoarytenoid muscle, *7* cricothyroid space (*red arrow*), *8* thyroid cartilage, *9* cricoid cartilage, *10* inferior paraglottic space

divided into two small groups with a sagittal arrangement: anterolateral group ventrally anchored to the perichondrium of the thyroid cartilage and anteromedial group with fibres arranged medially and dorsally with respect to Morgagni's ventricle.

At the glottosubglottic level [22], due to the anatomical complexity of the region, the submucosal membrane is reduced due to the presence of the vocal muscle. At this level, the conus elasticus adhering to the perimysium of the vocal muscle briefly interrupts the submucosal membrane which at this level is known as the inferior paraglottic space and is reduced to a thin layer of fibroadipose tissue in contact with the perichondrium of the thyroid lamina [23] (Fig. 11.9). Superiorly, at the free edge of the cord, the conus elasticus thickens forming the vocal ligament, extending between the vocal process of the arytenoid and the dihedral angle of the thyroid. The anterior convergence of the two vocal ligaments forms the anterior commissure tendon, or Broyle's tendon, closely anchored to the internal surface of the dihedral angle of the thyroid.

Posteroinferiorly, the adipose tissue of the paraglottic space is continuous with the loose tissue located around the cricoarytenoid joint [24]. At this level the anterolateral surface of the arytenoid, starting from its apex, presents a depression, the triangular fossa over which is the colliculus of the arytenoid, an area rich in seromucous glands in contact

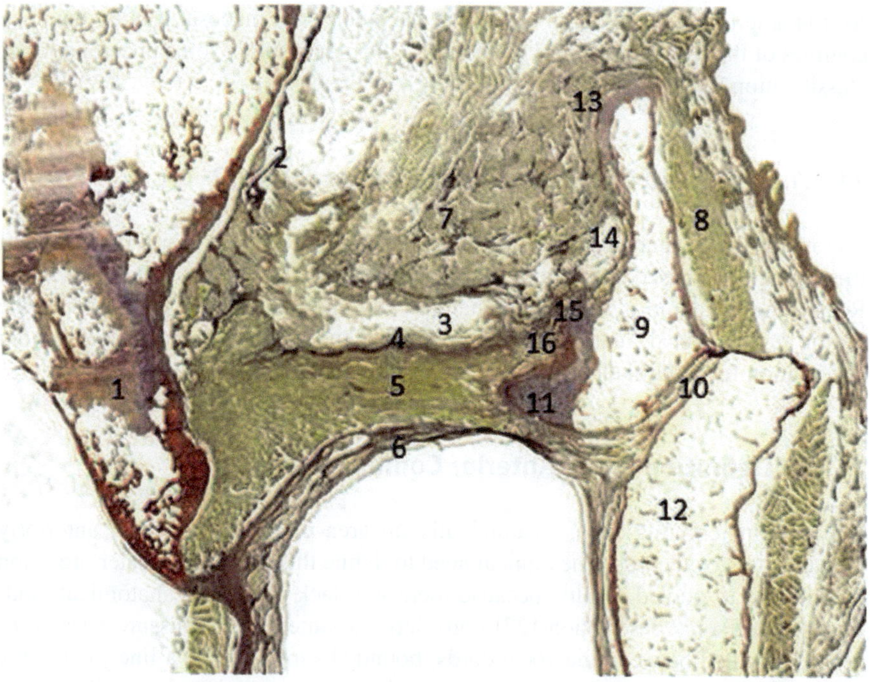

Fig. 11.10 Inferior paraglottic space (sagittal paramedian macrosection). *1* Thyroid cartilage, *2* thyroepiglottic ligament, *3* Morgagni's ventricle, *4* vocal ligament, *5* thyroarytenoid muscle, *6* conus elasticus, *7* vestibular ligament, *8* transverse arytenoid muscle, *9* arytenoid cartilage, *10* cricoarytenoid joint, *11* vocal process of arytenoid cartilage, *12* cricoid lamina, *13* colliculus, *14* triangular fossa, *15* crista arcuata, *16* oblong fossa

with the dorsal part of the laryngeal ventricle. Beneath are the crista arcuata, the area of insertion of the cranial fibres of the vocal ligament and of the vestibular ligament, and the oblong fovea with the insertion of the fibres of the thyroarytenoid muscle (Fig. 11.10).

The subglottic region of the larynx [25] corresponds to the cricoid area that is occupied laterally by the elastic cone, which at this level subdivides into a lateral band of greater consistency, that inserts in the superior edge of cricoid cartilage and bounds the circothyroid space from below. The medial band of the elastic cone continues at the bottom with the elastic membrane of the trachea, not before forming anteriorly in the hypoglottic cone a dense elastic web that reduces the submucosa to an almost virtual space.

11.4 Sites and Subsites

The clinical and postsurgical staging of malign tumours (TNM) has reached its seventh edition [26]. It is a fundamental knowledge basis for comparing data and the results of the various antineoplastic treatments. It is a clinically oriented classification, so it is able to "see" the organ only from the side of the mucosa, leaving the deep confines of the sites and subsites undefined. This may result in some uncertainties of classification [2].

11.4.1 Arytenoid Cartilage: Glottis or Supraglottis?

As regards the arytenoids, these are traditionally assigned to the marginal portion of the supraglottis, based on the criterion of continuity of the mucosa that covers them. Really, the arytenoid is a structure anatomically and functionally pertaining to the glottic region. Vocal ligament, vocal process of the arytenoid, and cricoarytenoid articulation form the skeletal framework of the vocal cord.

11.4.2 Definition of the Anterior Commissure

The anterior commissure is anatomically an area of mucosa situated anteriorly between the vocal cords. The clinical need to define this area with greater precision has not led to pooled results, because there is a lack of precise anatomical landmarks. Oloffson's description [27] considers the anterior commissure an area of mucosa located between the vocal cords, bounded superiorly by a line joining the angle of the ventricles and extending inferiorly for not more than 3 mm. The AJCC and Ogura [28] consider the glottis, and therefore also the anterior commissure, as being included between the horizontal plane passing through the free margin of the vocal cord and a virtual plane located 1 cm more caudally. If we consider as the glottis only that which is covered by composite pavement epithelium, the matter becomes more complicated because the height of the epithelium reaches a maximum of 5 mm at the middle third of the cord and tends to get thinner, and in some cases to disappear, at the anterior level.

Stell and Morton [29] say that "the anterior commission is a point rather than an area, which tumours may pass through, but from which they cannot originate".

11.4.3 Boundary Between Glottis and Subglottis

All authors do not agree on this definition either. Oloffson [27] considers the subglottis the inferior face of the vocal cord. Carlon [1] reckons that the boundary could be set where the elastic cone separates into its two components, the one pertaining to the mucous membrane and the one pertaining to the cartilaginous skeleton (cricoid).

This limit corresponds to a plane about 1 cm from the free margin of the vocal cord, and this agrees with what Ogura has established concerning the caudal limits of the anterior commissure.

Take-Home Messages

- The larynx has a complex compartmental structure with the presence of muscle and ligament barriers that are able to condition and sometimes limit the growth and spread of neoplasias, at least in the initial phases.
- The areas of weakness are generally characterised by a reduced submucosal membrane (anterior commissure), by a particular abundance of seromucous glands (area under the anterior commissure, subglottis, cricoarytenoid articulation) or by the vicinity of bordering membrane structures (cricothyroid membrane, posterior portion of the inferior paraglottic space). The possible involvement of these areas must be carefully evaluated during preoperative staging.

References

1. Carlon G (1990) Il carcinoma della laringe. Dalla patologia alla clinica. Ed. Piccin, Padova
2. Carlon G, Della Libera D, Dei Tos AP (1990) A whole organ sectioning method for histologic examination of laryngeal and hypopharyngeal specimens. In: Sacristan T (ed) Otolaryngology, head and neck surgery. Kugler & Ghedini, Amsterdam, pp 2367–2371
3. Kirchner JA (1989) Fifteen Daniel C. Baker, Jr, memorial lecture. What have whole organ section contributed to the treatment of laryngeal cancer? Ann Otol Rhinol Laryngol 98(9):661–667
4. Blitzer A (1979) Mechanisms of spread of laryngeal carcinoma. Bull N Y Acad Med 55: 813–821
5. Glanz HK (1984) Carcinoma of the larynx. Adv Otorhinolaryngol 32:1–123
6. Carter DR, Meyers AD (1979) Anatomy of the subglottic larynx. Otolaryngol Head Neck Surg 87(2):203–206
7. Olofsson J, van Nostrand AWP (1973) Growth and spread of laryngeal and hypopharyngeal carcinoma with reflections on the effect of preoperative irradiation. Acta Otolaryngol Suppl (Stockh) 308:1–84
8. Turk LM, Hogg DA (1993) Age changes in the human laryngeal cartilages. Clin Anat 6:154–162
9. Carter RL, Tanner NSB (1979) Local invasion by laryngeal carcinoma the importance of focal (metaplastic) ossification within laryngeal cartilage. Clin Otolatyngol 7:283–290
10. Kirchner JA (1984) Invasion of the framework by laryngeal cancer. Acta Otolaryngol (Stockh) 27:392–397
11. Pressman JJ (1956) Submucosal compartmentation of the larynx. Ann Otol Rhinol Laryngol 65:766–771
12. Kirchner JA, Carter D (1987) Intralaryngeal barriers to the spread of cancer. Acta Otolaryngol (Stockh) 103:503–513
13. Kirchner JA (1996) Relative prognostic importance of histologic invasion of the laryngeal framework by hypopharyngeal cancer. Ann Otol Rhinol Laryngol 7(5):101–108
14. Reidenbach MM (1996) The peri-epiglottic space: topographic relations and histologic organisation. J Anat 188:173–182
15. Silver CE (1981) Surgery for cancer of the larynx and related structures. Churchill Livingstone, New York
16. Micheau C, Luboinski B, Sancho H, Cachin Y (1976) Modes of invasion of cancer of the larynx. Cancer 38:346–360

17. Kleinsasser O (1987) Tumoren des Larynx und des Hypopharynx. Thieme, Stuttgart
18. Tucker GF (1974) The anatomy of laryngeal cancer. Can J Otolaryngol 3:417–431
19. Kirchner JA, Carter D (1987) Intralaryngeal barriers to the spread of cancer. Acta Otolaryngol 103(5–6):503–513
20. Tucker GF, Smith HR (1962) A histological demonstration of the development of laryngeal connective tissue compartments. Trans Am Acad Ophthalmol Otolaryngol 66:308–318
21. Meyer-Breiting E, Burkhardt A (1988) Tumours of the larynx. Springer, Berlin/Heidelberg/New York
22. Stracchi R, De Pasquale V et al (1992) Particular structure of the anterior third of the human true vocal cord. Acta Anat 7(5):189–194
23. Mu L, Sanders I, Wu BL, Biller HF (1994) The intramuscular innervation of the human interarytenoid muscle. Laryngoscope 104:33–39
24. Welsh LW, Welsh JJ, Behlke FM (1962) Analysis of laryngeal compartments. Ann Otol Rhinol Laryngol 71:913–921
25. Welsh LW, Welsh JJ, Rizzo TA (1989) Internal anatomy of the larynx and the spread of cancer. Ann Otol Rhinol Laryngol 98:228–234
26. Union Internationale Contre le Cancer (2010) TNM atlas: illustrated guide to the TNM/pTNM classification of malignant tumours, VIth edn. Springer, Berlin
27. Oloffson J (1974) Growth and spread of laryngeal carcinoma. Can J Otol 3:446–59
28. AJCC (1988) Manual for staging of cancer, 3rd edn. Lippincott Company, Philadelphia
29. Stell PM, Morton RP (1984) What is the glottis? In: Wigand ME, Steiner W, Stell PM (eds) Laryngectomy: conservation surgery for carcinoma of the larynx. Springer, New York

Pattern of Spreading of Larynx Cancer 12

Duilio Della Libera

Core Messages

- The structural organisation of the larynx and some of its individual anatomical components determine the way in which the carcinoma spreads. This can be clearly seen from the studies of histological macrosections of the neoplastic larynx.

12.1 Laryngeal Structure and Neoplastic Spreading

The subdivision of the larynx into sites and subsites (supraglottis, glottis, and subglottis) is amply justified from the oncological viewpoint because it is useful for defining the prognosis and planning therapy for the laryngeal carcinoma. The means of neoplastic growth depend partly on the intrinsic properties of the tumour population but also on their site of origin and on the anatomical structures involved in their growth [1, 2]. In the process of neoplastic spread, assessed according to three-dimensional criteria [3], definite development vectors prevail which, at least in the initial phases, are conditioned by the anatomical permeability of the site of origin and by the presence and conformation of submucosal compartments and of ligamental and osteocartilaginous barriers [4]. The different means of growth also involve a different pattern of metastatic spread, for different access to the different laryngeal lymphatic networks.

For each laryngeal site, series of coronal and sagittal macrosections of pathological larynxes will be proposed, with the aim of identifying and discussing their tumoural growth patterns.

12.1.1 Infiltration of the Cartilages

Infiltration of the laryngeal cartilages is most frequent in their ossified portion. Bone cavities are occupied by adipose tissue and by well-vascularised hematopoietic tissue which do not offer any important mechanical resistance to neoplastic growth.

M. Lucioni, *Practical Guide to Neck Dissection*,
DOI 10.1007/978-3-642-33977-6_12, © Springer-Verlag Berlin Heidelberg 2013

The hyaline cartilage has a compact organisation and is without blood vessels, so it offers greater resistance [1].

The ossified areas are present, from the third decade of life, at the point of insertion of ligaments, membranes, and muscles, mostly as a response to mechanical stress stimuli. In these areas, there is an interruption of the perichondral barrier and the development of a rich vascular network [5].

The most important sites of invasion of the cartilage are (1) the dihedral angle of the thyroid cartilage where the tendon of the anterior commissure is inserted, (2) the insertions of the cricothyroid membrane in the corresponding cartilages, (3) the anterior portion of the thyroid lamina near the origin of the vocal muscle, (4) the posterior margin of the thyroid lamina adjacent to the piriform sinus, and (5) the cricoarytenoid articular capsule [9].

12.2 Supraglottic Carcinoma

The definition of the site of origin of the supraglottic carcinoma is not easy because by the time of diagnosis, the tumour is frequently already evolved, involves a broad surface of mucosa, and may develop asymmetrically in both surface and depth. In less than a quarter of cases of supraglottic carcinoma, only one laryngeal subsite is involved at the moment of diagnosis, while in the remaining three quarters of cases, two or more subsites are involved. There are various classifications of supraglottic carcinoma, depending on the site of origin. We shall adopt a simple classification which subdivides supraglottic carcinomas into median (with symmetrical and asymmetrical development), lateral, and marginal.

Supraglottic carcinomas tend to grow for a long time within the submucosal spaces (preepiglottic space and superior paraglottic space), determining the late spread to the underlying glottic region as though there were a kind of barrier between the two laryngeal sites. Really, in the initial phases of their development, vestibular neoplasias are impeded medially by the vocal ligament and laterally by the ventricle. They tend to infiltrate the paraglottic spaces locally which, at this level, behave like real expansion vessels of neoplastic growth. The glottic extension is therefore a late event with the exception of median neoplasias which involve the region of anterior supracommissural region at the level of the pedicle of the epiglottis or which grow laterally to Morgagni's ventricle.

12.2.1 Median Supraglottic Carcinoma with Symmetrical Development

It originates from the central region of the laryngeal side of the epiglottis (Fig. 12.1) and extends symmetrically in all directions [6]. In the advanced phases, the ventricular bands and the aryepiglottic folds may be involved. In over two thirds of cases, the neoplasia prematurely invades the preepiglottic

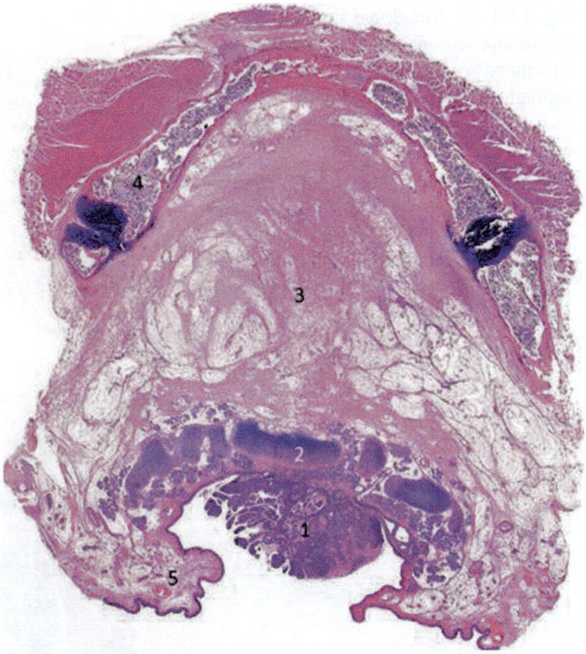

Fig. 12.1 Medial symmetrical supraglottic carcinoma (axial macrosection). *1* Superficial carcinoma of laryngeal surface of epiglottis, *2* epiglottis, *3* preepiglottic space, *4* hyoid bone, *5* aryepiglottic fold

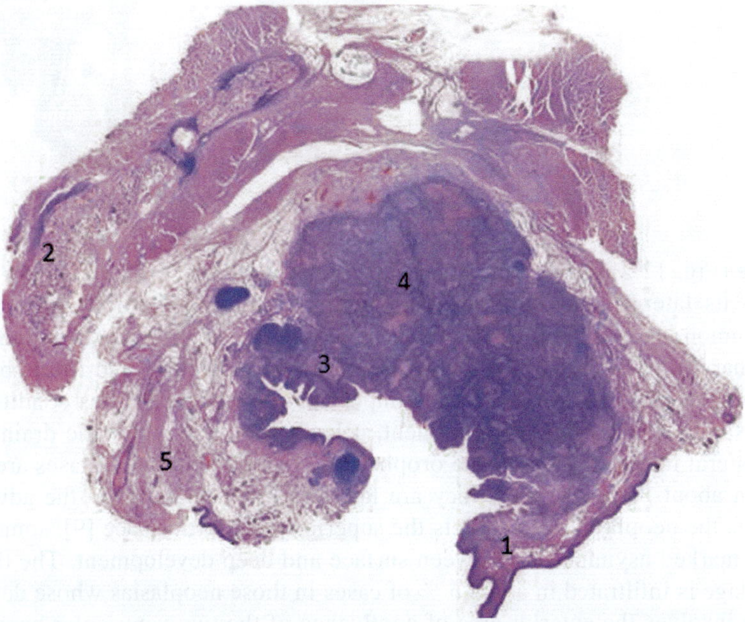

Fig. 12.2 Advanced medial supraglottic carcinoma (axial macrosection). *1* Aryepiglottic fold, *2* hyoid bone, *3* epiglottis, *4* neoplastic invasion of preepiglottic space, *5* superior paraglottic space

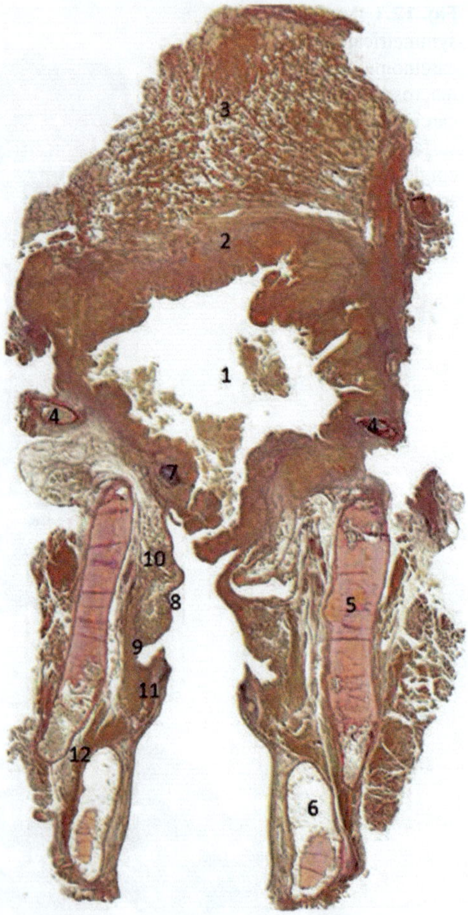

Fig. 12.3 Multiregional supraglottic carcinoma (coronal macrosection). *1* Advanced spread of multiregional supraglottic carcinoma with invasion of preepiglottic and paraglottic spaces and oropharyngeal region through hyoepiglottic membrane, *2* retrolingual region, *3* tongue, *4* hyoid bone, *5* thyroid cartilage, *6* cricoid cartilage, *7* epiglottis, *8* ventricular fold, *9* Morgagni's ventricle, *10* superior paraglottic space, *11* vocal cord, *12* cricothyroid space

space (Fig. 12.2) through the glandular foramina of the epiglottis, either encircling its lateral edges or destroying it. From the preepiglottic space [7, 8], the tumour may extend through the hyoepiglottic membrane to the adjacent oropharynx in the region of the glossoepiglottic valleculae and to the base of the tongue (Fig. 12.3). This fact is important because, as well as conditioning a less conservative surgical treatment, it leads to a new lymphatic drainage in the lateral lingual system of the oropharynx. Lymph nodal metastases are present in about 25 % of cases; they are late and often bilateral. In the advanced forms, the neoplasia often affects the superior paraglottic space [9], sometimes with marked asymmetries between surface and deep development. The thyroid cartilage is infiltrated in about 5 % of cases in those neoplasias whose development involves the anterior area of confluence of the two ventricular bands that

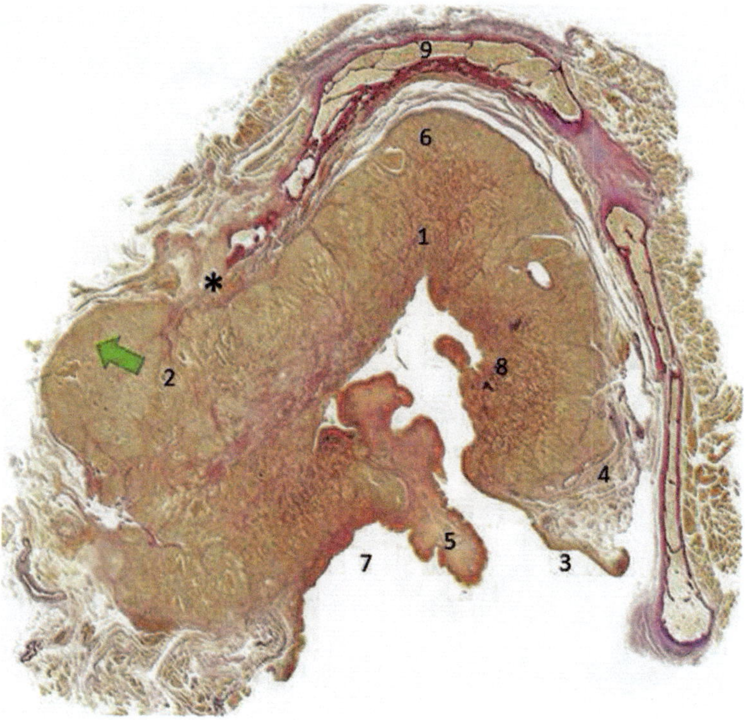

Fig. 12.4 Multiregional supraglottic carcinoma (axial macrosection). *1* Extensive vestibular spread of vestibular carcinoma with invasion of hyoid bone (*) and lateral lamina of thyroid cartilage, *2* lateral extralaryngeal diffusion, *3* ventricular fold, *4* superior paraglottic space, *5* aryepiglottic fold, *6* preepiglottic space, *7* piriform sinus, *8* epiglottis, *9* hyoid bone

correspond to the pedicle of the epiglottis and the underlying thyroepiglottic ligament (Fig. 12.4).

12.2.2 Median Supraglottic Carcinoma with Asymmetrical Development

Carcinoma of the larynx corner (Fig. 12.5) involves the join between the median portion of the epiglottis and a ventricular fold [10]. It is the most common supraglottic neoplasia (50 % of cases). In over 90 % of cases, the neoplasia makes an early infiltration of the submucosal membrane of the laryngeal vestibule, often with a marked asymmetry between the superficial mucosal component and the infiltrating component, which more frequently prevails over the former (Fig. 12.6). Due to their particular position on the border between the preepiglottic space and the superior paraglottic space, these neoplasias can grow both caudally and ventrally. In the latter

Fig. 12.5 Ventrolateral supraglottic carcinoma (larynx corner carcinoma) (axial macrosection). *1* Angular region, between ventricular fold and epiglottis, *2* epiglottis, *3* neoplastic invasion of preepiglottic space, *4* preepiglottic space, *5* aryepiglottic fold, *6* hyoid bone

case, anterosuperior growth leads to the invasion of the preepiglottic space (Fig. 12.7), often with destruction of the lateral margin of the epiglottis (Fig. 12.8). In some cases, through the preepiglottic space, they reach the adjoining oropharyngeal region of the glossoepiglottic valleculae and the base of the tongue. The second preferential direction of growth is posteroinferior, with infiltration of the superior paraglottic space and of the arytenoid region. The extension of the neoplasia beyond the lateral angle of Morgagni's ventricle is often accompanied by infiltration of the lateral lamina of the thyroid cartilage (Fig. 12.9), sometimes with extension to the medial wall of the piriform sinus. In this neoplasia, lymph node metastases are present in about half of the cases and are generally homolateral with the neoplasia.

12.2.3 Lateral Supraglottic Carcinoma

It arises in the mucosal surface of the ventricular folds in the area corresponding to the quadrangular membrane and in Morgagni's ventricle (Fig. 12.10). It often has a carpet-like surface growth pattern, and in half the cases, it infiltrates the superior paraglottic space [11]. The median vestibular regions of the larynx are involved by

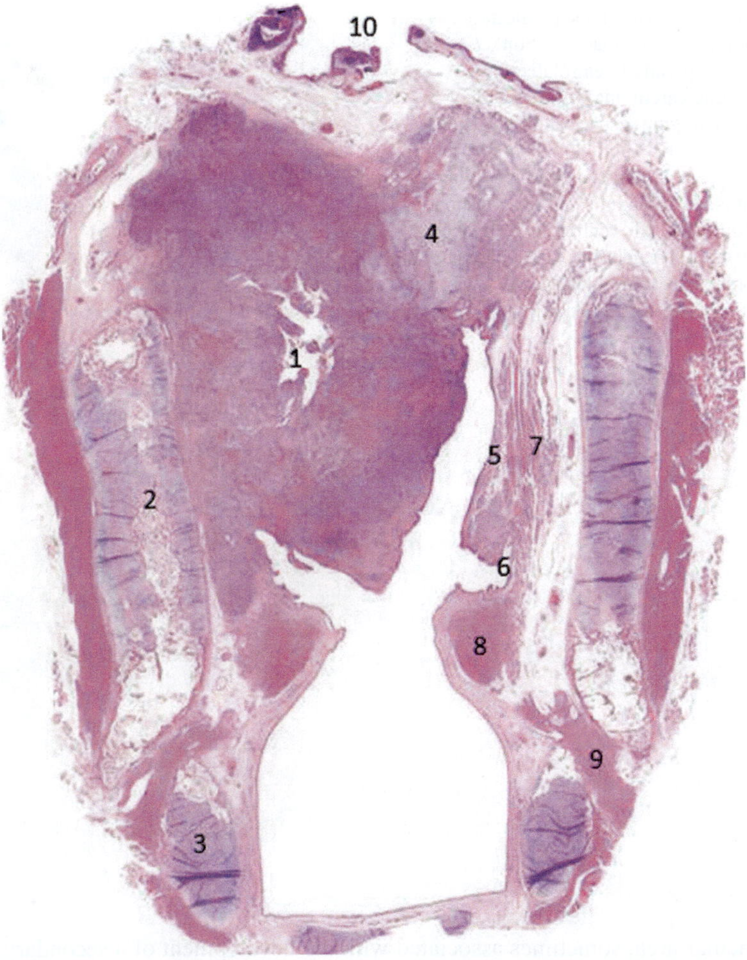

Fig. 12.6 Ventrolateral supraglottic carcinoma (larynx corner carcinoma) (coronal macrosection). *1* Advanced spread of lateral supraglottic carcinoma with invasion of paraglottic spaces through quadrangular membrane and preepiglottic space, *2* thyroid cartilage, *3* cricoid cartilage, *4* epiglottis, *5* ventricular fold, *6* Morgagni's ventricle, *7* superior paraglottic space, *8* vocal cord, *9* cricothyroid space, *10* glossoepiglottic valleculae

neoplastic growth at a late stage. Lymph nodal metastases are present in about one third of the cases and are generally homolateral.

As regards carcinoma of the ventricle [12], there is no general agreement on its definition. Some authors in fact use the term ventriculosaccular as a synonym for transglottic, while others consider it an independent neoplasia originating from the mucosa that covers the ventricle (Fig. 12.11). The primitive origin of the neoplasia of the ventricular mucosa is a rare event, considering the rarity of finding squamous metaplasia of the cylindrical ciliated epithelium of a respiratory type which habitually covers the ventricle. It is thought that the rare neoplasias confined only to the

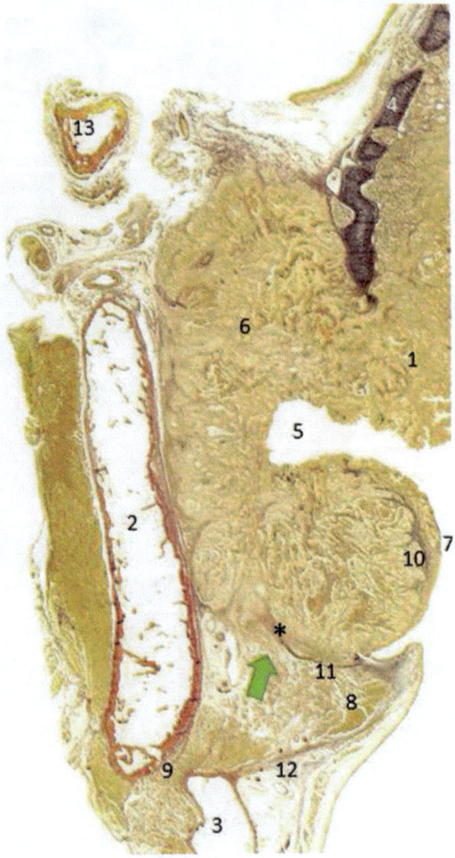

Fig. 12.7 Ventrolateral supraglottic carcinoma (coronal macrosection). *1* Advanced spread of ventrolateral supraglottic carcinoma with invasion of superior paraglottic space and Morgagni's ventricle; glottic–supraglottic boundaries (*), *2* thyroid cartilage, *3* cricoid cartilage, *4* epiglottis, *5* Morgagni's ventricle, *6* superior paraglottic space, *7* ventricular fold, *8* vocal cord, *9* cricothyroid space, *10* quadrangular membrane, *11* ventricular ligament, *12* elastic cone, *13* hyoid bone

ventricular area, sometimes associated with the development of a secondary laryngocele due to ectasia of the ventricular saccule, may derive from metaplastic areas of the ventricular mucosa or from the join between the superior edge of the vocal cord and the ventricle floor.

12.2.4 Marginal Supraglottic Carcinoma

It arises in the marginal region of the laryngeal vestibule [13, 14], that is, at the level of the free margin of the epiglottis, the aryepiglottic folds, and the apex of the arytenoid. In the first case, it may spread to the lingual side of the epiglottis, to the oropharyngeal region of the glossoepiglottic valleculae, and to the laryngeal side of the epiglottis.

The neoplasia of the aryepiglottic fold (Figs. 12.12 and 12.13) generally presents a surface growth on the medial wall of the piriform sinus or in the region of the

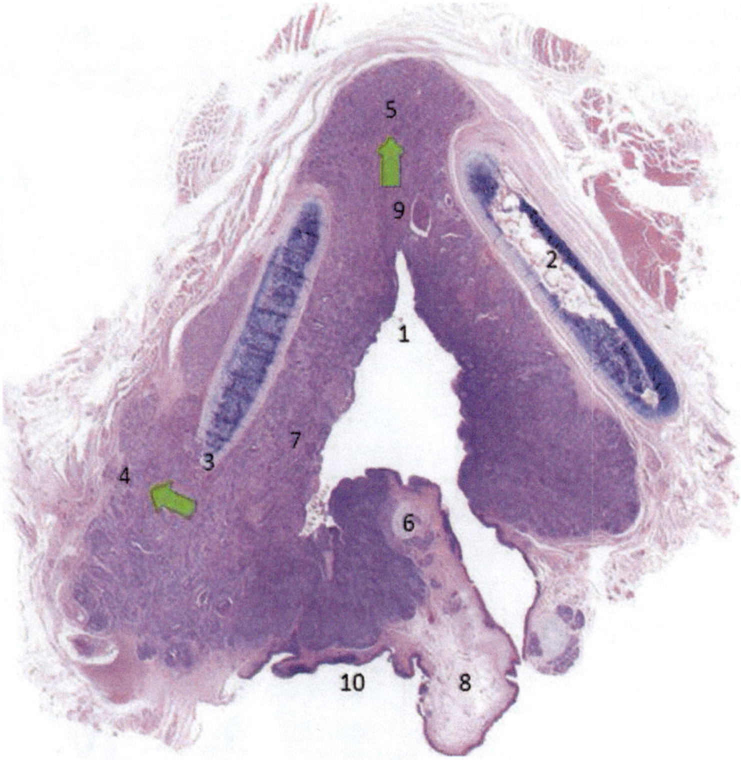

Fig. 12.8 Multiregional supraglottic carcinoma (axial macrosection). *1* Extensive vestibular spread of supraglottic carcinoma with invasion of paraglottic space, *2* thyroid cartilage, *3* infiltration of lateral lamina of thyroid cartilage, *4* lateral extralaryngeal diffusion, *5* medial extralaryngeal diffusion through anterior thyroid cartilage incisure, *6* ventricular fold, *7* superior paraglottic space, *8* aryepiglottic fold, *9* preepiglottic space, *10* piriform sinus

threefolds, the anatomical confluence of the aryepiglottic, glossoepiglottic, and pharyngoepiglottic folds.

Carcinoma of the apex of the arytenoid may extend to the medial wall of the piriform sinus, to the retroarytenoid area, and less commonly to the ventricular band and the aryepiglottic fold.

12.3 Glottic Carcinoma

Glottic carcinoma (Fig. 12.14) originates in most cases from the anterior portion of the vocal cord. In the initial phases, the neoplasia involves the free margin of the vocal cord which, due to the low supply of lymphatic vessels, is associated in this phase with a low risk of lymph nodal metastases. Afterwards, the tumour infiltrates

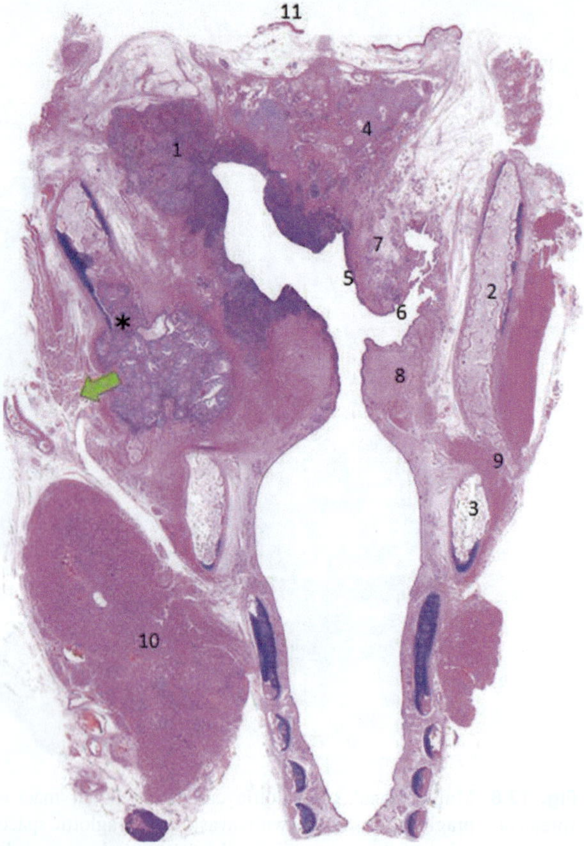

Fig. 12.9 Ventrolateral supraglottic carcinoma (coronal macrosection). *1* Advanced spread of ventrolateral supraglottic carcinoma with invasion of superior paraglottic space and lateral lamina of thyroid cartilage (*), *2* thyroid cartilage, *3* cricoid cartilage, *4* epiglottis, *5* ventricular fold, *6* Morgagni's ventricle, *7* superior paraglottic space, *8* vocal cord, *9* cricothyroid space, *10* thyroid, *11* glossoepiglottic valleculae

the vocal (thyroarytenoid) muscle and the inferior paraglottic space, and it may spread in a craniocaudal and lateral direction [15, 16].

12.3.1 Glottic Carcinomas Limited to the Glottic Site

In the initial phases, the carcinoma of the medio-anterior third of the vocal cord is characterised by an infiltration limited to the mucosa of the vocal ligament, to Reinke's space, and to the muscular fibres of the underlying thyroarytenoid muscle [17, 18]. Less frequently the glottic neoplasia arises in the posterior third of the cord (Fig. 12.15), with early involvement of the arytenoid cartilage, its vocal process, and sometimes the cricoarytenoid articulation. There are often peritumoural areas of dysplasia and carcinoma in situ. The initial glottic tumour of the anterior commissural area (Fig. 12.16) prematurely involves the vocal cords bilaterally and presents a high risk of infiltration of the dihedral angle of the thyroid cartilage (Fig. 12.17), even in the case of initial neoplasias with small dimensions (Fig. 12.18). Afterwards, the neoplasia infiltrates the thyroarytenoid muscle in depth and tends to involve the

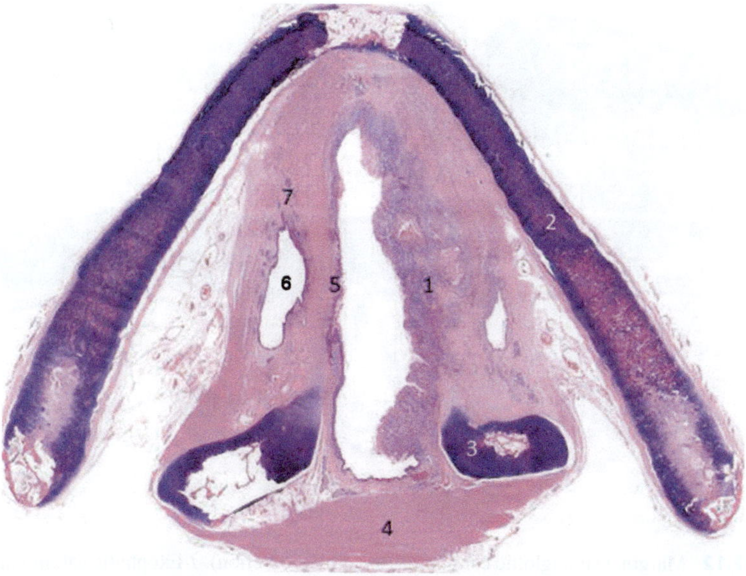

Fig. 12.10 Lateral supraglottic carcinoma (axial macrosection). *1* Superficial spread of lateral supraglottic carcinoma with minimal invasion of superior paraglottic space through quadrangular membrane, *2* thyroid cartilage, *3* arytenoid cartilage, *4* interarytenoid muscle, *5* ventricular fold, *6* Morgagni's ventricle, *7* superior paraglottic space

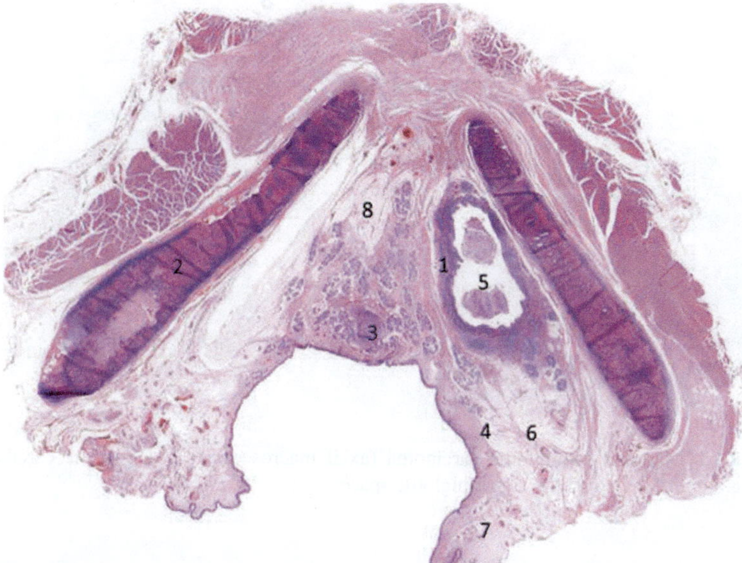

Fig. 12.11 Ventricular supraglottic carcinoma (ventriculosaccular carcinoma) (axial macrosection). *1* Superficial spread of ventricular carcinoma with minimal invasion of paraglottic space, *2* thyroid cartilage, *3* petiole of epiglottis, *4* ventricular fold, *5* Morgagni's ventricle, *6* superior paraglottic space, *7* aryepiglottic fold, *8* preepiglottic space

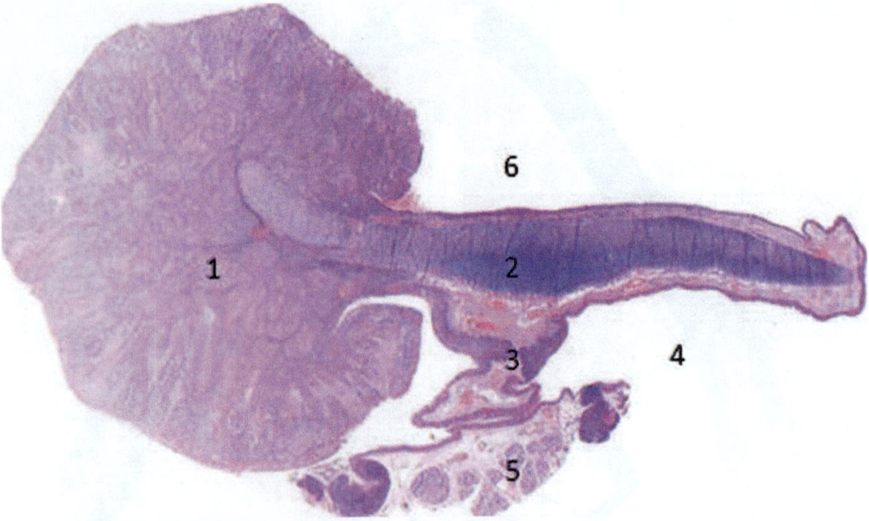

Fig. 12.12 Marginal supraglottic carcinoma (axial macrosection). *1* Exophytic carcinoma of free margin of epiglottis and aryepiglottic fold, *2* epiglottis, *3* median glossoepiglottic fold, *4* glossoepiglottic vallecula, *5* retrolingual region, *6* laryngeal surface of epiglottis

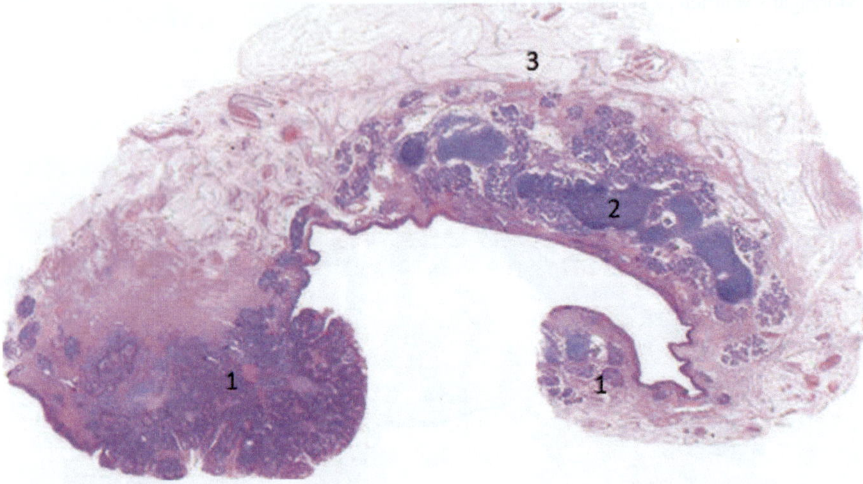

Fig. 12.13 Marginal supraglottic carcinoma (axial macrosection). *1* Exophytic carcinoma of aryepiglottic fold, *2* epiglottis, *3* preepiglottic space

inferior paraglottic space (Fig. 12.19), often with an ample exophytic intraluminal component. In over 85 % of cases, only one vocal cord is involved with extension to the anterior commissure [19] in about 10 % of cases and to the contralateral cord in less than 5 % of cases.

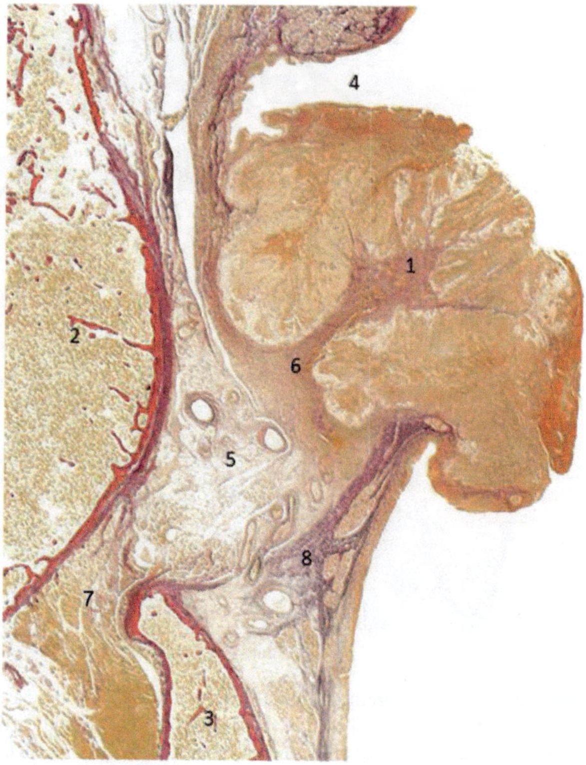

Fig. 12.14 Glottic carcinoma (coronal macrosection). *1* Local spread of glottic carcinoma arising from the free edge of vocal cord, invasion of cricothyroid muscle, *2* thyroid cartilage, *3* cricoid cartilage, *4* Morgagni's ventricle, *5* inferior paraglottic space, *6* vocal cord, *7* cricothyroid space, *8* elastic cone

12.3.2 Glottic Carcinoma with Subglottic Extension

The neoplasia presents a caudal diffusion to the vocal ligament and to the inferior face of the vocal cord towards the anterolateral subglottic region (Fig. 12.20). In its initial phases, the neoplasia spreads within the inferior paraglottic space which at this level is bounded anterolaterally by the perichondrium of the thyroid cartilage and inferomedially by the conus elasticus. In this phase of its growth, the tumour is pushed downwards and tends to infiltrate the subglottic site [20]. Afterwards, the neoplasia reaches the cricothyroid space with possible infiltration of the cricothyroid membrane and extralaryngeal extension. In most cases, the extension to the cricothyroid space coincides with the infiltration of the inferior margin of the lateral lamina of the thyroid cartilage. In about half the cases, the neoplasia extends to the anterior commissure [21] and to the subglottic region under the anterior commissure [22], an event which is accompanied by a high percentage of infiltration of the cartilage and of the cricothyroid membrane with possible anterior extralaryngeal extension (Fig. 12.21).

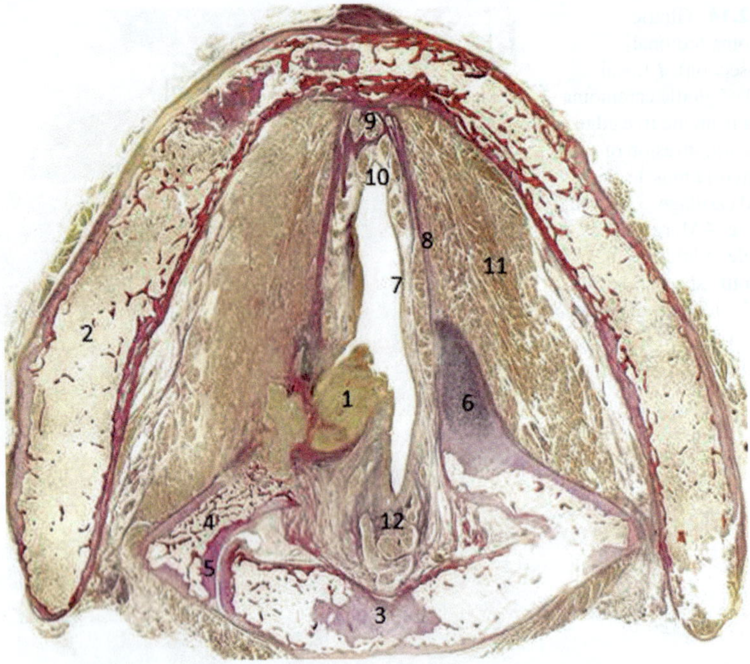

Fig. 12.15 Glottic carcinoma (axial macrosection). *1* Glottic carcinoma arising from the free edge of posterior third of vocal cord, *2* thyroid cartilage, *3* cricoid cartilage, *4* arytenoid cartilage, *5* cricoarytenoid joint, *6* vocal process of arytenoid cartilage, *7* vocal cord, *8* vocal ligament, *9* tendon of anterior commissure, *10* anterior commissure, *11* thyroarytenoid muscle, *12* posterior commissure

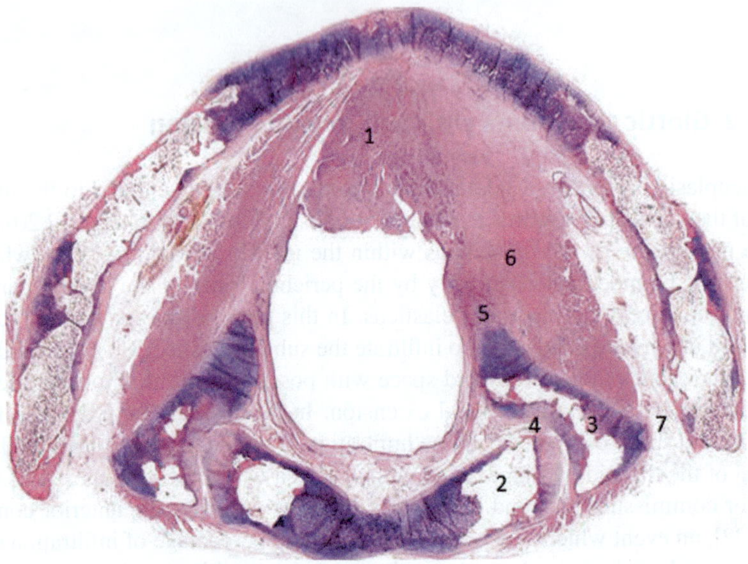

Fig. 12.16 Glottic carcinoma (axial macrosection). *1* Anterior commissure carcinoma, *2* cricoid cartilage, *3* arytenoid cartilage, *4* cricoarytenoid joint, *5* vocal process of arytenoid cartilage, *6* vocal cord, *7* cricothyroid space

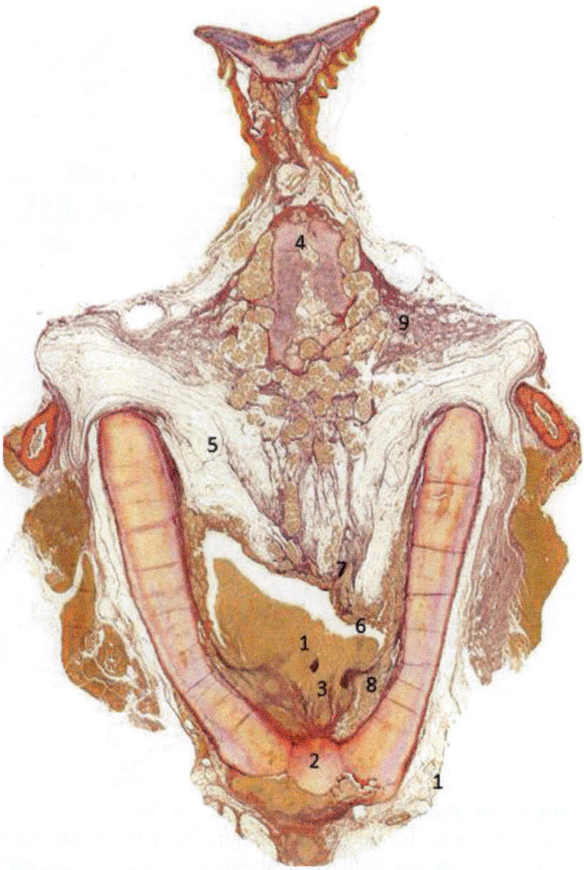

Fig. 12.17 Primary anterior commissure carcinoma (coronal macrosection). *1* Advanced spread of anterior commissure glottic carcinoma, *2* thyroid cartilage, *3* anterior commissure tendon, *4* epiglottis, *5* superior paraglottic space, *6* ventricular fold, *7* quadrangular membrane, *8* conus elasticus, *9* preepiglottic space

12.3.3 Glottic Carcinoma with Supraglottic Extension

These are subdivided into neoplasias with limited or prevalent vertical extension, depending on the manner of growth.

Glottic carcinomas with limited vertical extension to the supraglottis are characterised by a prevalently mucosal spread along the tunica propria with minimum infiltration of the paraglottic space. They generally extend to the wall of Morgagni's ventricle and to its saccule and less frequently to the ventricular band along the arytenoid region and to the epiglottis along its pedicle. The anterior commissure [23] is involved in a low percentage of cases, sometimes with superficial spread to the contralateral vocal cord.

Glottic carcinomas with prevalent vertical extension to the supraglottis present a cranial diffusion along two preferential routes consisting of the anterior commissure and the paraglottic space. In the first case, the neoplasia spreads along the anterior commissural region [24] with infiltration of the thyroepiglottic ligament and of the caudal part of the preepiglottic and superior paraglottic

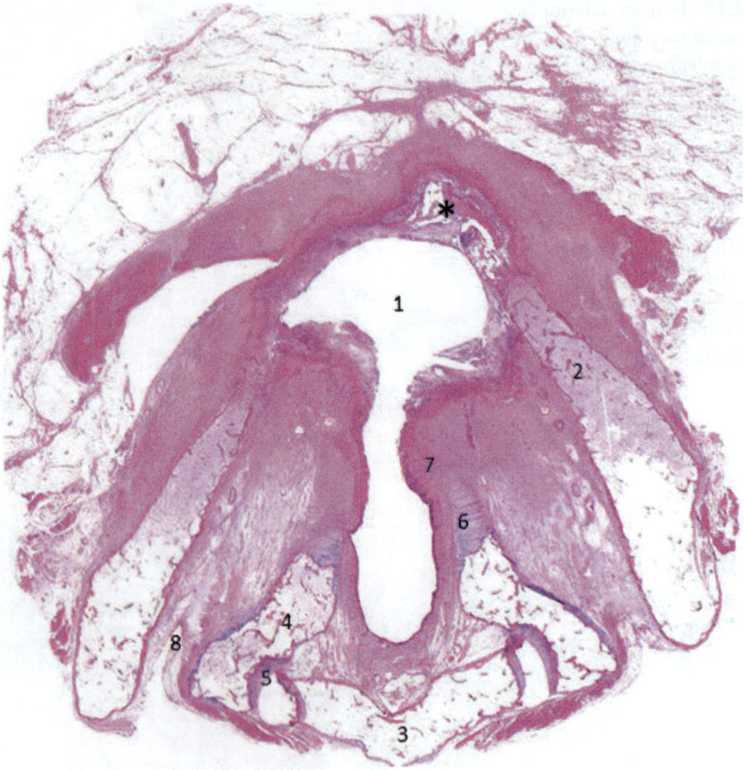

Fig. 12.18 Primary anterior commissure carcinoma (axial macrosection). *1* Anterior commissure carcinoma with thyroid cartilage infiltration and extralaryngeal tumour diffusion (*), *2* thyroid cartilage, *3* cricoid cartilage, *4* arytenoid cartilage, *5* cricoarytenoid joint, *6* vocal process of arytenoid, *7* vocal cord, *8* cricothyroid space

spaces [25]. In most cases, the neoplasias that originate from the medio-anterior third of the vocal cord rapidly reach the tunica propria of the ventricle and the inferior paraglottic space which at this level is largely occupied by the fibres of the vocal muscle. In these cases, the neoplastic growth (Fig. 12.22) is deviated upwards by the lateral lamina of the thyroid cartilage with infiltration of the superior paraglottic space (Fig. 12.23) [26], sometimes without involving the mucosal plane of the ventricular band. The role of Morgagni's ventricle in the cranial spread of the tumour may be decisive in some cases, especially in neoplasias originating from the superior edge of the vocal cord which at an early stage involve the floor of the ventricle and its aditus (Fig. 12.24). The endoventricular growth of the neoplasia generally involves the destruction of the ventricular walls and its spread into the superior paraglottic space. In some cases, the neoplasia enclosing the initial portion of the ventricle may produce a secondary laryngocele through a valve-type mechanism.

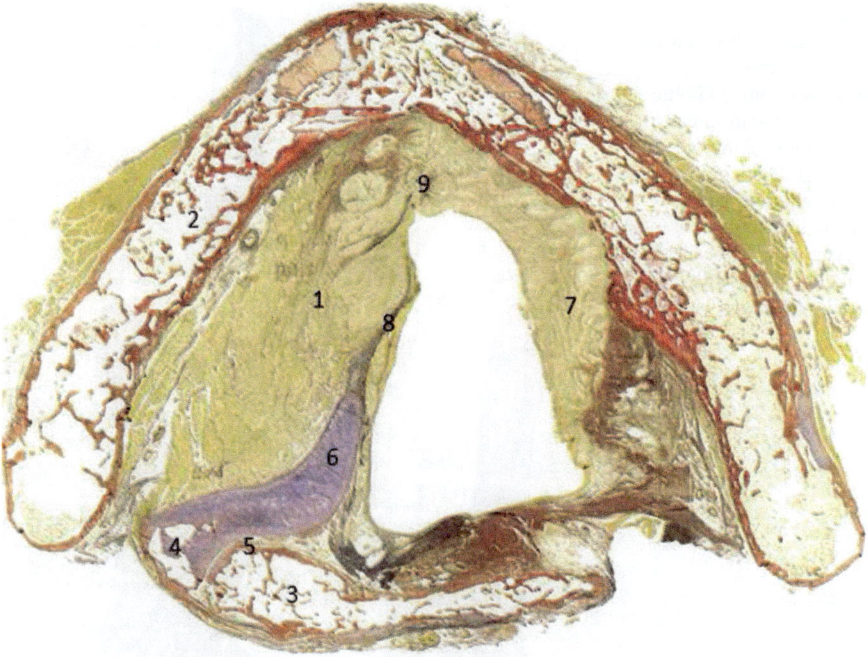

Fig. 12.19 Glottic carcinoma (axial macrosection). *1* Advanced glottic carcinoma, *2* thyroid cartilage, *3* cricoid cartilage, *4* arytenoid cartilage, *5* cricoarytenoid joint, *6* vocal process of arytenoid cartilage, *7* vocal cord, *8* vocal ligament, *9* anterior commissure

12.3.4 Transglottic Carcinoma

These are advanced forms of glottic neoplasias which extend in a craniocaudal direction along the axis of Morgagni's ventricle (Figs. 12.25 and 12.26). This term is often used to indicate voluminous tumours for which the seat of origin cannot be established with certainty or glottic neoplasias with supraglottic extension.

12.4 Subglottic Carcinoma

The primitive tumour of the subglottis [27] is a neoplasia in search of identity. In fact, if we consider neoplasias of the inferior face of the vocal cord as glottic tumours and exclude glottic neoplasias with a subglottic diffusion, primitive subglottic tumours [28] are extremely rare (Fig. 12.27) and account for less than 1 % of the most credible case histories [29]. In general, they present a growth and metastasisation pattern similar to tracheal tumours. They are characterised by early infiltration of the cricoid cartilage and of the cricothyroid membrane with possible extralaryngeal extension [30].

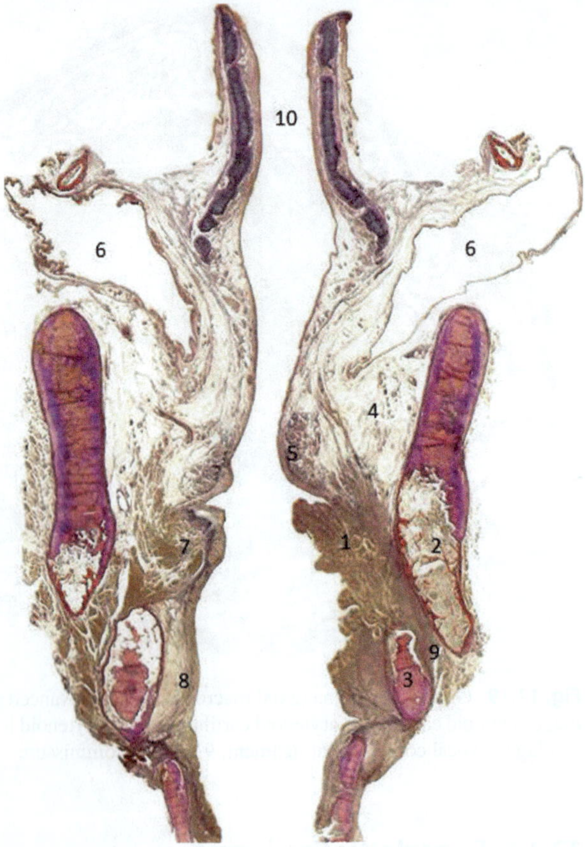

Fig. 12.20 Glottic carcinoma with subglottic spread (coronal macrosection). *1* Glottic carcinoma with spread to hypoglottis, *2* thyroid cartilage, *3* cricoid cartilage, *4* paraglottic space, *5* ventricular fold, *6* laryngocele, *7* vocal cord, *8* subglottis, *9* cricothyroid space, *10* epiglottis

12.5 External Laryngeal Carcinomas

The definition of external laryngeal carcinomas includes neoplasias of the hypopharynx and of the inferior regions of the oropharynx (glossoepiglottic valleculae, retrolingual area, caudal lateral oropharyngeal walls including the region of the threefolds) which in their growth involve the different laryngeal regions and whose surgical treatment therefore includes laryngectomy.

12.5.1 Hypopharynx Carcinoma

The hypopharynx is divided into three subsites: piriform sinuses, posterior wall of the hypopharynx, and retrocricoid region.

Carcinoma of the piriform sinus (Figs. 12.28 and 12.29) is the most frequent hypopharyngeal neoplasia. In an early stage of its development, it involves the

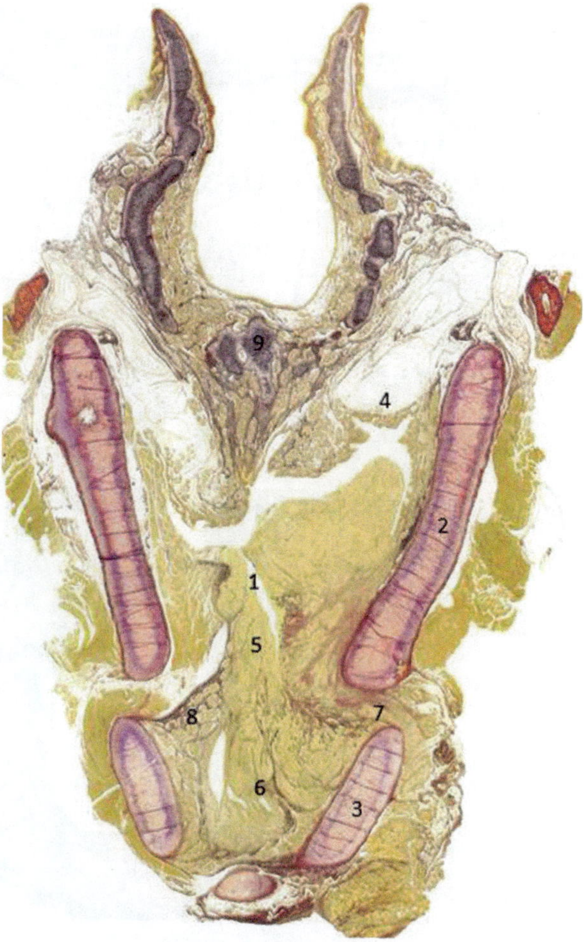

Fig. 12.21 Glottic carcinoma with subglottic spread (coronal macrosection). *1* Glottic carcinoma with prevalent spread to anterior subglottis, *2* thyroid cartilage, *3* cricoid cartilage, *4* superior paraglottic space, *5* anterior commissure, *6* hypoglottis, *7* cricothyroid space, *8* conus elasticus, *9* epiglottis

laryngeal structures along three preferential routes: (a) anterolaterally first to the paraglottic space and then to the hyothyroepiglottic cavity; (b) spreading in a dorsocaudal direction towards the cricothyroid joint, the rear wall of the larynx, and the aditus of the oesophagus; and finally (c) spreading to the lateral vestibular walls. In the case of neoplasias of the piriform sinus, the presence of fixity of the vocal cord indicates infiltration of the cricoarytenoid complex, of the interarytenoid muscles, and, less frequently, of the recurrent nerve.

From the retrocricoid area (Fig. 12.30), the neoplasias spread craniocaudally towards the paraglottic spaces and the aditus of the oesophagus. From the rear wall of the hypopharynx, the neoplastic growth involves first the lateral wall of the piriform sinus, the lateral lamina of the thyroid cartilage, and the submucosal compartments of the larynx.

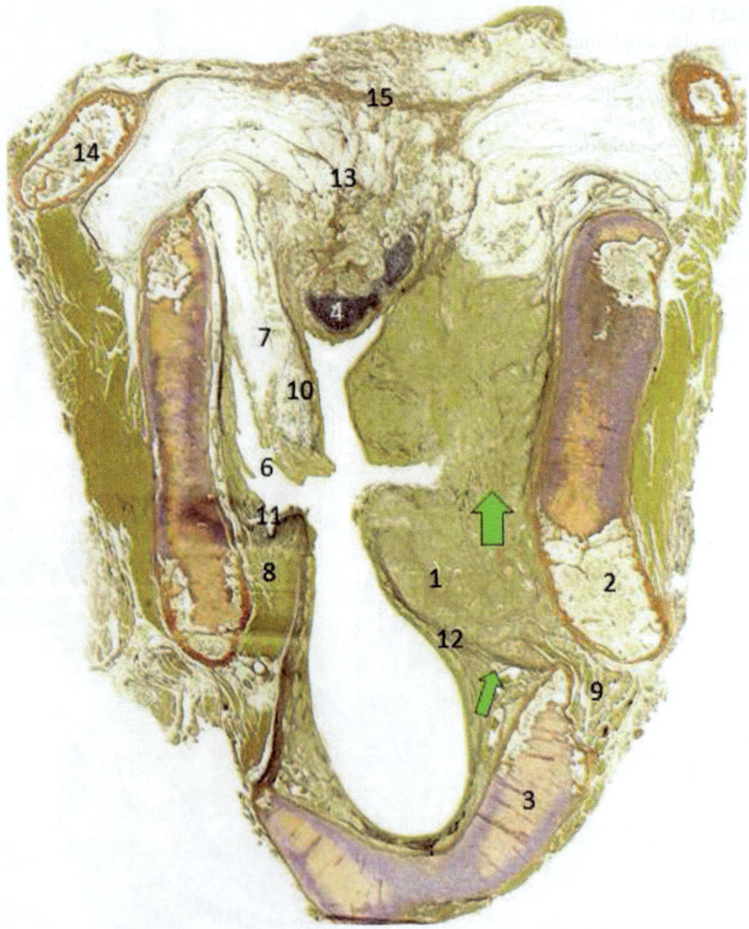

Fig. 12.22 Glottic carcinoma with supraglottic spread (coronal macrosection). *1* Advanced glottic carcinoma involving the superior paraglottic space, the ventricular fold (so-called transglottic carcinoma), and Morgagni's ventricle, *2* thyroid cartilage, *3* cricoid cartilage, *4* epiglottis, *5* Morgagni's ventricle, *6* superior paraglottic space, *7* ventricular fold, *8* vocal cord, *9* cricothyroid space, *10* quadrangular membrane, *11* ventricular ligament, *12* conus elasticus, *13* preepiglottic space, *14* hyoid bone, *15* hyoepiglottic ligament

12.5.2 Glossoepiglottic Region Carcinoma

The glossoepiglottic region includes (a) the glossoepiglottic valleculae, (b) the tongue base, and (c) the region of the threefolds.

The neoplasias of the vallecular region (Fig. 12.31) are limited in the first stages of their development by the hyoepiglottic membrane which favours their extension anteriorly towards the tongue base and posteriorly towards the lingual side of the epiglottis. Then, infiltrating the membrane, the neoplasia infiltrates the preepiglottic space, involving the laryngeal spaces and possibly destroying the epiglottis.

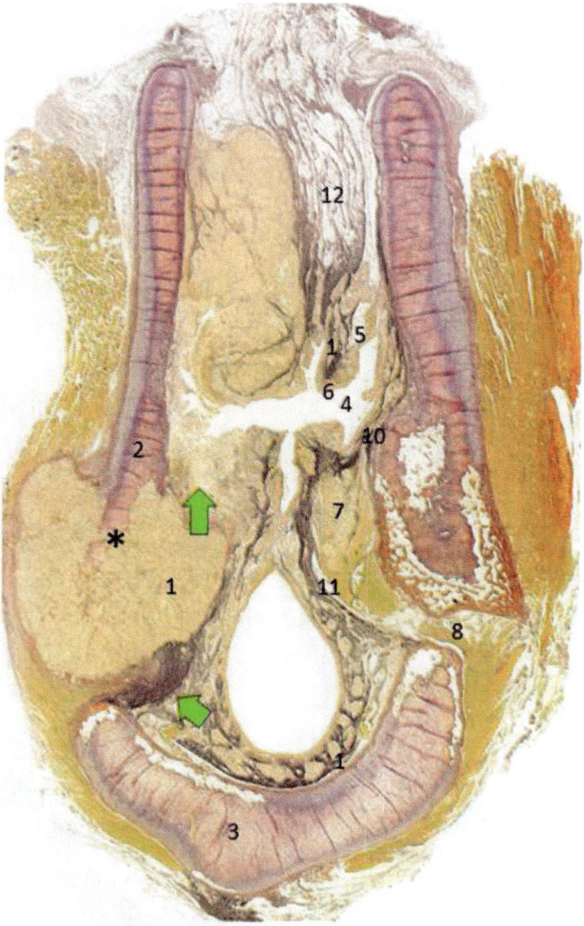

Fig. 12.23 Glottic carcinoma with supraglottic spread (coronal macrosection). *1* Advanced glottic carcinoma involving the superior paraglottic space, the ventricular fold (so-called transglottic carcinoma), and Morgagni's ventricle; invasion of cricothyroid space and thyroid cartilage, *2* thyroid cartilage (*), *3* cricoid cartilage, *4* Morgagni's ventricle, *5* superior paraglottic space, *6* ventricular fold, *7* vocal cord, *8* cricothyroid space, *9* quadrangular membrane, *10* ventricular ligament, *11* conus elasticus, *12* preepiglottic space

Carcinomas of the tongue base (Fig. 12.32) infiltrate the intrinsic and extrinsic muscles of the tongue at an early stage and then extend to the muscle of the buccal pelvis. Infiltration of the genioglossus muscle, inserted in fan formation between the mental spine of the mandible, the body of the hyoid bone, and the epiglottis, involves early neoplastic spreading to the body of the tongue and the floor of the mouth. Lastly, the absence of anatomical boundaries leads to constant involvement of the vallecular region with oropharyngeal–laryngeal extension. In both cases, the neoplasias of the glossoepiglottic region are characterised by early lymph nodal metastases in over 50 % of cases at the time of diagnosis, often bilateral.

Carcinomas of the region of the threefolds tend to extend to the adjoining regions of the larynx and the pharynx (Fig. 12.33).

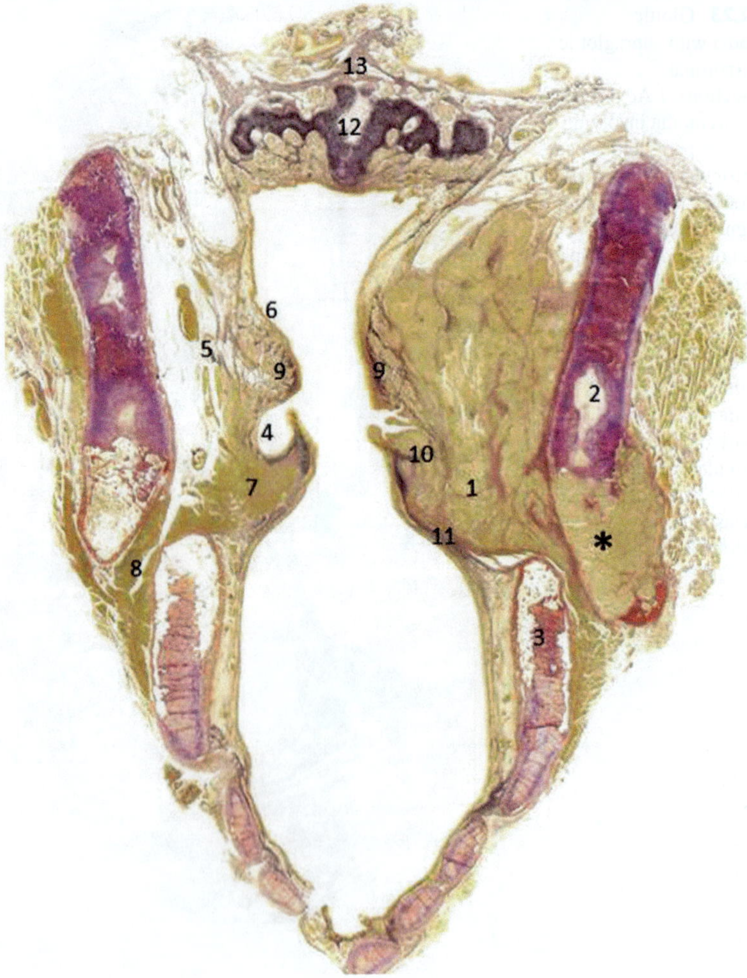

Fig. 12.24 Glottic carcinoma with supraglottic spread (coronal macrosection). *1* Advanced glottic carcinoma involving the superior paraglottic space with mucosal integrity of ventricular fold; invasion of cricothyroid space, thyroid cartilage (*), and extralaryngeal extension, *2* thyroid cartilage, *3* cricoid cartilage, *4* Morgagni's ventricle, *5* superior paraglottic space, *6* ventricular fold, *7* vocal cord, *8* cricothyroid space, *9* quadrangular membrane, *10* ventricular ligament, *11* conus elasticus, *12* epiglottis, *13* preepiglottic space

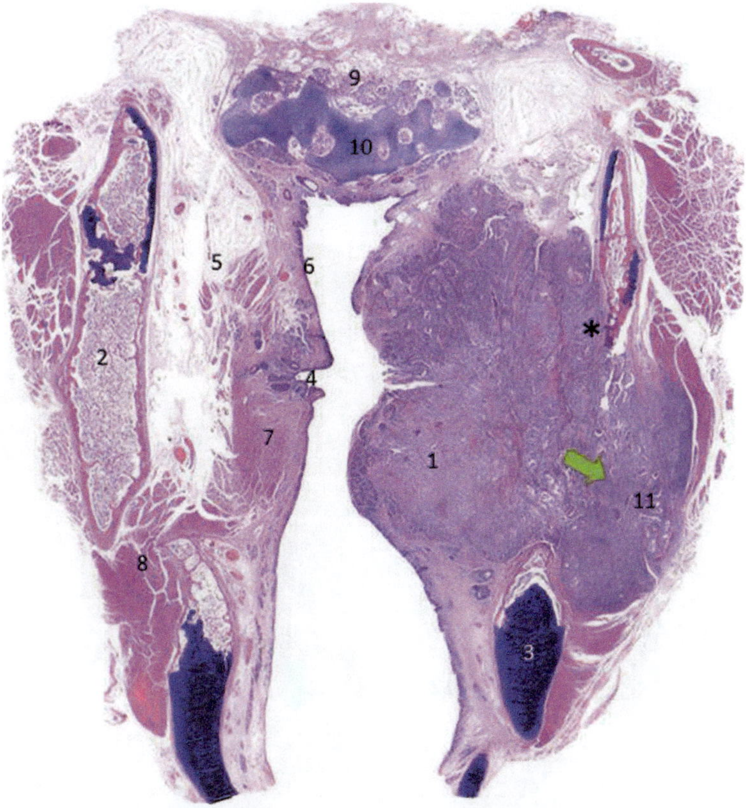

Fig. 12.25 Transglottic carcinoma (coronal macrosection). *1* Advanced glottic carcinoma involving the superior paraglottic space, the ventricular fold, and Morgagni's ventricle; invasion of cricothyroid space and thyroid cartilage (*), *2* thyroid cartilage, *3* cricoid cartilage, *4* Morgagni's ventricle, *5* superior paraglottic space, *6* ventricular fold, *7* vocal cord, *8* cricothyroid space, *9* preepiglottic space, *10* epiglottis, *11* lateral extralaryngeal diffusion

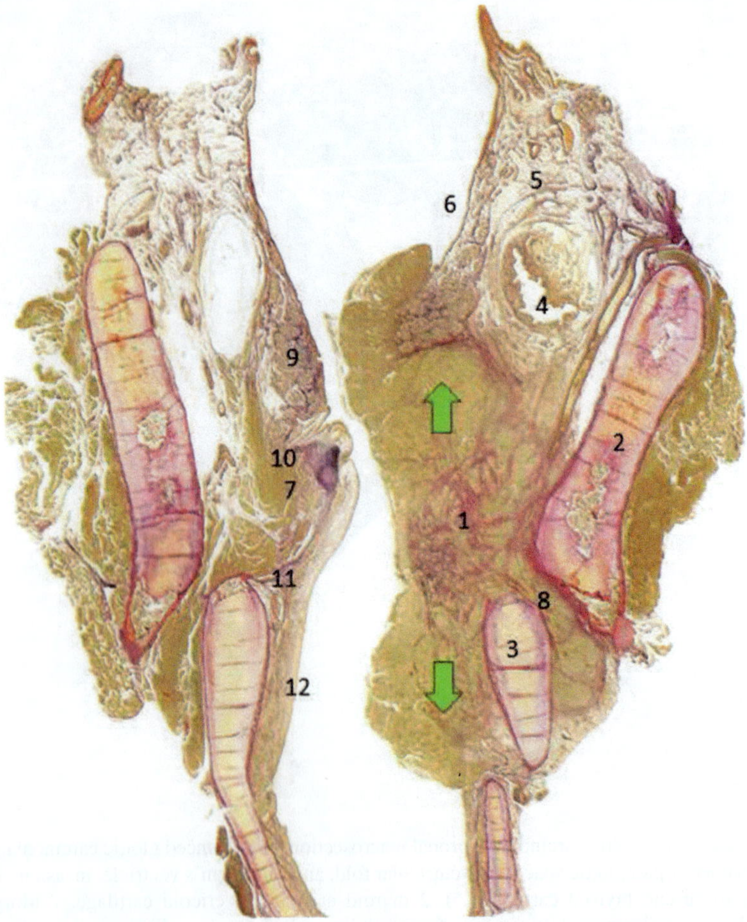

Fig. 12.26 Transglottic carcinoma (coronal macrosection). *1* Advanced glottic carcinoma involving glottis, superior paraglottic space, ventricular fold, Morgagni's ventricle, and hypoglottis, *2* thyroid cartilage, *3* cricoid cartilage, *4* Morgagni's ventricle, *5* superior paraglottic space, *6* ventricular fold, *7* vocal cord, *8* cricothyroid space, *9* quadrangular membrane, *10* ventricular ligament, *11* conus elasticus, *12* subglottis

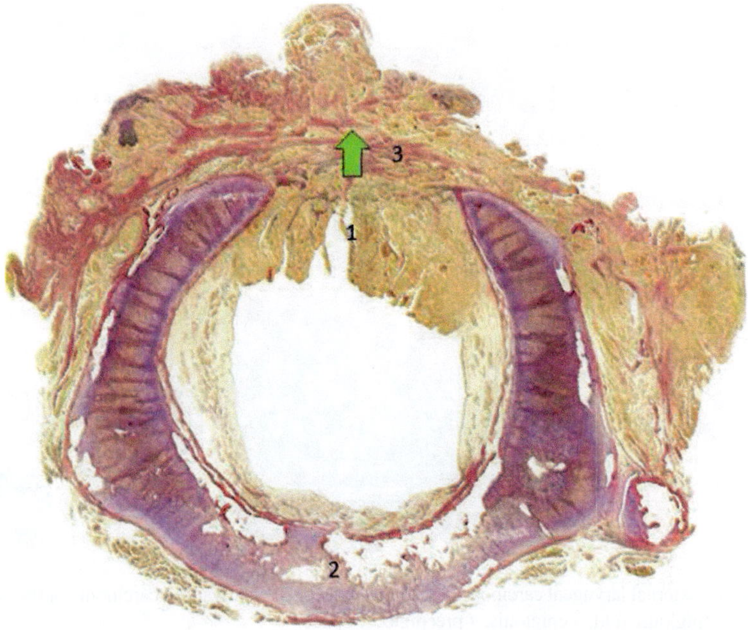

Fig. 12.27 Subglottic carcinoma (axial macrosection). *1* Anterior subglottic carcinoma, *2* cricoid cartilage, *3* cricothyroid membrane

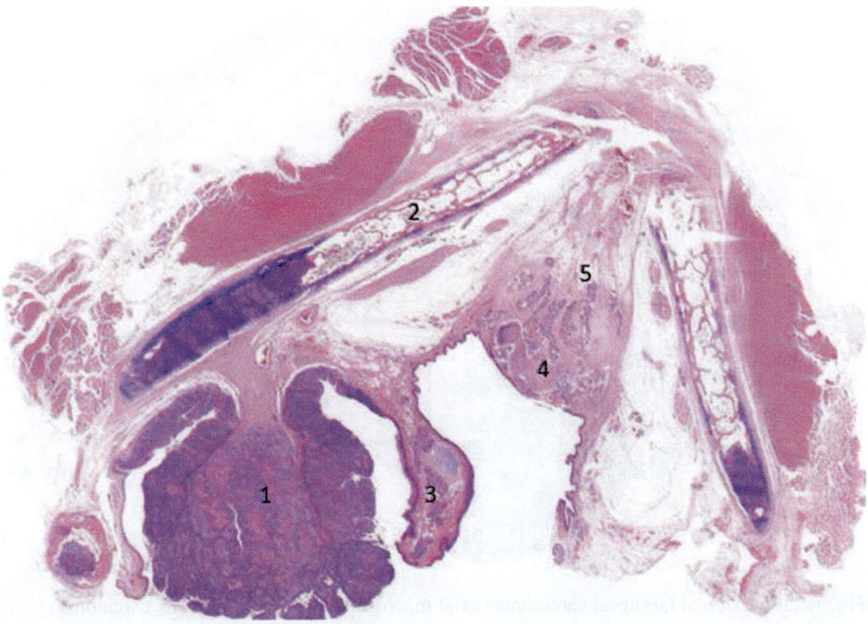

Fig. 12.28 External laryngeal carcinoma (axial macrosection). *1* Exophytic carcinoma of the piriform sinus, *2* thyroid cartilage, *3* aryepiglottic fold, *4* epiglottis, *5* preepiglottic space

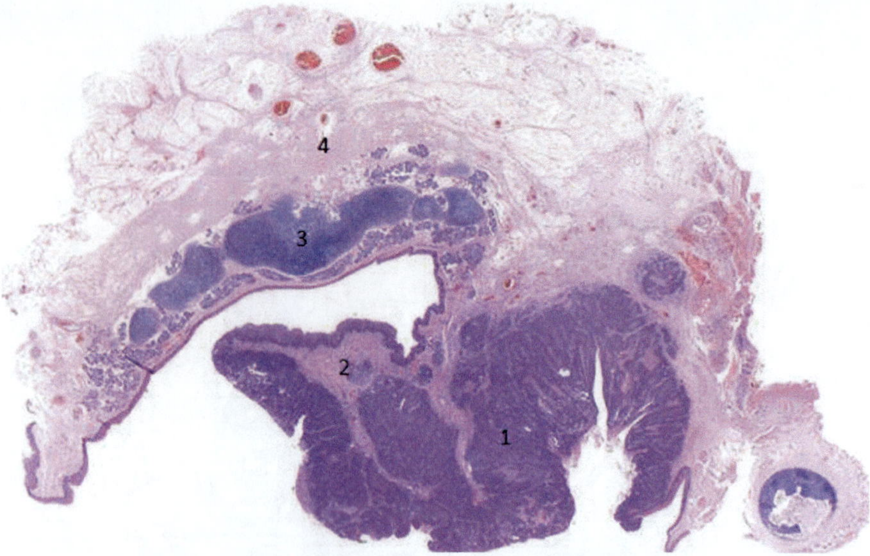

Fig. 12.29 External laryngeal carcinoma (axial macrosection). *1* Advanced carcinoma of the piriform sinus, *2* aryepiglottic fold, *3* epiglottis, *4* preepiglottic space

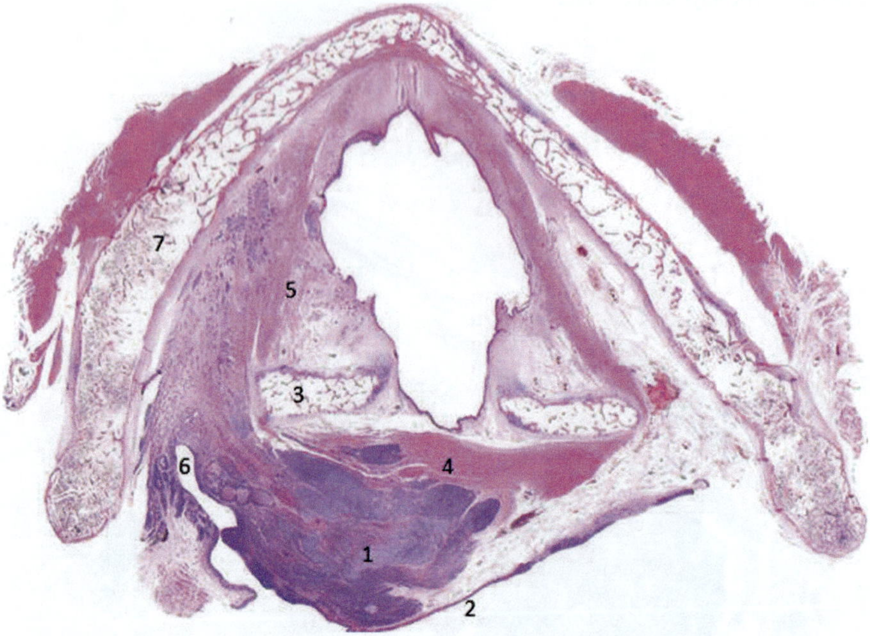

Fig. 12.30 External laryngeal carcinoma (axial macrosection). *1* Hypopharynx carcinoma (postcricoid area), *2* postcricoid area, *3* arytenoid cartilage, *4* interarytenoid muscle, *5* inferior paraglottic space, *6* piriform sinus, *7* thyroid cartilage

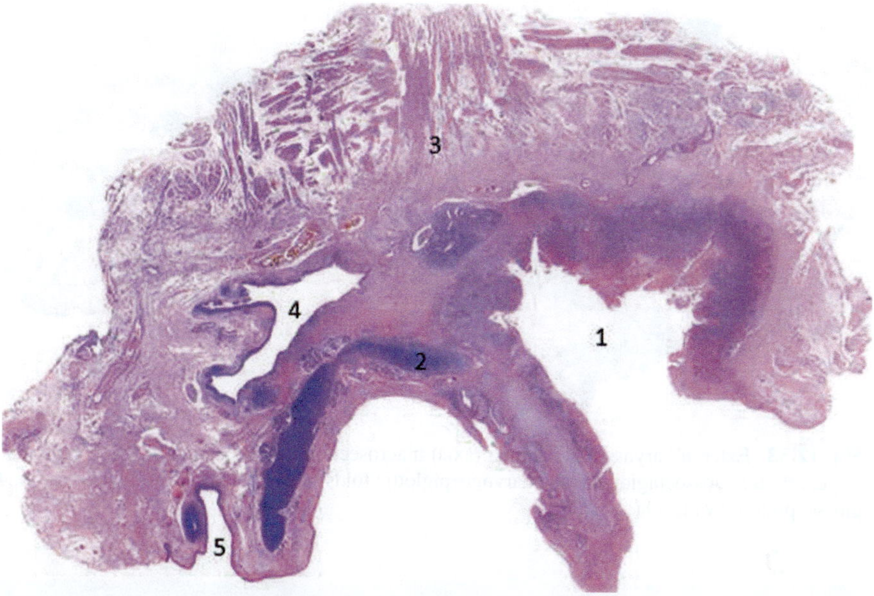

Fig. 12.31 External laryngeal carcinoma (axial macrosection). *1* Carcinoma of the glossoepiglottic vallecula (lateral), *2* epiglottis, *3* retrolingual region, *4* glossoepiglottic vallecula, *5* piriform sinus

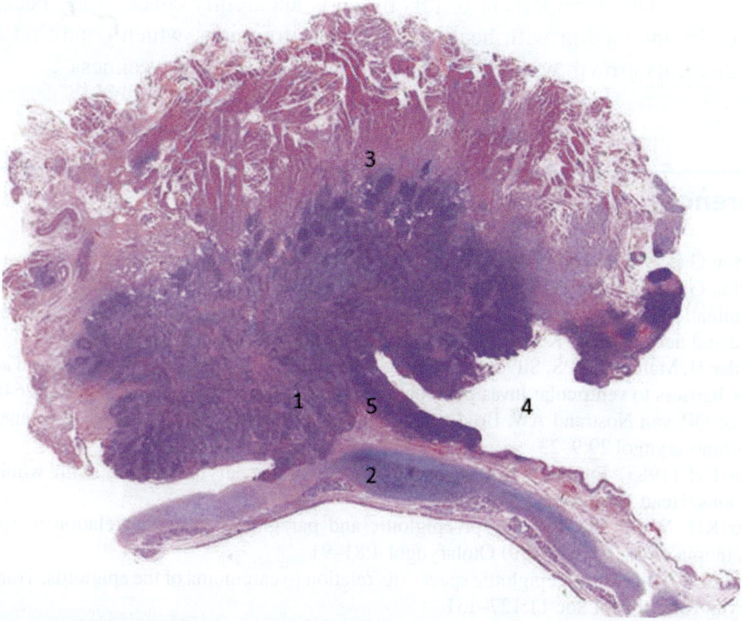

Fig. 12.32 External laryngeal carcinoma (axial macrosection). *1* Carcinoma of the glossoepiglottic vallecula (medial), *2* epiglottis, *3* retrolingual region, *4* glossoepiglottic vallecula, *5* glossoepiglottic medial fold

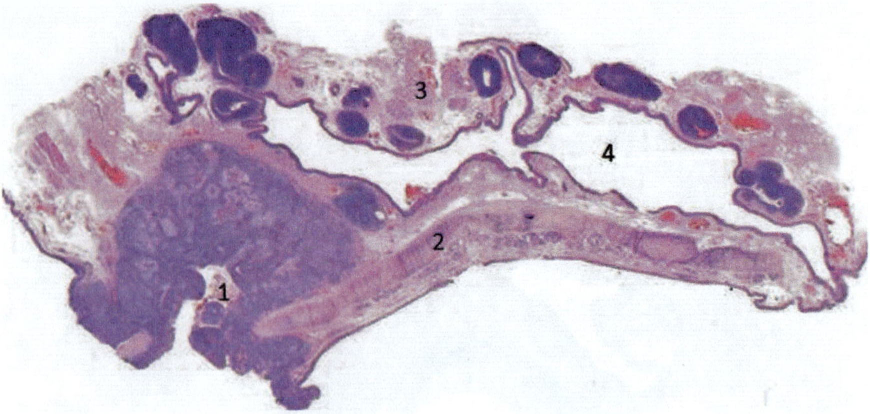

Fig. 12.33 External laryngeal carcinoma (axial macrosection). *1* Carcinoma of the "threefolds" (aryepiglottic, glossoepiglottic, and pharyngoepiglottic folds), *2* epiglottis, *3* retrolingual region, *4* glossoepiglottic vallecula

Take-Home Messages

- In laryngeal carcinoma, the proliferating cells grow in number forming complex neoplastic agglomerates but still with a certain structural organisation. The development of this tissue is apparently chaotic only because of its interaction with healthy laryngeal structures, which condition and direct its growth with "lines of strength" and "lines of weakness".

References

1. Carlon G (1990) Il carcinoma della laringe. Dalla patologia alla clinica. Piccin, Padova
2. Carlon G, Della Libera D, Dei Tos AP (1990) A whole organ sectioning method for histologic examination of laryngeal and hypopharyngeal specimens. In: Sacristan T (ed) Otolaryngology, head and neck surgery. Kugler & Ghedini, Amsterdam, pp 2367–2371
3. Beitler JJ, Mahadevia PS, Silver CE, Wadler S, Rubin JS, Bello JA, Mitnik RJ, Vikram B (1994) New barriers to ventricular invasion in paraglottic laryngeal cancer. Cancer 15(73):2648–2652
4. Bryce DP, van Nostrand AW, Brodarec I (1983) Growth and spread of laryngeal cancer. Adv Otorhinolaryngol 29:9–23
5. Lam KH (1983) Extralaryngeal spread of cancer of the larynx: a study with whole-organ sections. Head Neck Surg 5:410–424
6. Lam KH, Wong J (1983) The preepiglottic and paraglottic spaces in relation to spread of carcinoma of the larynx. Am J Otolaryngol 4:81–91
7. Cleri LH (1944) The preepiglottic space – its relation to carcinoma of the epiglottis. Trans Amer Laryng Rhinol Otol Soc 11:127–131
8. Dayal VS, Bahri H, Stone PC (1972) Pre-epiglottic space. An anatomic study. Arch Otolaryngol 95:130–133
9. Kirchner JA, Som ML (1971) Clinical and histological observations on supraglottic cancer. Ann Otol (St Louis) 80:638–646

10. McDonald TJ, Desanto LW, Weiland LH (1976) Supraglottic larynx and its pathology as studied by whole laryngeal sections. Laryngoscope (St Louis) 86:635–648
11. Meyer-Breiting E (1984) Squamous cell carcinomas of the anterior wall of the larynx. In: Wigand ME, Steiner W, Steli PM (eds) Functional partial laryngectomy conservation surgery for carcinoma of the larynx. Springer, Berlin, pp 140–143
12. Micheau C, Luboinski B, Lanchi P, Cachin Y (1978) Relationship between laryngocele and laryngeal carcinomas. Laryngoscope (St Louis) 88:680–688
13. Micheau C, Leonardelli GB, Gérard-Marchant R, Cachin Y (1973) Modalités d'envahissement des tumeurs du vestibule larynge: aspects histopathologiques et statistiques. Nuovo Arch Ital Otol 1:279–291
14. Sessions DG, Ogura JH (1976) Classification of laryngeal cancer. In: Alberti PW, Bryce DP (eds) Workshops from the centennial conference on laryngeal cancer. Appleton-Century-Crofts, New York, pp 83–89
15. Olofsson J (1976) Growth and spread of laryngeal carcinoma. In: Alberti WP, Bryce DP (eds) Workshops from the centennial conference on laryngeal cancer. Appleton-Century-Crofts, New York, pp 40–53
16. Kirchner JA (1977) Two hundred laryngeal cancers: patterns of growth and spread as seen in serial sections. Laryngoscope (St Louis) 87:474–482
17. Broyles EN (1943) The anterior commissure tendon. Ann Otol Rhinol Laryngol 52:342–345
18. Kleinsasser O, Glanz H (1982) Microcarcinoma and microinvasive carcinoma of the vocal cords. Clin Oncol 1:479–487
19. Andrea M (1981) Vasculature of the anterior commissure. Ann Otol Rhinol Laryngol 90:18–20
20. Freelanci AP, Van Nostrand P (1976) The applied anatomy of the anterior commissure and subglottis. In: Alberti PW, Bryce DP (eds) Workshops from the centennial conference on laryngeal cancer. Appleton-Century-Crofts, New York
21. Bagatella F, Bignardi L (1981) Morphological study of the laryngeal anterior commissure with regard to the spread of cancer. Acta Otolaryngol 92:167–171
22. Kirchner JA, Fischer JJ (1976) Anterior commissure cancer – a clinical and laboratory study of 39 cases. In: Alberti PW, Bryce DP (eds) Workshops from the centennial conference on laryngeal cancer. Appleton-Century-Crofts, New York, pp 645–651
23. Barbosa MM, Araujo VJ, Boasquevisque E, Carvalho R, Romano S, Lima RA, Dias FL, Salviano SK (2005) Anterior vocal commissure invasion in laryngeal carcinoma diagnosis. Laryngoscope 115(4):724–730
24. Andrea M, Guerrier Y (1981) The anterior commissure of the larynx. Clin Otolaryngol 6:259–264
25. Bagatella F, Bignardi L (1983) Behavior of cancer at the anterior commissure of the larynx. Laryngoscope (St Louis) 93:353–356
26. Della Libera D, Bittesini L, Falconieri G (1997) Anterior commissure involvement in glottic laryngeal carcinoma. In: Advances in laryngology in Europe. Proceedings of the first scientific conference of the European Laryngological Society. Elsevier Science B.V, pp 355–357
27. Harrison DF (1971) The pathology and management of subglottic cancer. Ann Otol 80:6–12
28. Steli PM, Gregory I, Watt J (1980) Morphology of the human larynx. II. The subglottis. Clin Otolaryngol Allied Sci 5:389–395
29. Olofsson J (1976) Specific features of laryngeal carcinoma involving the anterior commissure and subglottic region. In: Alberti PW, Bryce DP (eds) Workshops from the centennial conference on laryngeal cancer. Appleton-Century-Crofts, New York, pp 626–644
30. Lund WS (1976) Classification of subglottic tumors and discussion of their growth and spread. In: Alberti WP, Bryce DP (eds) Workshops from the centennial conference on laryngeal cancer. Appleton-Century-Crofts, New York

Concerning Endoscopy, Imaging, and Surgical Options in Larynx Cancer

13

Core Messages

- The growth of laryngeal tumours depends a great deal on the site of onset and takes place along preferential routes. Some structures, such as tendons and cartilages, within certain limits "divert" the tumour, which instead easily colonises the epithelium, and the adipose and glandular tissue. The knowledge of the anatomy of the larynx and the study of the spread of tumours are the basis of the concepts of functional laryngeal surgery.

Significant anatomical structures: quadrangular membrane, conus elasticus, Broyle's tendon, endoscopic laser surgery, horizontal supraglottic laryngectomy, supracricoid laryngectomy, total laryngectomy, thyroid cartilage, anterior commissure, Delphian limph node, preepiglottic space, superior paraglottic spaces, inferior paraglottic space, cricoarytenoid joint, lateral cricoarytenoid space, anterior cricoarytenoid space, subglottis, cricoarynenoid unit.

13.1 Topographic Anatomy of the Endolarynx*

From the point of view of its topographic anatomy, the larynx is composed of an external cartilaginous skeleton, connected by fibrous ligaments and articulations, and an internal elastic skeleton, which follows the profile of the mucosal surface and which, at supraglottic level, takes the name quadrangular membrane and, at glottic–subglottic level, conus elasticus. The elastic skeleton chondrifies in some regions and forms the epiglottis, the cuneiform and corniculate cartilages, and the vocal process of the arytenoids. The laryngeal mucosa, consisting of a ciliated epithelium

*All the larynx MRI images in this chapter came from the Department of Radiology of University of Brescia (R. Maroldi, M. Ravanelli)

M. Lucioni, *Practical Guide to Neck Dissection*,
DOI 10.1007/978-3-642-33977-6_13, © Springer-Verlag Berlin Heidelberg 2013

and a tunica propria, lies on the surface of the elastic skeleton, while the laryngeal submucosa is between the elastic and the cartilaginous skeleton. The internal and the external skeleton are very close together in some sites (anterior commissure, vocal folds), making the presence of submucosa minimum or virtual (Fig. 13.1a, b).

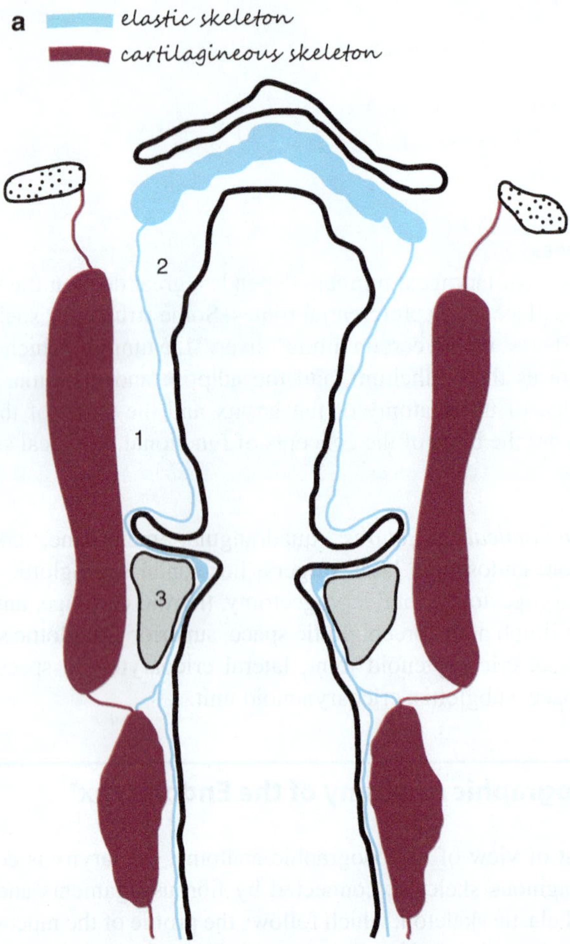

Fig. 13.1 (**a**) Schematic coronal section of the larynx: quadrangular membrane and conus elasticus, which make up the elastic skeleton of the larynx. On its surface is the mucosa (*1*). Between the elastic skeleton and the cartilagineous skeleton is the submucosa (*2*). Vocal muscle (*3*). (**b**) Larynx MRI: T2-weighted sequence on coronal plane. The paraglottic fat space is shown by a *dotted line*. The inferior paraglottic space reaches a door (*curved arrow*) that leads outside the larynx between the opposite surfaces of the cricoid and thyroid cartilage; conus elasticus (*arrowheads*) extending from the inferior aspect of the true vocal cord to the upper margin of the cricoid cartilage; cricothyroid muscle (*straight arrows*)

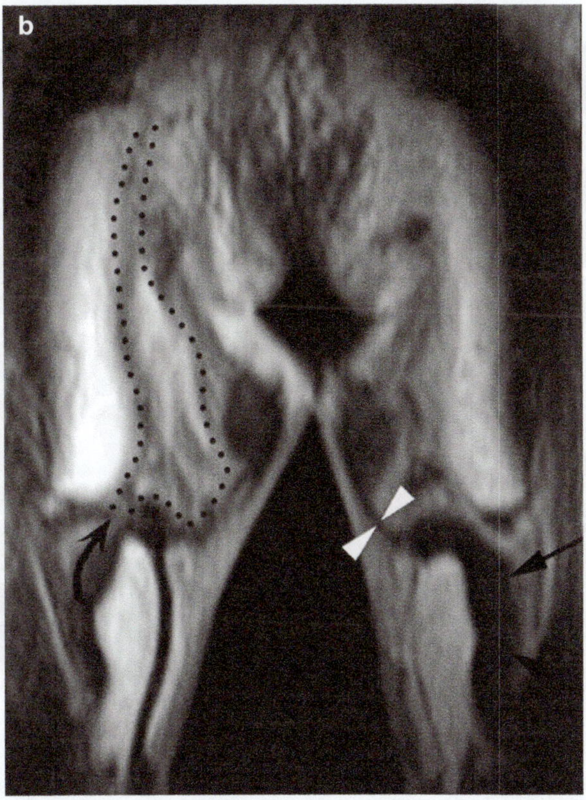

Fig. 13.1 (continued)

In coronal macrosections, the submucosa appears in the shape of an hourglass, abundant in the supraglottis and in the subglottis, scarce at the glottic level. It corresponds to what we call paraglottic spaces, rich in loose connective tissue, glands and lymphatic vessels. The paraglottic spaces are:

1. The preepiglottic space (anterior to the epiglottis)
2. The superior paraglottic space (at the level of the ventricular folds)
3. The inferior paraglottic space (at the level of the vocal cord)

These anatomical spaces are in communication with each other, and the "bottlenecks" or the presence of fibroelastic thickening, for example, Broyle's tendon, condition the routes by which the tumour spreads [1] (Fig. 13.2a–d).

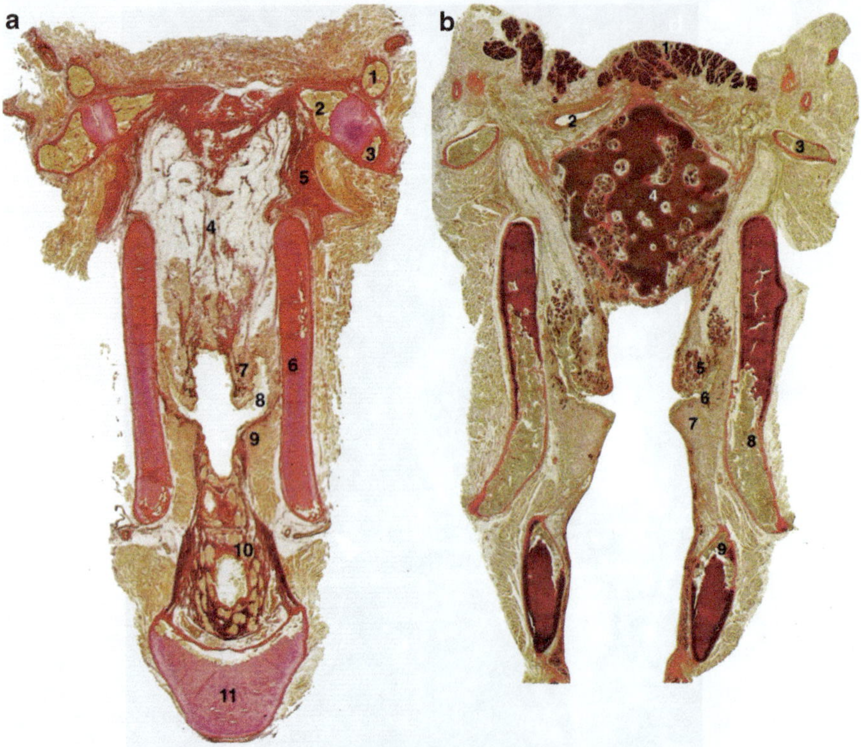

Fig. 13.2 (**a**) Coronal macrosection of the larynx: preepiglottic space. *1* Lesser cornu of hyoid bone, *2* corpus of hyoid bone, *3* greater cornu of hyoid bone, *4* preepiglottic space, *5* thyrohyoid membrane, *6* thyroid cartilage, *7* ventricular fold, *8* Morgagni's ventricle, *9* vocal cord, *10* conus elasticus, *11* cricoid ring. (**b**) Coronal macrosection of the larynx: epiglottis. *1* Tongue base, *2* glossoepiglottic vallecula, *3* greater cornu of hyoid bone, *4* epiglottis, *5* ventricular fold, *6* Morgagni's ventricle, *7* vocal cord, *8* thyroid cartilage, *9* cricoid cartilage. (**c**) Coronal macrosection of the larynx: glottic and paraglottic spaces. *1* Epiglottis, *2* quadrangular membrane, *3* superior paraglottic space, *4* inferior paraglottic space, *5* conus elasticus, *6* Morgagni's ventricle, *7* vocal ligament. (**d**) Coronal macrosection of the larynx: arytenoid cartilages and posterior commissure. *1* Epiglottis, *2* interarytenoid muscles, *3* aryepiglottic fold, *4* piriform sinus, *5* thyroid cartilage, *6* arytenoid cartilage, *7* posterior commissure, *8* cricoarytenoid joint, *9* cricoid cartilage

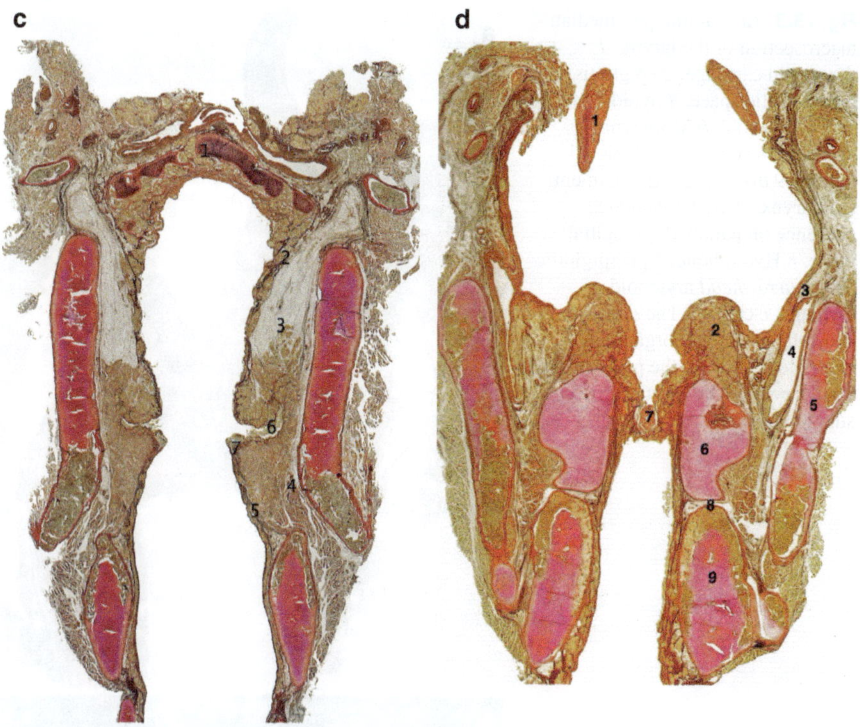

Fig. 13.2 (continued)

In sagittal macrosections, it may be noted how the epiglottis, which is an elastic cartilage, actually represents an area of chondrification of the quadrangular lamina, which inserts in it (Fig. 13.3a, b).

At glottic level, the presence of the vocal muscle may create confusion concerning the definition of the inferior paraglottic space. Tucker [2, 3] considers it the space bounded anterolaterally by the thyroid cartilage, inferomedially by the elastic cone and dorsally by the piriform sinus; it is therefore implied that the vocal muscle is contained within the inferior paraglottic space. Instead, Carlon [1] defines it "the space situated laterally to the even structures of the larynx (ventricular bands and vocal cords)". Other authors [4–7] also consider the inferior paraglottic space the thin layer of loose connective tissue that lies laterally to the vocal muscle. This is currently the most accepted interpretation. It also agrees with pT3 glottic staging, which is thus established when the tumour infiltrates this space, as well as by fixity of the vocal cords [8] (Fig. 13.4a, b).

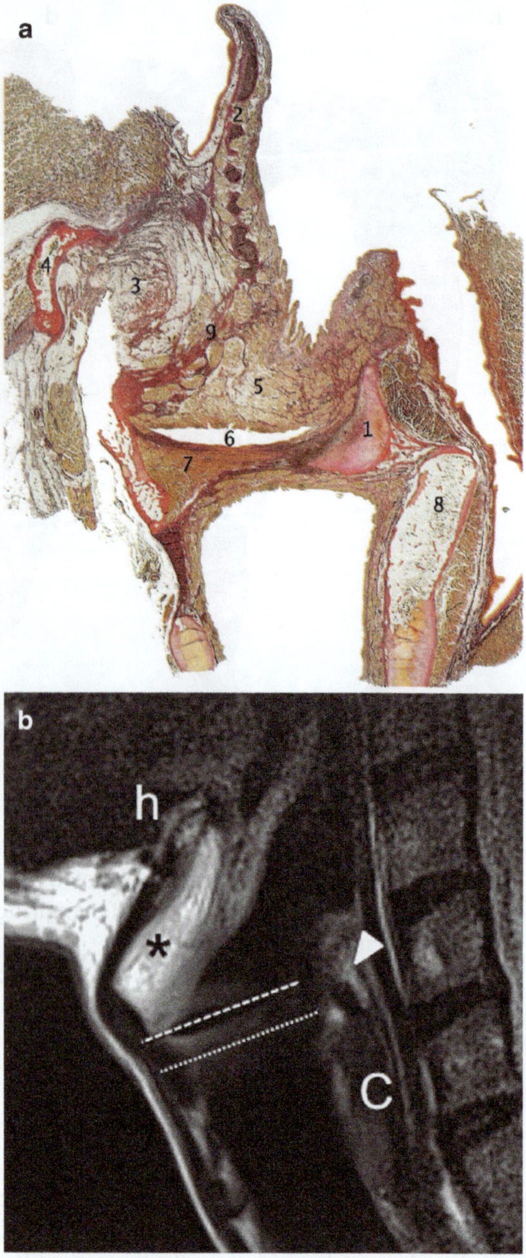

Fig. 13.3 (**a**) Sagittal paramedian macrosection of the larynx. *1* Arytenoid cartilage, *2* epiglottis, *3* preepiglottic space, *4* hyoid bone, *5* ventricular fold, *6* Morgagni's ventricle, *7* vocal cord, *8* cricoid lamina, *9* thyroepiglottic ligament. (**b**) Larynx MRI: T2-weighted sequence on paramedian sagittal plane. *h* Hyoid bone, * preepiglottic space, *arrowhead* arytenoid cartilage, *c* cricoid. The *dotted lines* show the plane of Morgagni's ventricle (-----). An the plane of the boundary between glottic and subglottic sites (.........)

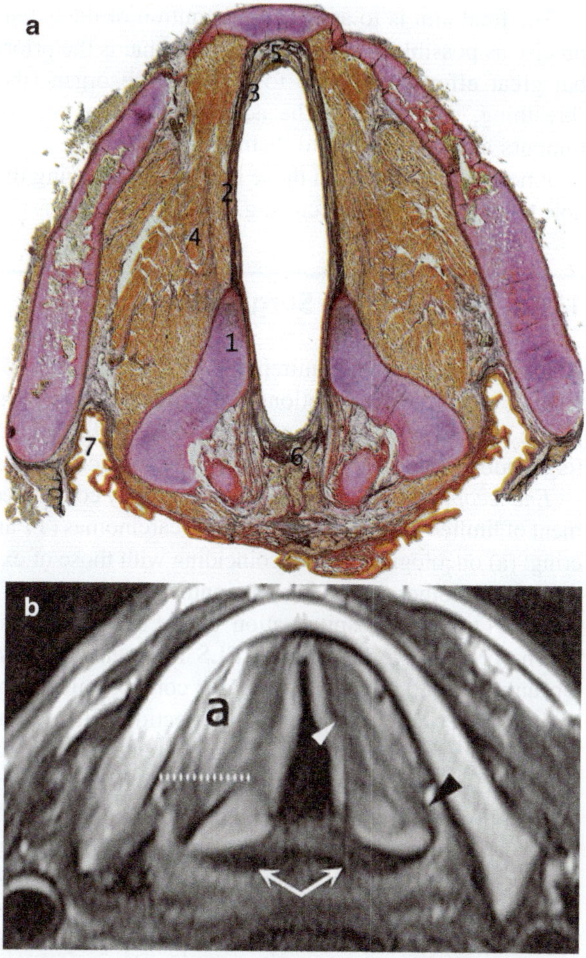

Fig. 13.4 (**a**) Axial macrosection of the larynx on glottic plane. *1* Vocal process of arytenoid cartilage, *2* vocal ligament, *3* epithelial layer, *4* vocal muscle, *5* anterior commessure, *6* posterior commessure, *7* piriform sinus. (**b**) Larynx MRI: T2-weighted sequence on axial plane. Vocal muscle (*white arrowhead*) and vocal cord mucosa (hyperintense layer on the luminal surface of the muscle); insertion of the lateral cricoarytenoid muscle on the muscular process of arytenoid cartilage (*dark arrowhead*); paraglottic space (*a*); interarytenoid muscle (*arrows*)

13.2 Oncological Staging

In the current TNM staging, the larynx is divided into three sites: supraglottis, glottis, and subglottis. Other subsites identify more detailed topographic areas [8].

The diagnostic work-up consists of numerous steps. The endoscopic examination may be carried out with various modes (video-flexible endoscopy, videolaryngostroboscopy, narrow band imaging (NBI), supravital staining with 2 % toluidine blue, saline infusion of Reinke's space, autofluorescence diagnosis, microlaryngoscopy, intraoperative rigid endoscopy by 0°, 30°, 70°, and 120° telescope) [9, 10]. Radiological imaging is based on computed tomography (CT) and on magnetic resonance imaging (MRI) [11, 12].

The final aim is to arrive at a definition of the extent of the neoplasia that is as precise as possible. Curing the patient remains the priority objective of the therapy, but great efforts are made to preserve the organ (the larynx) and the function (breathing, deglutition). The accurate study of the local extension of laryngeal tumours aims to be able to distinguish better those cases in which a conservative treatment is possible from those in which conserving the organ brings an unacceptable risk of the disease recurring.

13.3 What are the Surgical Options?

Another indispensable requirement for this mission is to possess an in-depth knowledge of all the surgical options of laryngeal oncology, so as not to force the therapeutic indication due to a poor knowledge of the various possibilities of treatment (e.g. endoscopic laser surgery as against supracricoid laryngectomies).

Endoscopic laser surgery (ELS) is correctly considered the first choice in the treatment of limited glottic and supraglottic carcinomas (T1 and some T2). That is considering: (a) oncological results coinciding with those of external approach surgery and those of radiotherapy; and (b) a significant reduction of postoperative morbidity, of treatment times, of hospitalisation, and of the overall costs of the treatment [13].

The progressive increase of ELS in recent years is due to the following fact. Up until the end of the 1990s, laser cordectomies were systematically performed with a subperichondrial plane of dissection (Fig. 13.5a, b). Postoperative dysphonia was therefore consistent, certainly greater than that resulting from radiotherapy and from cordectomy in laryngofissure. Considering that the great majority of glottic T1a do not go deeper than the plane of the vocal ligament [14], in 2000, the European Laryngological Society introduced the concept of functional cordectomies for glottic tumours. Having established five types of laser cordectomy, from the most superficial to the deepest and most extensive, the choice is made at the end of the diagnostic work-up of the patient [15]. The oncological results have not substantially changed, but the functional results are decidedly better.

Horizontal supraglottic laryngectomy (HSL) has a solid tradition. Introduced in clinical practice in the 1950s, it continues to be indicated in T3 cases and in some supraglottic T4 because it gives good oncological and functional results.

Supracricoid laryngectomy (SCL) (or reconstructive subtotal), in its numerous variations, is a relatively more recent acquisition. The extent of surgical exeresis may vary from the most conservative situation (crico-hyoido-epiglottopexy, CHEP) to the most destructive (tracheo-hyoido-epiglottopexy, THEP) [16]. The indications are for some T2, some T3, and some T4. In order to propose one of these conservative operations, it is indispensable to save at least one arytenoid cartilage.

In our case history, the indication for *total laryngectomy* (TL), which is an incapacitating operation due to the necessity for a permanent tracheostoma and the consequent aphonia, is applied in about 10 % of patients with a pure laryngeal tumour [17].

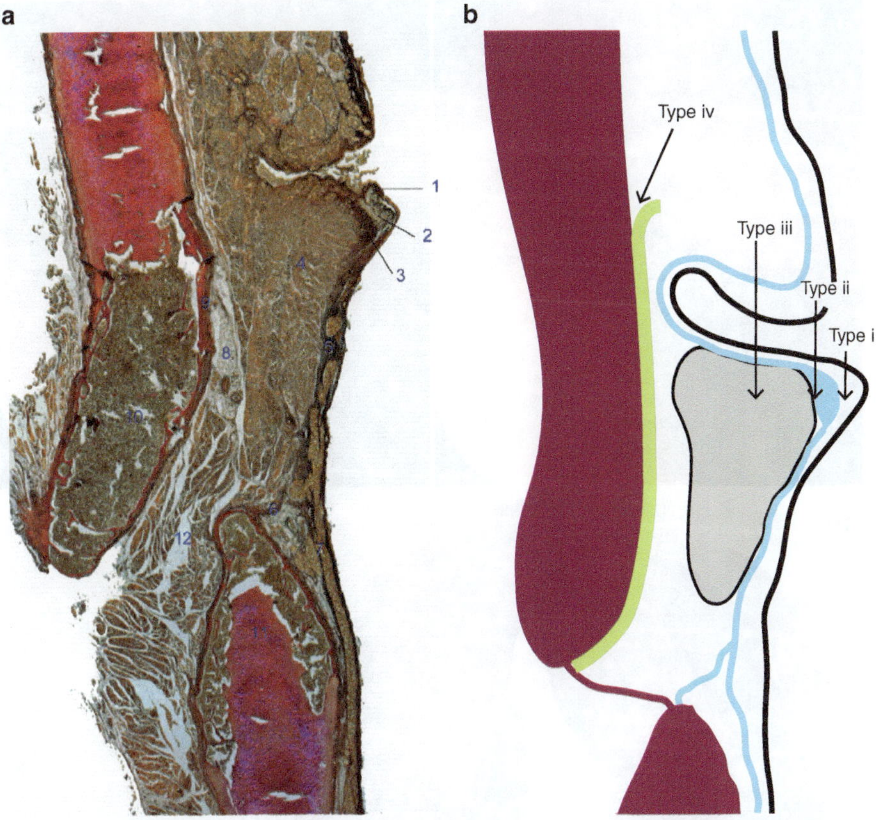

Fig. 13.5 (**a**) Coronal macrosection of the larynx: stratification from epithelium to cartilage. *1* Epithelial layer, *2* Reinke's space, *3* vocal ligament, *4* vocal muscle, *5* elastic conus, *6* elastic conus (deep layer), *7* conus elasticus (superficial layer), *8* inferior paraglottic space, *9* perichondrium, *10* thyroid cartilage, *11* cricoid cartilage, *12* cricothyroid space. (**b**) Types of cordectomies with relation to the depth of dissection

13.4 Choice of Surgery Based on the Assessment of the Critical Areas

The first thing the surgeon looks at is usually the endoscopic image (appearance, site, superficial extension). His second thought goes to the possibility of deep infiltration of the neoplasia (cord mobility). Radiological imaging (CT and MRI) is of great assistance for evaluating some anatomical areas considered "critical" in view of the natural routes for the spread of the laryngeal tumour. Whether or not these areas are involved will lead to the surgical proposal that must be as conservative as possible with acceptable oncological certainty.

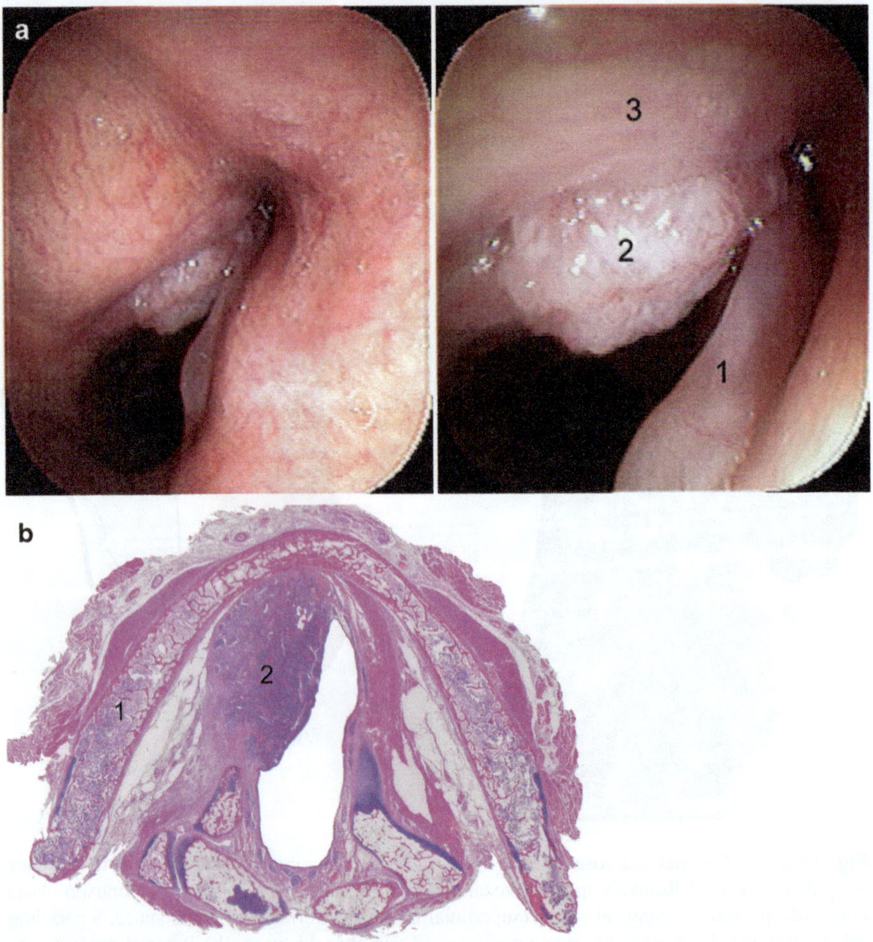

Fig. 13.6 Carcinoma of the left vocal cord reaching the anterior commissure. (**a**) Video laryngoscopy. *1* Right vocal cord, *2* neoplasia, *3* left ventricular fold. (**b**) Axial macrosection. *1* Thyroid cartilage, *2* neoplasia

Critical Area 1: The *Thyroid Cartilage*: In glottic and transglottic tumours, its infiltration is a contraindication for ELS and points towards supracricoid laryngectomy, not necessarily towards total laryngectomy.

The *anterior commissure* is a critical anatomical site for the laryngeal surgeon. The epithelium is only 1–2 mm from the cartilage, there is no perichondrium and, of the whole larynx, it is the site most difficult to expose. The T1a tumour of the vocal cord which arrives at the anterior is a fairly frequent situation and easy to control. CT is rarely required (Fig. 13.6a, b).

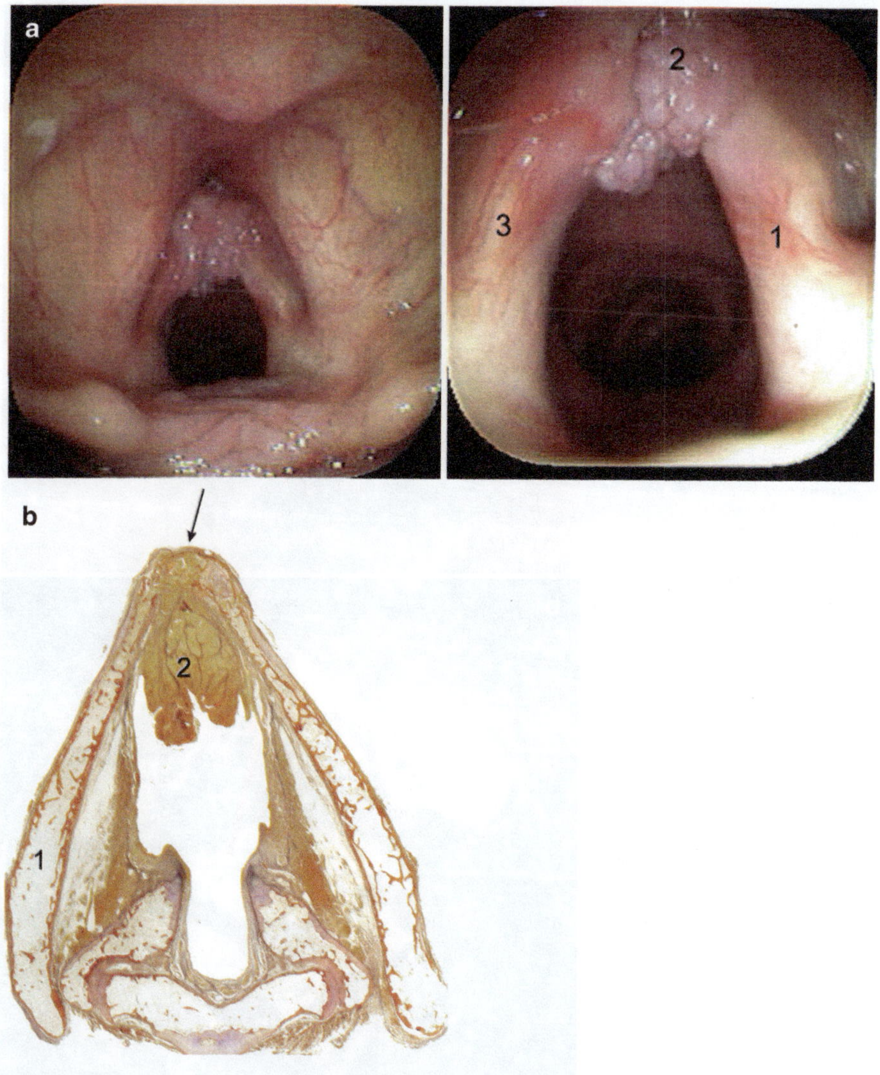

Fig. 13.7 Primary carcinoma of the anterior commissure. (**a**) Video laryngoscopy. *1* Right vocal cord, *2* neoplasia, *3* left vocal cord. (**b**) Axial macrosection showing cartilaginous infiltration at the level of the anterior commissure (*arrow*) *1* thyroid cartilage, *2* neoplasia

Instead, the primitive carcinoma of the anterior commissure, which is fortunately much rarer, is a more risky situation due to its frequent transglottic and transcartilaginous extension. In these cases, it is obligatory to perform a CT (Fig. 13.7a, b).

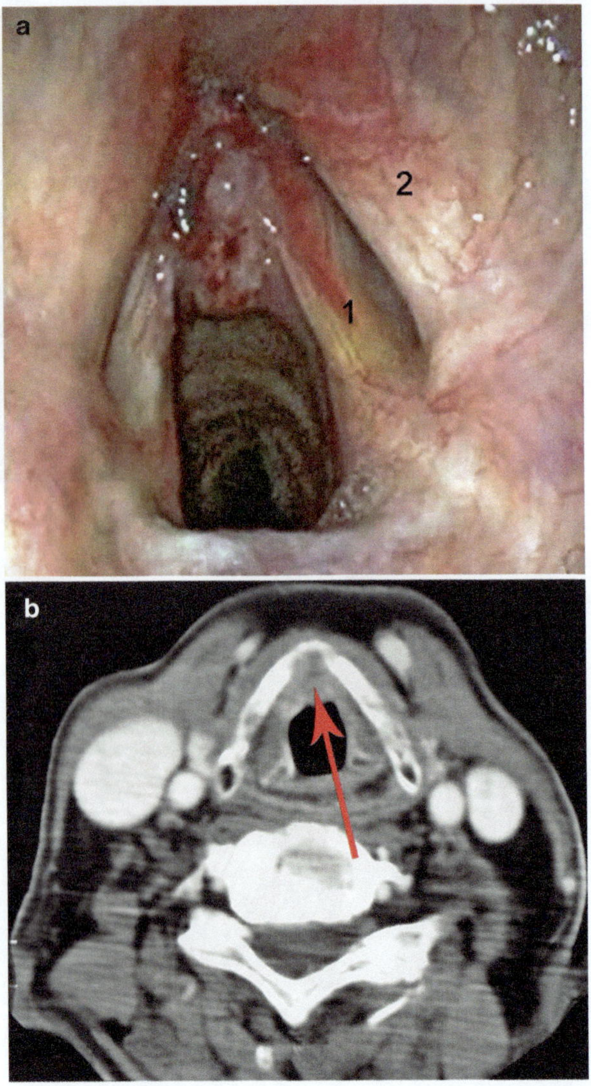

Fig. 13.8 Primary carcinoma of the anterior commissure. (**a**) Video laryngoscopy. Median lesion of the anterior glottis. *1* Right vocal cord, *2* right ventricular fold (**b**) the TC shows erosion of the thyroid cartilage

In the case of the patient in Fig. 13.8b, a primitive tumour of the anterior commissure can be seen, with an infiltrating aspect. In fact, the CT demonstrates the erosion of the thyroid cartilage (cT3), though without going beyond it. A laser cordectomy, even if performed with a subperichondrial dissection plane, would certainly not be radical. In this case, it was correctly decided to perform supracricoid laryngectomy (CHEP).

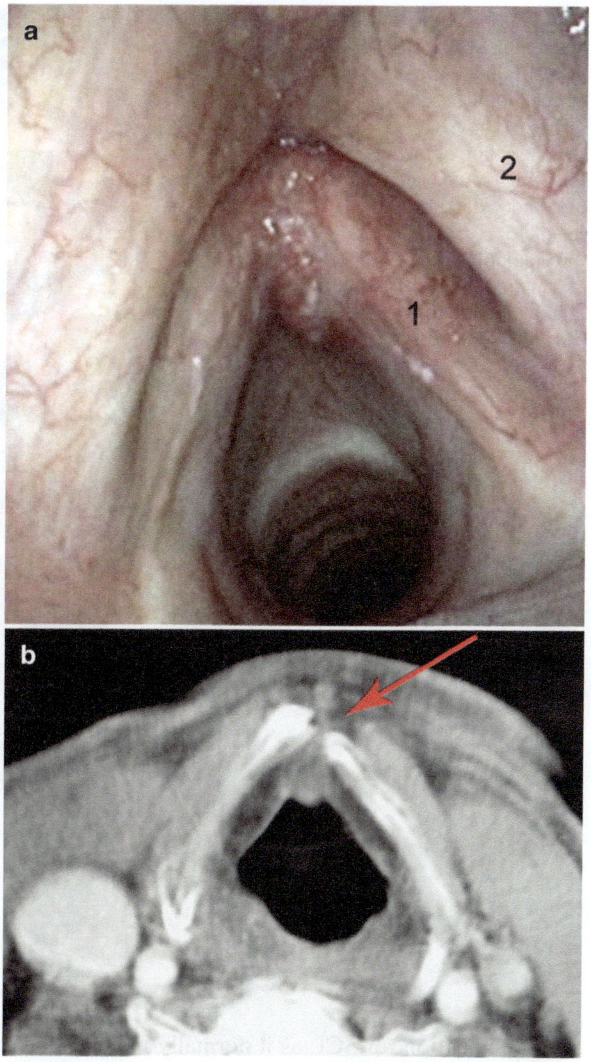

Fig. 13.9 Primitive carcinoma of the anterior commissure. (**a**) Video laryngoscopy. *1* Right vocal cord, *2* right ventricular fold. (**b**) The CT shows an anterior extension of the tumour through the thyroid cartilage, with colonization of Delphian lymph node (*red arrow*)

The small primitive tumour of the anterior commissure affecting the patient in Fig. 13.9 would seem, at first sight, easy to control with ELS. The CT result is startling: the spread of the tumour through the cartilage can be clearly seen (whereby passing from T1a to T4), going so far as to involve the Delphian lymph node of Robbins level VI (pN1) [18] (Fig. 13.9b).

ELS is of course contraindicated in this case. Instead, the patient underwent supracricoid laryngectomy (CHEP, with preservation of both arytenoids). It is important to consider that the anterior extension of the tumour is not a

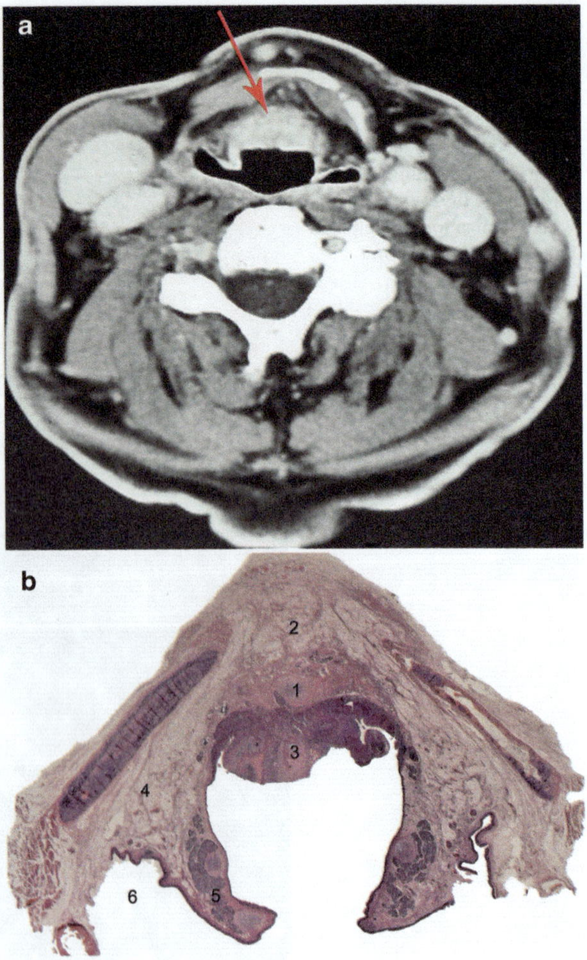

Fig. 13.10 Supraglottic carcinoma (pT2). (**a**) The CT excludes involvement of the preepiglottic space (*red arrow*). (**b**) Axial macrosection of the larynx showing that the neoplasia does not extend beyond the epiglottis. *1* Epiglottis, *2* preepiglottic space, *3* tumour, *4* superior paraglottic space, *5* aryepiglottic fold, *6* piriform sinus

contraindication for SCL as it normally includes the complete ablation of the subhyoid muscles, just as in total laryngectomy. In this case, clearance of level VI is mandatory.

Critical Area 2: The *Preepiglottic Space* and the *Superior Paraglottic Spaces*: CHEP is contraindicated in transglottic tumours that infiltrate the preepiglottic space and the superior paraglottic spaces.

The preservation of the suprahyoid portion of the epiglottis normally allows better phonation because the preserved arytenoid adapts to it, thus forming an efficacious neoglottis. Whether or not the preepiglottic space is involved must be carefully assessed before deciding to preserve the infrahyoid epiglottis.

The imaging of the patient in Fig. 13.10a, b shows that the preepiglottic space is intact.

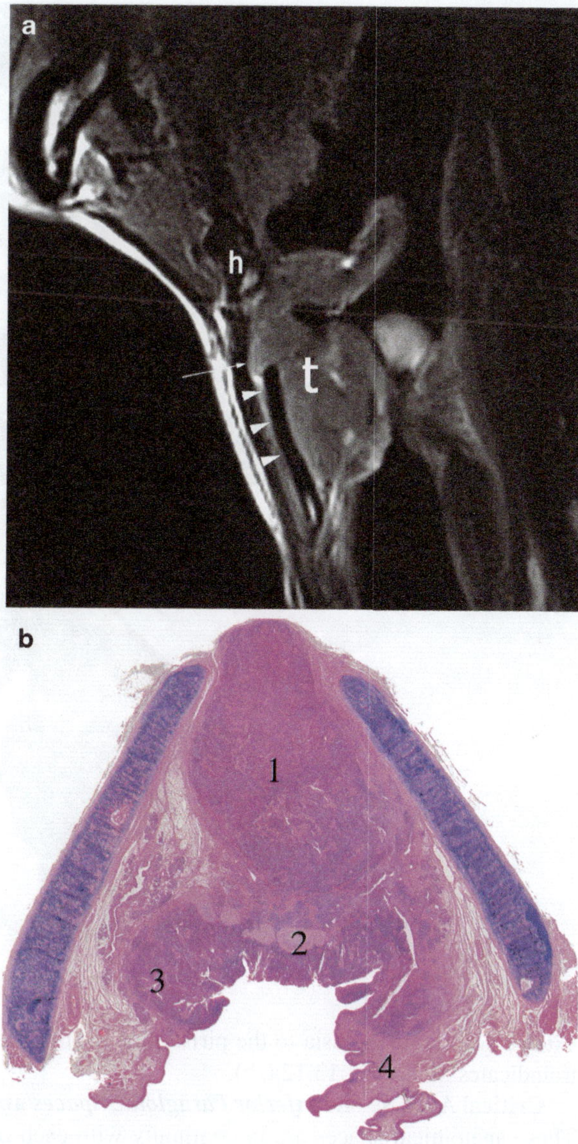

Fig. 13.11 Supraglottic carcinoma (pT3). (**a**) Larynx MRI: T2-weighted sequence on parasagittal plane (orthogonal to the thyroid lamina). Massive infiltration of the preepiglottic space (*t*). Focal infiltration of the thyrohyoid membrane (*arrow*) above the upper margin of the non ossified thyroid lamina (*arrowheads*). The hyoid bone (*h*) is not infiltrated. (**b**) Axial macrosection passing through the preepiglottic space. *1* Tumour, *2* epiglottis, *3* superior paraglottic space (infiltrated by tumour), *4* ventricular fold

Instead, the imaging of the patient in Fig. 13.11a, b shows that the tumour has massively infiltrated the preepiglottic space.

Also, the involvement of the superior paraglottic spaces imposes the sacrifice of the whole epiglottis and the clearance of the preepiglottic space. The possible

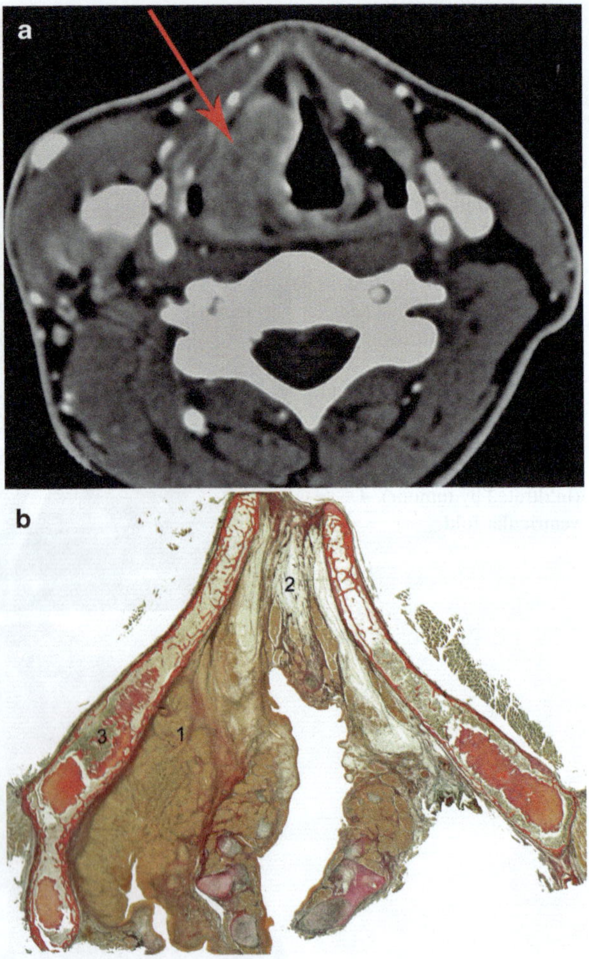

Fig. 13.12 Supraglottic carcinoma (pT3). (**a**) The CT shows the involvement of the superior paraglottic space (*red arrow*). (**b**) Axial macrosection of the larynx showing the involvement of the superior paraglottic space with extension to the piriform sinus. *1* Superior paraglottic space (infiltrated by tumour), *2* petiole of the epiglottis, *3* thyroid cartilage

extension of the neoplasia to the piriform sinus must also be assessed, which contraindicates SCL (Fig. 13.12a, b).

Critical Area 3: The *Inferior Paraglottic Spaces* and the *Cricoarytenoid Joint*: These anatomical spaces are in continuity with each other. Their infiltration indicates or excludes resorting to ELS, SCL, and TL.

In patients with glottic T1a who, when submitted to the diagnostic work-up, demonstrate clearly superficial lesions, CT is not normally carried out. The primitive tumour of the anterior commissure is an exception, as we have already said. When there is a suspicion of profound infiltration or involvement of the ventricle, CT is obligatory, because the infiltration of the inferior paraglottic space is a risky situation for oncological radicality (Fig. 13.13a, b).

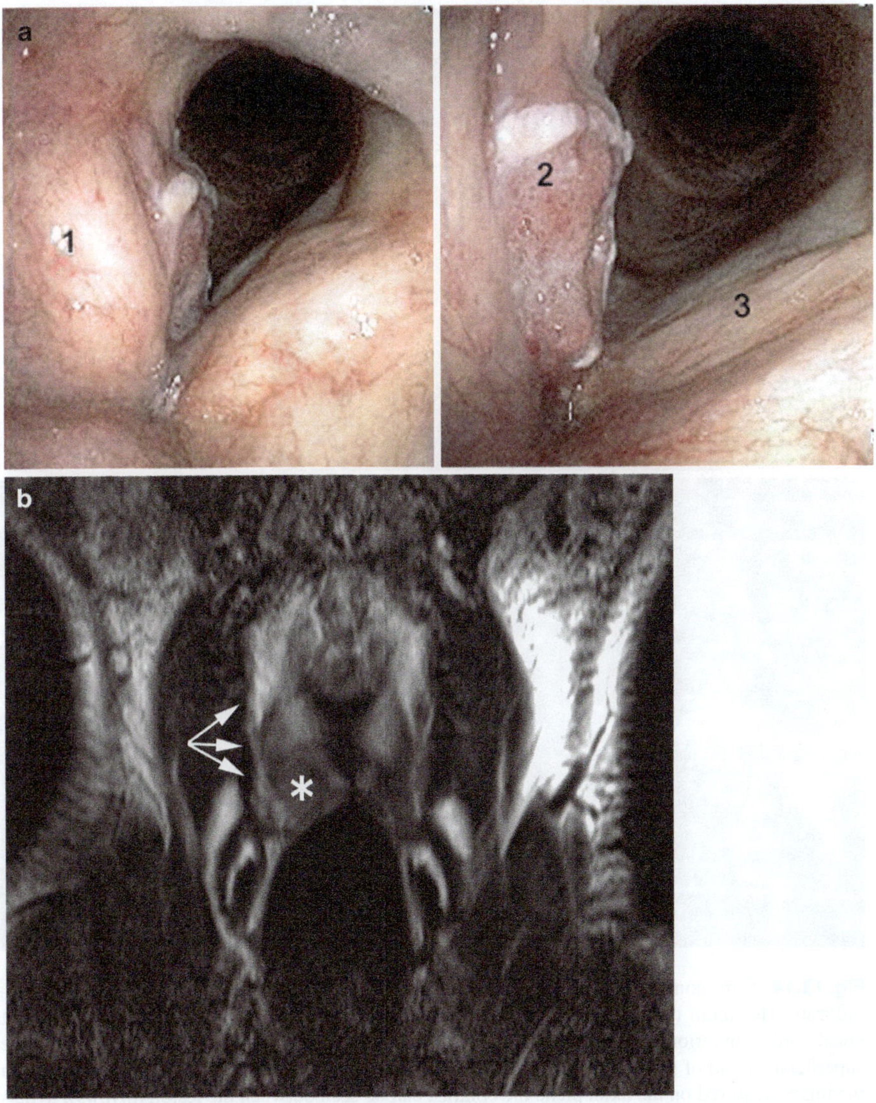

Fig. 13.13 Carcinoma of the right vocal cord with vocal fold hypomobility (cT2). (**a**) Video laryngoscopy: vegeto-infiltrating lesion of the right vocal cord. *1* Right ventricular band, *2* tumour, *3* left vocal cord. (**b**) Larynx MRI: T2-weighted sequence on coronal plane. The lesion (*) infiltrates the vocal muscle and spreads within the fibres of the lateral thyroarytenoid muscle through the ventricle till the false vocal cord level. The paraglottic fat space (hyperintense stripe pointed by *arrows*) is not infiltrated because the lesion is contained by the thyroarytenoid muscle

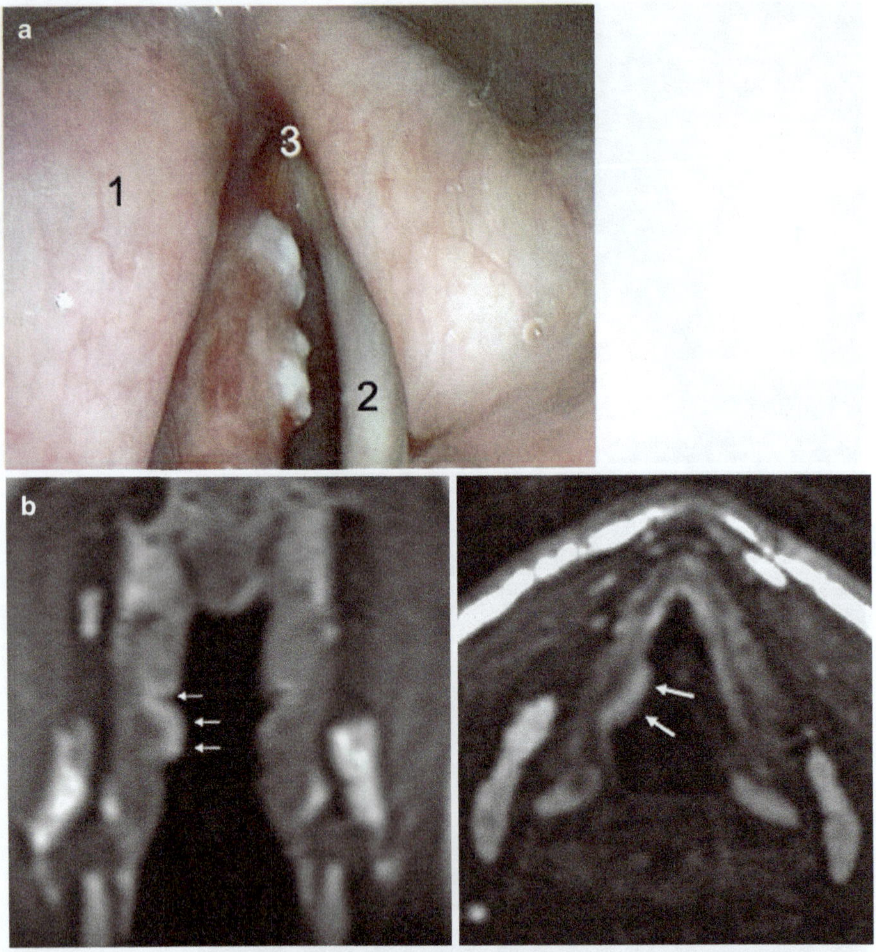

Fig. 13.14 Carcinoma of the left vocal cord. (**a**) Endoscopic view with rigid optics at 30°: vegeto-infiltrating lesion of the left vocal cord; the ventricle is not shown. *1* Left ventricular fold, *2* right vocal cord, *3* anterior commissure. (**b**) Larynx MRI: T2-weight sequence on coronal plane: the superficial spread of the neoplasia is clearly shown, extending as far as the ventricle (*arrows*). In the image acquired on the axial plane the contrast can be seen between the lesion (*arrows*), hyper-intense, and the underlying muscular layer, hypointense

MRI can give very detailed information about the possible deep extension of the neoplasia, and in some cases can even exclude it, as can be seen in Fig. 13.14a, b.

Posterior spread towards the cricoarytenoid joint can not only not be tackled with ELS but not even with SCL because, in this operation, the line of dissection passes right through the articular capsule (Fig. 13.15a, b). In this case, radical oncological

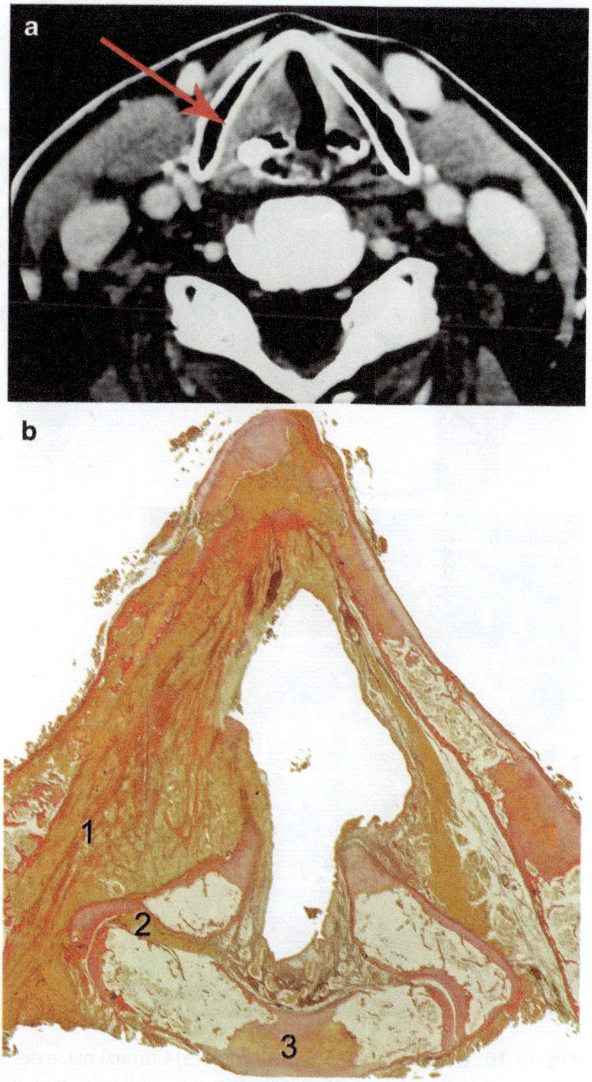

Fig. 13.15 Right transglottic carcinoma with fixity of the vocal cords. (**a**) The CT shows the involvement of the inferior paraglottic space (*red arrow*). (**b**) Axial macrosection of the larynx showing the involvement of the inferior paraglottic space and of the cricoarytenoid articulation. *1* Inferior paraglottic space, *2* cricoarytenoid articulation, *3* cricoid cartilage

treatment is still possible only with a conservative operation such as THEP, which contemplates the ablation of the entire cricoarytenoid articulation of the affected side.

Critical Area 4: The *Lateral Cricothyroid Space*: From the inferior paraglottic space the neoplasia can laterally penetrate the cricothyroid space and be expressed in the larynx (Fig. 13.16a–c).

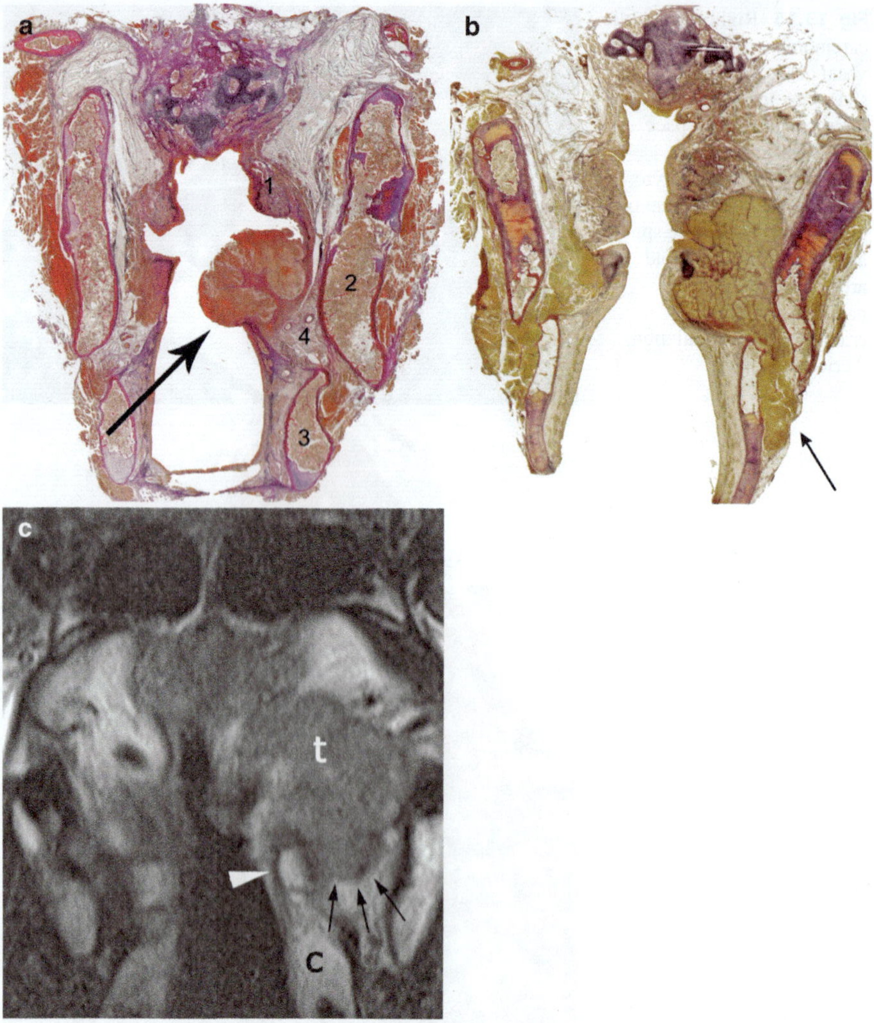

Fig. 13.16 Carcinomas of the vocal cord. (**a**) Coronal macrosection of the larynx showing that the tumour infiltrates the vocal muscle (*arrow*), while the paraglottic space is intact (cT1 or T2, depending on cord mobility). *1* Ventricular fold, *2* thyroid cartilage, *3* cricoid cartilage, *4* inferior paraglottic space. (**b**) Coronal macrosection of the larynx showing infiltration of the inferior paraglottic space (*arrow*) (pT3) and, laterally, extension from the larynx through the cricothyroid space (pT4). (**c**) Larynx MRI: T2-weighted sequence on coronal plane. This tumour (*t*) spreads towards the infero-posterior part of the paraglottic space, between the arytenoid (*arrowhead*) and the thyroid cartilages. The tumour does not extend beyond the larynx, and the inferior boundary between tumour and paraglottic space is visible (*arrows*)

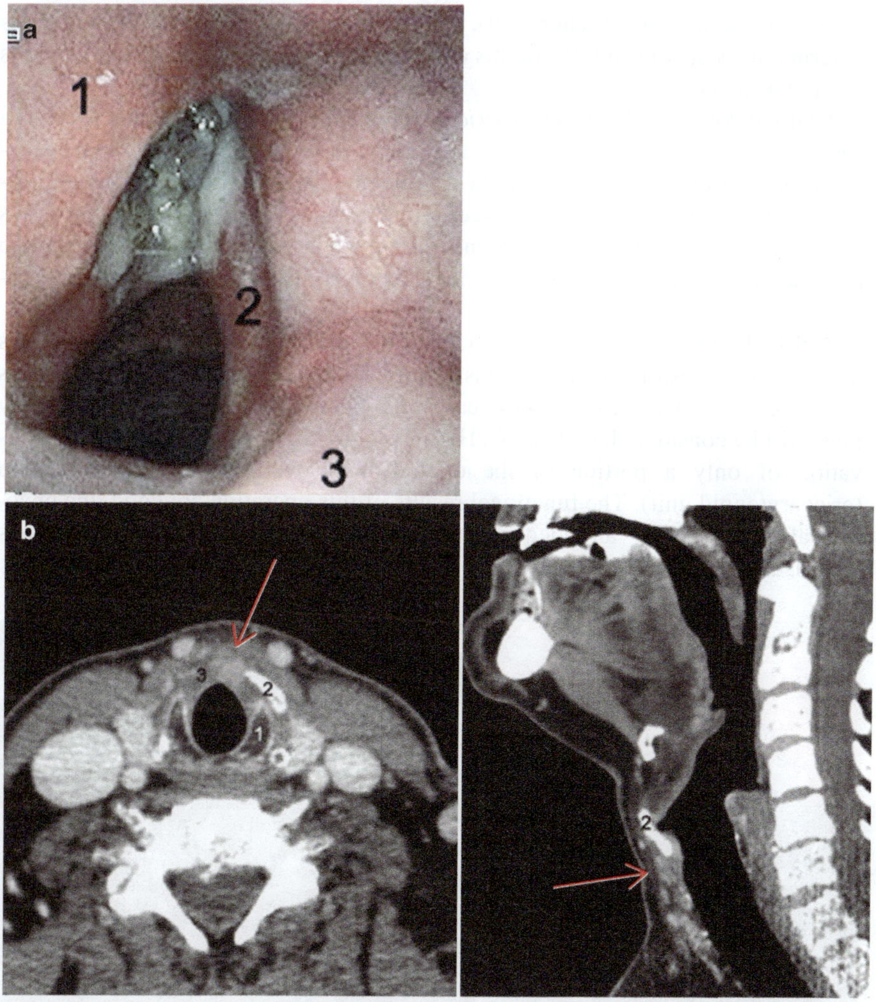

Fig. 13.17 Anterior glottic carcinoma. (**a**) Video laryngoscopy: the lesion is to the anterior glottis, with prevalence on the left. *1* Left ventricular fold, *2* right vocal cord, *3* right arytenoid. (**b**) Larynx TC: on the axial plane passing between cricoid and thyroid cartilages the outspreading of the tumour can be seen (*arrows*). On the sagittal plane there is a suspected pathologic Delphian node. *1* Cricoid cartilage, *2* thyroid cartilage, *3* cricothyroid space

The invasion of the inferior paraglottic space, in its rear portion, is typical of tumours classified clinically as T2 and after the operation as T3 (cT2/pT3). It is often difficult to pinpoint the cause of vocal fold hypomobility. Is it a "mass effect" due to the volume of the tumour? Is it an infiltration of the vocal muscle or, deeper down, of the inferior paraglottic space? Or is even the cricoarytenoid articulation involved? Anyway, a general rule must be kept in mind: while the profound anterior infiltration of the tumour can be easily controlled by type IV laser cordectomy (subperichondrial), the posterior infiltration gives worse results in terms of local control [19]. In this case, the surgical indication generally shifts from SCL to TL.

Critical Area 5: The *Anterior Cricothyroid Space*: The cricothyroid space is a very important anatomical site. It is easily identifiable by palpation along the anterior profile of the larynx, between the thyroid and cricoid cartilages. Digital pressure from the outside can be clearly seen with the endoscope. Any infiltration of this space means that the tumour has extended from the larynx and it will be difficult to propose the classic SCL due to very poor safety margins at the moment of supracricoid resection (Fig. 13.17a,b).

Critical Area 6: The *Subglottis*: One of the main contraindications for SCL is the caudal extension of the neoplasia towards the subglottis since exeresis is supracricoid by definition. In these cases, the possibility of conservative surgery must still be considered, that is, of THEP. This operation contemplates the preservation of only a portion of the cricoid with overlying healthy arytenoid (cricoarytenoid unit). The functional results of this operation are good, the oncological results need an adequate follow-up to obtain reliable conclusions (Fig. 13.18a–c).

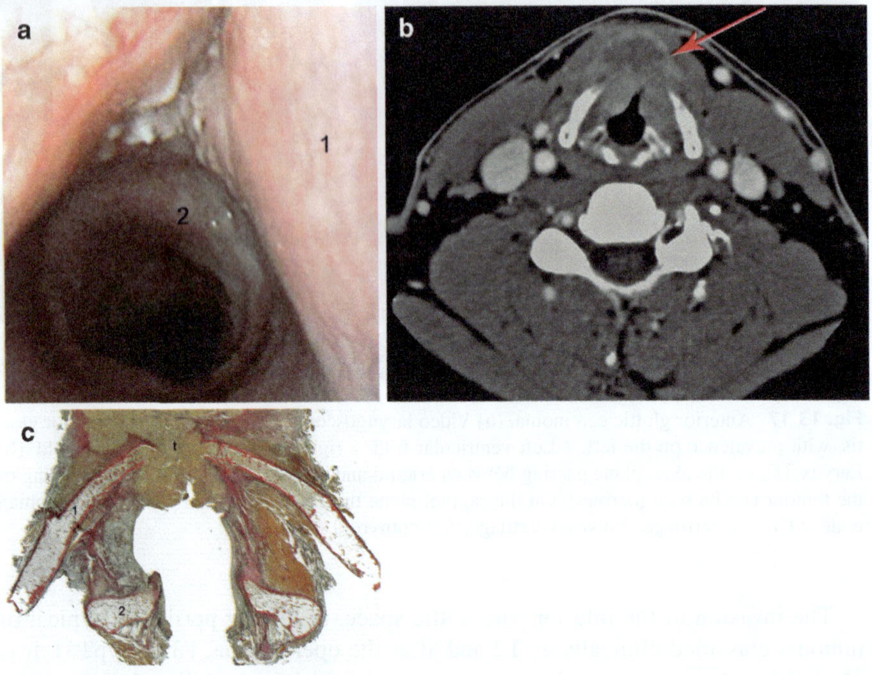

Fig. 13.18 Carcinoma of the anterior commissure, with subglottic extension. (**a**) On video laryngoscopy the lesion is apparently superficial. *1* Right vocal cord, *2* first tracheal ring. (**b**) the CT shows a massive anterior extension through the cricothyroid space and the thyroid cartilage (*arrow*). (**c**) the macrosection, taken on a more cranial plane, shows the same details. *1* Thyroid cartilage, *2* arytenoid cartilage, *t* tumor through thyroid cartilage

Take-Home Messages

- The correct staging of the tumour is fundamental for a correct therapy and, consequently, for an optimum oncological result. At the origin of local recurrences, there is very probably an error in staging which fails to recognise the real extent of the tumour and leaves some residue after surgical exeresis.
- The surgeon and the radiologist must focus their attention on the study (endoscopy, CT, MRI) of some anatomical areas that we have listed and defined as "critical".
- The surgeon must know all the possible options and be able to manage them; the possibility of a conservative approach with acceptable oncological results must always be considered.

References

1. Carlon G (1990) Il carcinoma della laringe. Dalla patologia alla clinica. Piccin, Padova
2. Tucker GF Jr, Smith HR (1962) A histological demonstration of the development of laryngeal connective tissue compartments. Trans Am Acad Ophthalmol Otolaryngol 66:308–318
3. Tucker GF Jr (1963) Some clinical inferences from the study of serial laryngeal sections. Laryngoscope 73:728–748
4. Maguire A, Dayal VS (1974) Supraglottic anatomy: the pre- or the peri-epiglottic space? Can J Otolaryngol 3:432–445
5. Sato K, Kurita S, Hirano M (1993) Location of preepiglottic space and its relationship to the paraglottic space. Ann Otol Rhinol Laryngol 102:930–934
6. Reidenbach MM (1996) The paraglottic space and transglottic cancer: anatomical considerations. Clin Anat 9:244–251
7. Reidenbach MM (1997) Borders and topographic relationships of the paraglottic space. Eur Arch Otorhinolaryngol 254:193–195
8. Union Internationale Contre le Cancer (2011) TNM classification of malignant tumors, 7th edn. Wiley-Liss, New York
9. Colden D, Zeitels SM, Hillman RE, Jarboe J, Bunting G, Spanou K (2001) Stroboscopic assessment of vocal fold keratosis and glottic cancer. Ann Otol Rhinol Laryngol 110: 293–298
10. Zbaren P, Becker M, Lang H (1996) Pretherapeutic staging of laryngeal carcinoma. Clinical findings, computed tomography, and magnetic resonance imaging compared with histopathology. Cancer 77:1263–1273
11. Barbera L, Groome PA, Mackillop WJ et al (2001) The role of computed tomography in the T classification of laryngeal carcinoma. Cancer 91:394–407
12. Becker M (2001) Malignant lesions of the larynx and hypopharynx. In: Hermans R (ed) Imaging of the larynx. Springer, Berlin, pp 57–84
13. Lucioni M, Marioni G, Bertolin A, Giacomelli L, Rizzotto G (2011) Glottic laser surgery: outcomes according to 2007 ELS classification. Eur Arch Otorhinolaryngol 268:1771–1778
14. Pittore B, Ismail-Koch H, Davis A et al (2009) Thyroarytenoid muscle invasion in T1 glottic carcinoma. Eur Arch Otorhinolaryngol 266:1787–1791
15. Remacle M, Eckel HE, Antonelli A et al (2000) Endoscopic cordectomy: a proposal for a classification by the Working Committee, European Laryngology Society. Eur Arch Otorhinolaryngol 257:227–231

16. Rizzotto G, Succo G, Lucioni M, Pazzaia T (2006) Subtotal laryngectomy with tracheohyoidopexy: a possible alternative to total laryngectomy. Laryngoscope 116(10):1907–1917
17. Lucioni M, Marioni G, Mangialaio M, Rizzotto G (2006) CO2 laser treatment of laryngeal stenoses after reconstructive laryngectomies with cricohyoidopexy, cricohyoidoepiglottopexy or tracheohyoidoepiglottopexy. Eur Arch Otorhinolaryngol 264:175–180
18. Robbins KT, Clayman G, Levine PA, Medina J, Sessions R, Shaha A, Som P, Wolf GT, American Head and Neck Society; American Academy of Otolaryngology-Head and Neck Surgery (2002) Neck dissection classification update: revision proposed by the American Head and Neck Society and the American Academy of Otolaryngology-Head and Neck Surgery. Arch Otolaryngol Head Neck Surg 128(7):751–758
19. Peretti G, Piazza C, Mensi MC, Magnoni L, Bolzoni A (2005) Endoscopic treatment of cT2 glottic carcinoma: prognostic impact of different pT subcategories. Ann Otol Rhinol Laryngol 114:579–586

Notes on Laryngeal Surgical Technique 14

Core Messages

- This chapter concludes the in-depth study of the larynx and of oncological laryngology in the second edition of the *Practical Guide to Neck Dissection.*
- It represents the application in the surgical field of all that has been learned from anatomical studies and of the knowledge of the preferential routes for the spread of laryngeal neoplasia.
- Only the principal surgical operations are illustrated here, with no claim of completeness. But we believe they represent about 80–90 % of all the operations carried out in the laryngeal oncology theatre.

14.1 Laser Cordectomy (Type II)

Endoscopic laser surgery has had a great development in recent years, especially in Europe thanks to the insight and ability of Wolfgang Steiner. In order to standardise the nomenclature of the various operations and thus be able to compare the results of the various case histories, the European Laryngological Society has drawn up a code classifying endoscopic cordectomies of various types, according to the depth and extension of exeresis [1–4]. The operation we are going to discuss in this paragraph is the type most frequently performed.

1 Principles of Operation: It consists of removing the vocal cord tumour, following a dissection plane that lies below the vocal ligament. The aim is both therapeutic and diagnostic, since it is quite rightly becoming less frequent to carry out a biopsy prior to operation in initial lesions of the cords [5].

2 Indications: dysplasias, carcinomas "in situ", microinfiltrating glottic T1a and T1b

3 Surgical Technique: A satisfactory surgical exposure is fundamental for endoscopic laser surgery. The study of the neoplasia with rigid optics before starting the operation is essential for defining its limits (Fig. 14.1).

M. Lucioni, *Practical Guide to Neck Dissection*,
DOI 10.1007/978-3-642-33977-6_14, © Springer-Verlag Berlin Heidelberg 2013

The removal of the ventricular band allows excellent control of the glottis. The cordectomy normally includes all the mucosa of the floor of the ventricle (Fig. 14.2).

Once we have found the plane of the vocal ligament and exposed the surface of the vocal muscle, these are dissected until we reach the caudal limit of the vocal

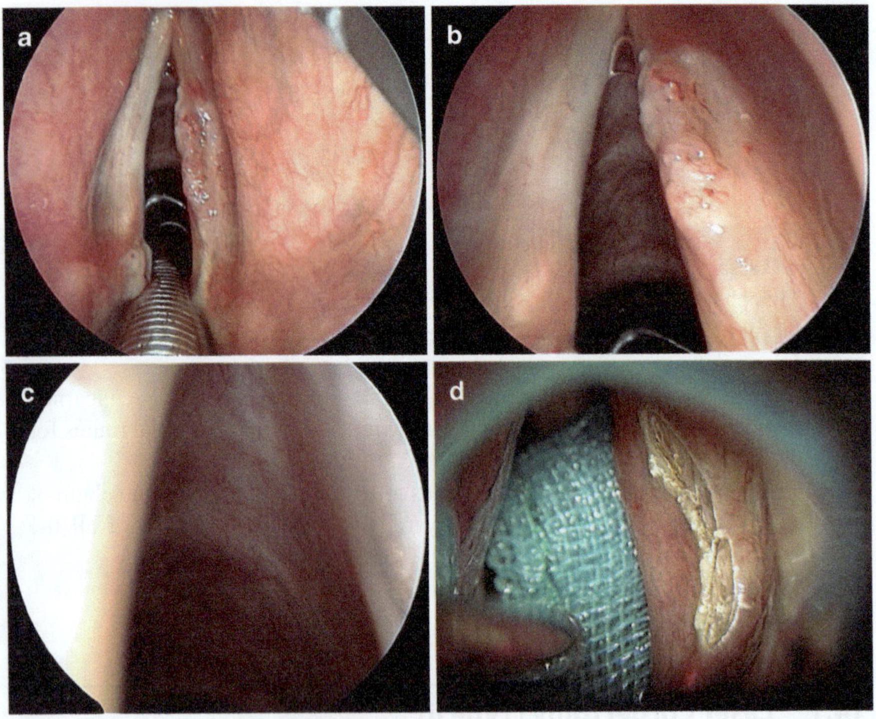

Fig. 14.1 Laser Cordectomy (Type II) T1a epidermoid carcinoma of the right vocal cord (**a**); the floor of the ventricle is free from disease (**b**); as is the subglottis (**c**); applying a protection on the cord plane, the ventricular fold is resected so as to see and work better on the glottis below (**d**)

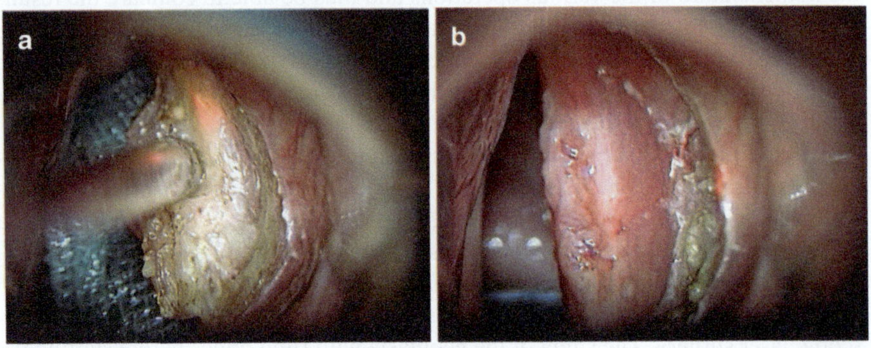

Fig. 14.2 Laser Cordectomy (Type II) The ventricular band is completely removed (**a, b**); the ventricle mucosa is pulled medially and dissected from the vocal muscle below (**c, d**)

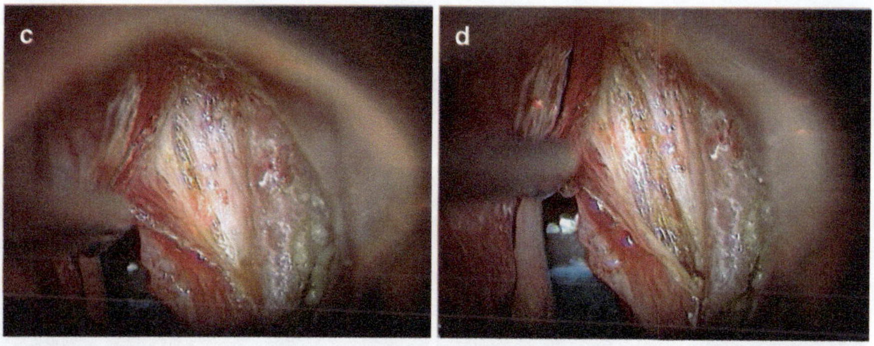

Fig. 14.2 (continued)

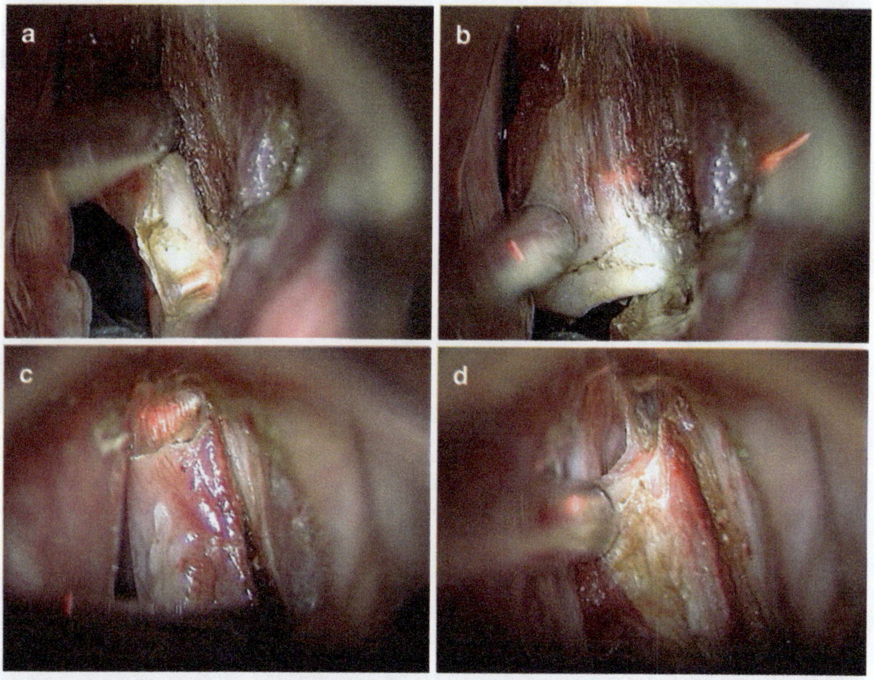

Fig. 14.3 Laser Cordectomy (Type II) The vocal ligament is sectioned immediately in front of the vocal process of the arytenoid (**a**, **b**); the anterior section is close to the anterior commissure (**c**, **d**)

muscle. The ligament is then resected at its two ends. The resection of the ligament is carried out only at this point so as to keep the cord well taut during dissection of the muscle and ligament (Fig. 14.3).

The vocal ligament is separated from the vocal muscle, paying attention to the resection margins which must be in healthy tissue (Fig. 14.4).

Exeresis is also completed anteriorly and the specimen is removed and oriented (Fig. 14.5).

Photocoagulation of the mucosal resection margins is then carried out and the field of operation is checked again with rigid optics at the end of the operation (Fig. 14.6).

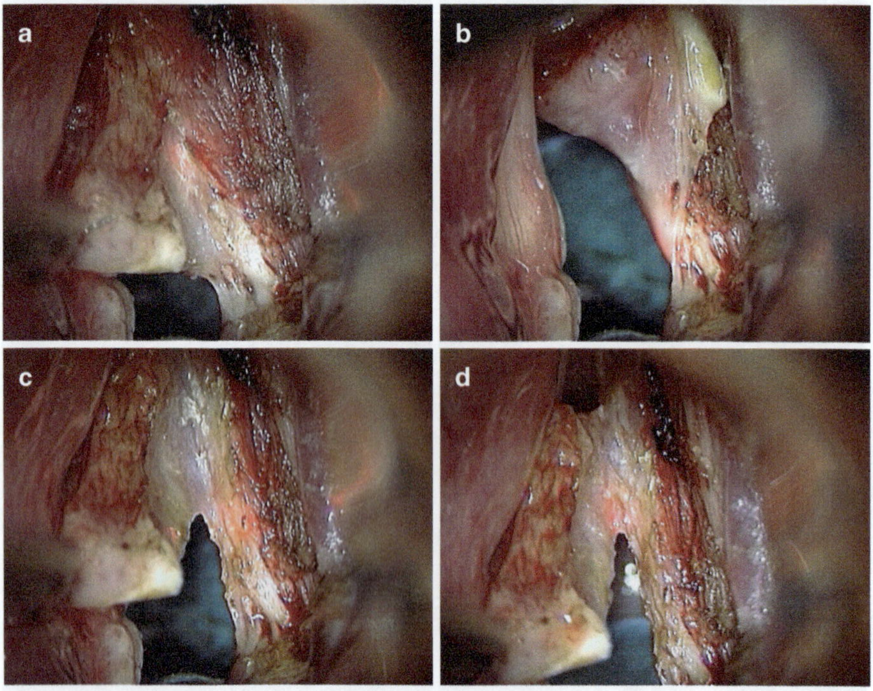

Fig. 14.4 Laser Cordectomy (Type II) The muscle fibres remain on the right while the vocal ligament is on the left; the microforceps grip the caudal end of the specimen and medialise it during dissection (**a**, **c**, **d**); the flap is lateralised at intervals so as to be able to check that the dissection is at a safe distance from the neoplasia (**b**)

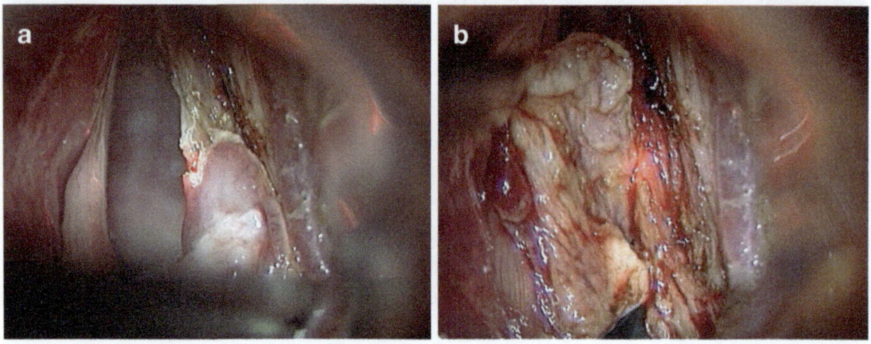

Fig. 14.5 Laser Cordectomy (Type II) Also the anterior margin is at a satisfactory distance from the neoplasia (**a**); the specimen is detached (**b**); it is put back in place to check its margins and orientation (**c**); before sending it for anatomo-pathological testing, the posterior margin of exeresis is marked with Indian ink (**d**)

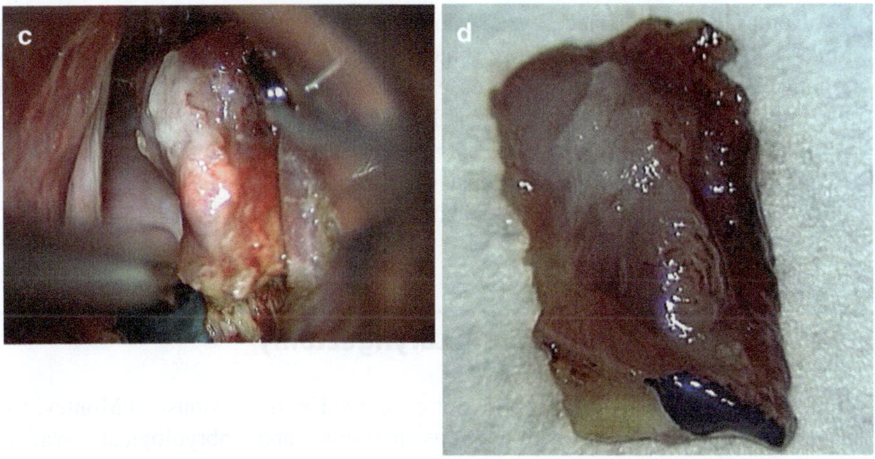

Fig. 14.5 (continued)

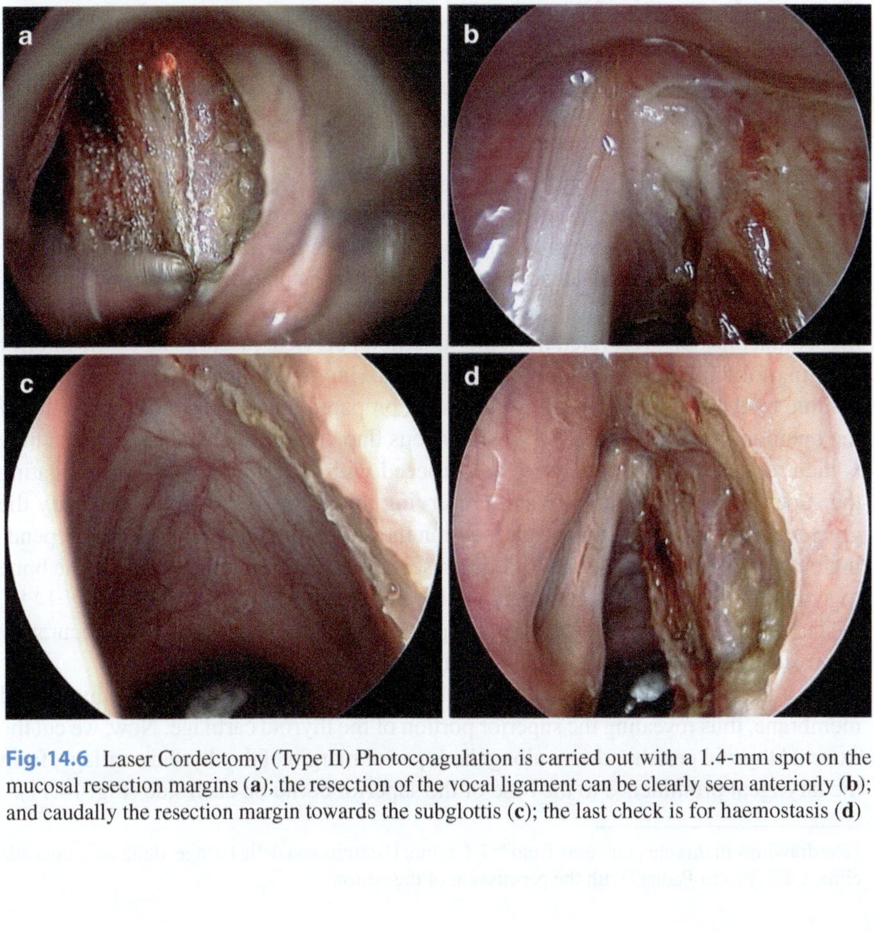

Fig. 14.6 Laser Cordectomy (Type II) Photocoagulation is carried out with a 1.4-mm spot on the mucosal resection margins (**a**); the resection of the vocal ligament can be clearly seen anteriorly (**b**); and caudally the resection margin towards the subglottis (**c**); the last check is for haemostasis (**d**)

4 Remarks: At a distance, type II cordectomy shows excellent functional outcomes, indistinguishable from those achieved with radiotherapy [6]. Laser photocoagulation seems to be effective in disease local control in case of close or superficial margin positivity at postoperative histological examination [7]. Positivity of the profound resection margins requires surgical revision, which in most cases consists of performing a deeper or more extensive cordectomy. In the case of margins with massive involvement, surgery with an external approach is indicated [8].

14.2 Supraglottic Horizontal Laryngectomy

Supraglottic horizontal laryngectomy was conceived by Justo Alonso of Montevideo around 1940 [9]. It resulted from previous anatomical and embryological considerations (Rouvière, Baclesse) [10] that presumed a lymphatic watershed at the level of the laryngeal ventricle which forms an obstacle to the downward spread of vestibular tumours.

1 Principles of the Operation: (a) complete ablation of the laryngeal vestibule, that is, of the portion of the larynx located above the glottic plane (the arytenoids are preserved); (b) ablation of the preepiglottic space; (c) ablation or non-ablation of the hyoid bone; and (d) the exeresis may be extended to one arytenoid, to the piriform sinus or to the tongue base (extended supraglottic laryngectomies), or to one vocal cord (laryngectomy type ¾) (Fig. 14.7a–d)*

2 Indications: T1, T2, and some supraglottic T3

3 Surgical Technique: The incision of the skin preferred is André's bilateral incision and usually laterocervical lymph node dissection is carried out beforehand. The incision involves the cutis, subcutaneous tissue, and the platysma muscle and extends from one mastoid to the other passing a finger's width above the jugulum. The large myocutaneous flap is elevated from the superficial cervical fascia as far as the inferior margin of the mandible. Before proceeding with laryngectomy, we perform the subisthmic tracheostomy. Three silk stitches are prepared at the top which, at the end of the operation, will be sutured to the cutaneous flap. The orotracheal anaesthesia tube is then replaced. At this point, we can proceed with the laryngectomy, after ligating the superior laryngeal pedicles and preserving the nerve (Fig. 14.8).We identify the greater cornu of the hyoid bone and section the supra- or subhyoid muscles (depending on whether the hyoid bone is to be preserved or not), remaining close to the bone to avoid damaging the lingual artery, which lies immediately underneath (Fig. 14.9).

This exposes the adipose tissue of the preepiglottic cavity which will be accurately removed with the specimen (Fig. 14.10).

We then section the subhyoid muscles horizontally at the level of the thyrohyoid membrane, thus revealing the superior portion of the thyroid cartilage. Now, we cut the external thyroid perichondrium along the superior margin of the thyroid cartilage, from the laryngeal prominence to the base of the superior cornu on each side (Fig. 14.11).

*The drawings in this chapter came from "G. Carlon, Il carcinoma della laringe: dalla patologia alla clinica, Ed. Piccin Padua" with the permission of the editor.

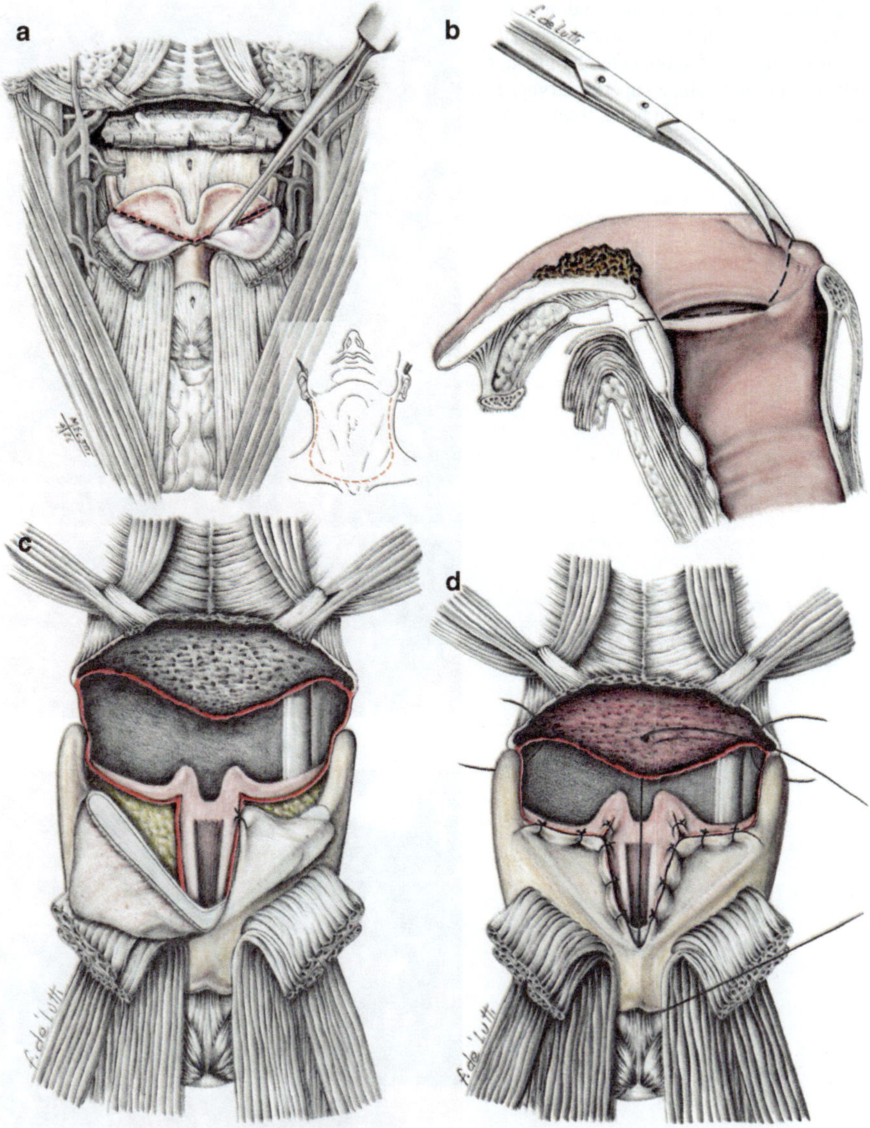

Fig. 14.7 Supraglottic Horizontal Laryngectomy: surgical steps (**a**) (1) large myocutaneous flap, (2) subisthmic tracheostomy, (3) ligating the superior laryngeal pedicles, (4) section the supra- or subhyoid muscles, (5) clearance of preepiglottic space, (6) resection and detaching of thyroid perichondrium, (7) resection of thyroid cartilage (**b**). (1) horizontal section in front of arytenoid cartilage, along the ventricle as far as anterior commissure (2) section line through the glossoepiglottic valleculae and the aditus of piriform sinus (**c**) reconstruction of the glottic plane with the perichondrium (**d**) (1) pexy between the base of tongue (the hyoid bone when preserved) and the residual thyroid cartilage (2) second suture layer between the subhyoid muscles and the suprahyoid cervical fascia

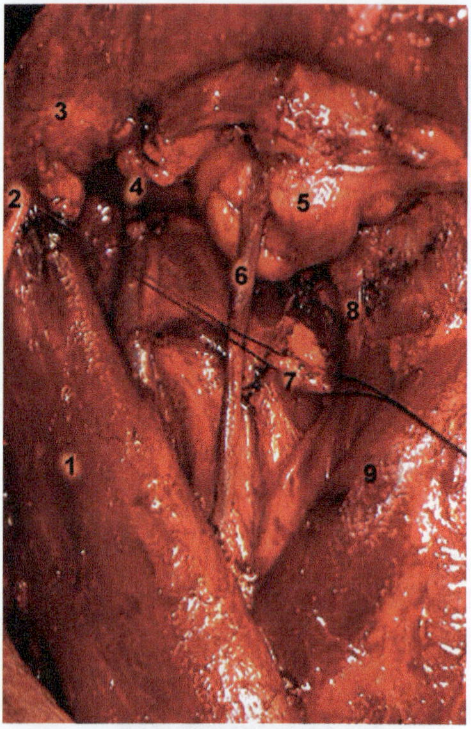

Fig. 14.8 *1* Sternocleidomastoid muscle, *2* great auricular nerve, *3* parotid gland, *4* digastric muscle, *5* submandibular gland, *6* retromandibular vein, *7* superior laryngeal artery and vein, *8* hyoid bone, *9* subhyoid muscles

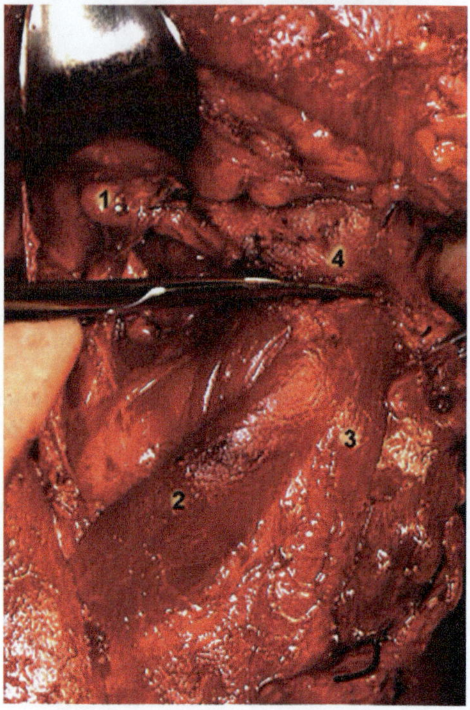

Fig. 14.9 *1* Apex of the greater cornu of the hyoid bone, *2* omohyoid muscle, *3* sternohyoid muscle, *4* body of the hyoid bone

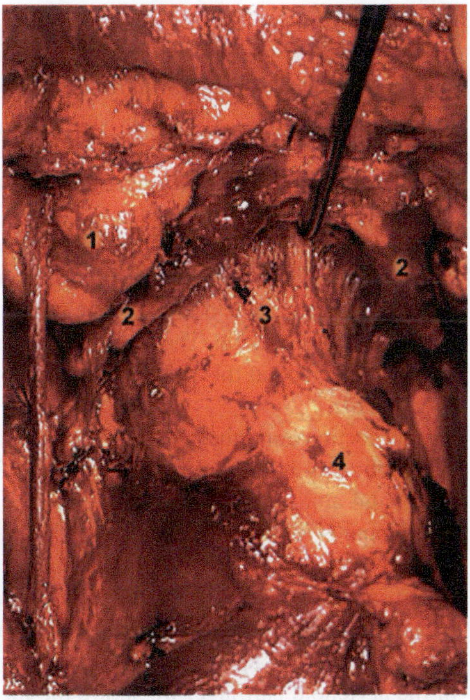

Fig. 14.10 *1* Submandibular gland, *2* greater cornu of the hyoid bone, *3* preepiglottic space (cleared), *4* adipose tissue of the preepiglottic space (lifted down)

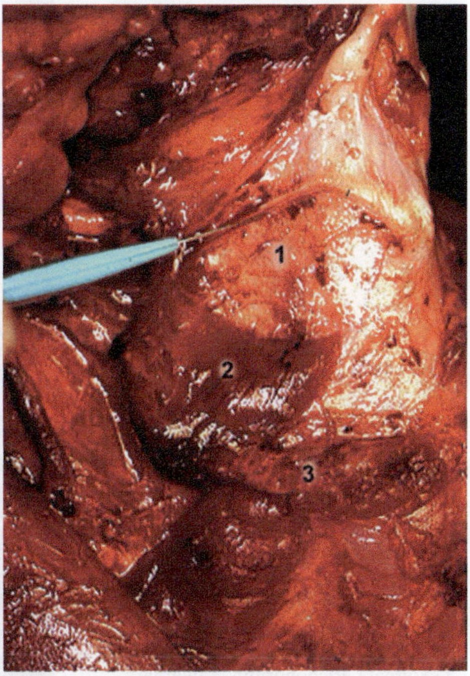

Fig. 14.11 *1* Right thyroid lamina, *2* thyrohyoid muscle (sectioned), *3* subhyoid muscles (sectioned)

First with an angled separator and then with a dry gauze swab, the perichondrium is elevated downwards on both sides, as far as the line that joins the base of the superior cornu to the small cartilage depression internally corresponding to the anterior commissure. Along this same ideal line, we then proceed with the bilateral elevation of the internal thyroid perichondrium, so as to create two subperichondrial supraglottic tunnels, converging above the anterior commissure (Fig. 14.12).

The thyroid cartilage is then sectioned along the same line with the Lister's scissors, inserting the distal blade in the subperichondrial tunnels (Fig. 14.13).

We proceed with supraglottic laryngectomy using as mode of access the pharyngostome made in the side wall of the piriform sinus, choosing the side less affected by the neoplastic lesion. The section is extended to the mucosa of the glossoepiglottic valleculae and at this point a good exposure of the neoplasia is usually possible (Fig. 14.14).

The section continues down towards the entrance to the piriform sinus and then runs obliquely along the anterior aspect of the arytenoid cartilage, then becomes horizontal along the floor of the ventricle, as far as the anterior commissure. The same route is followed on the other side (Fig. 14.15).

The glottic plane is reconstructed using the two triangles of external perichondrium, preserved during the demolitive phase of the operation, which are turned back to cover the section edges of the thyroid cartilage and the inferior paraglottic spaces. We fix the apex of the triangle to the mucosa that covers the vocal process of the arytenoid cartilage; we suture the medial side of the triangle with the lateral

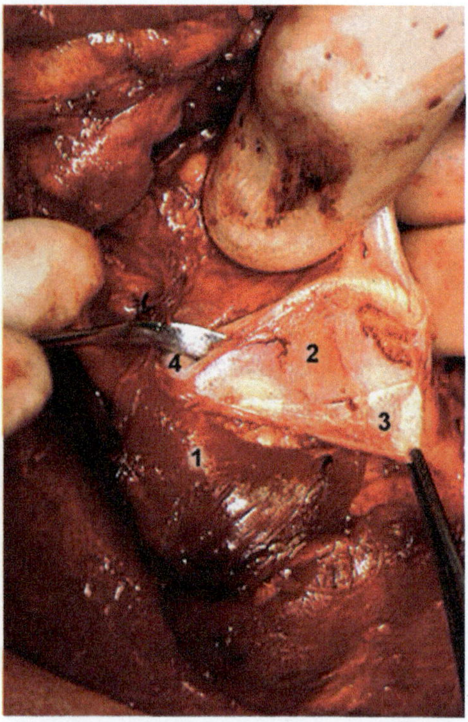

Fig. 14.12 *1* Thyrohyoid muscle (sectioned), *2* right thyroid lamina, *3* external perichondrium, *4* internal perichondrium

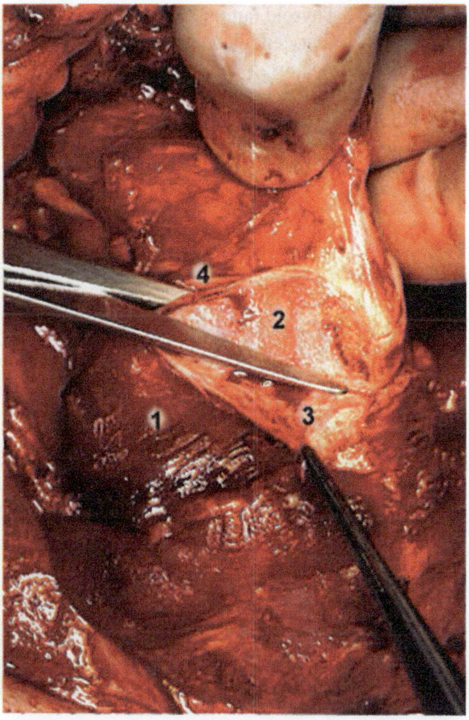

Fig. 14.13 *1* Thyrohyoid muscle (sectioned), *2* right thyroid lamina, *3* external perichondrium, *4* internal perichondrium

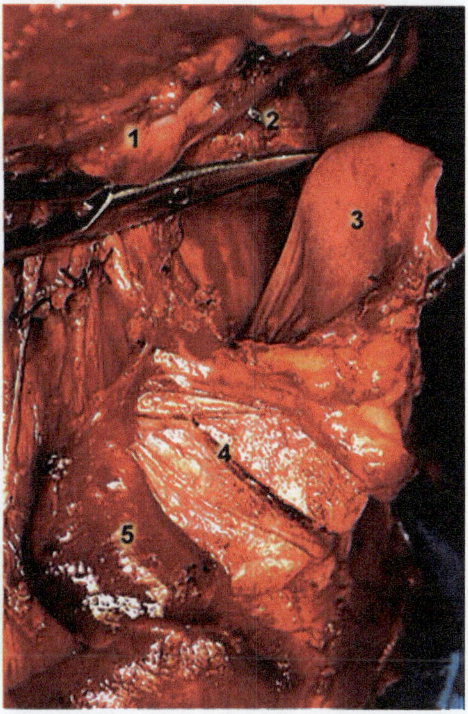

Fig. 14.14 *1* Greater cornu of the hyoid, *2* tongue base, *3* epiglottis, *4* thyroid lamina, *5* thyrohyoid muscle

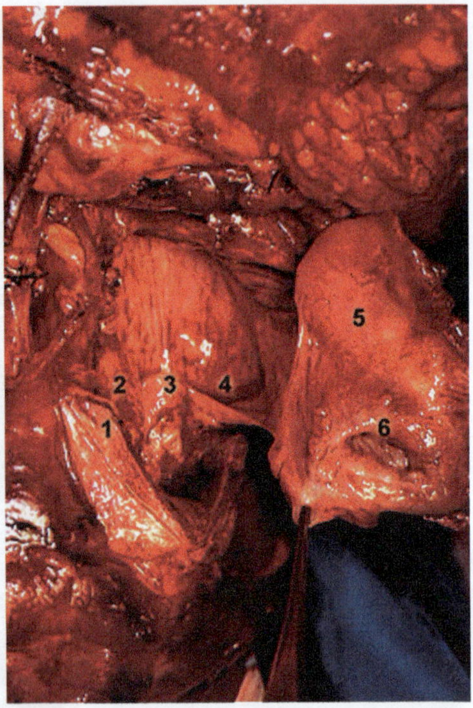

Fig. 14.15 *1* Thyroid lamina, *2* right arytenoid, *3* left arytenoid, *4* entrance to the left piriform sinus, *5* epiglottis, *6* carcinoma of the left vestibule (ventricular fold and infrahyoid epiglottis)

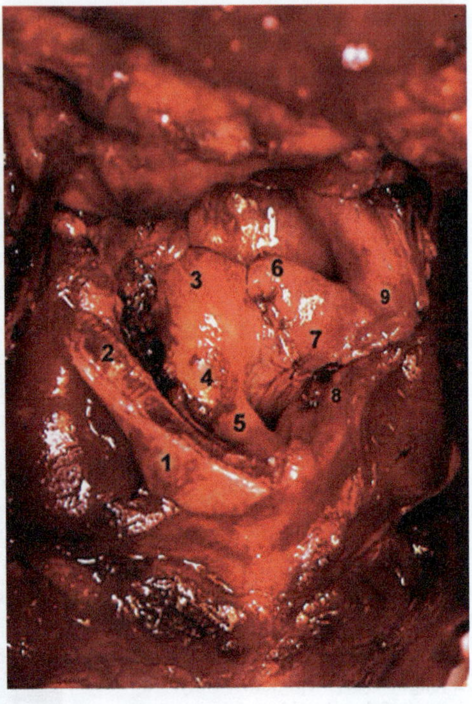

Fig. 14.16 *1* Right external perichondrium, *2* right thyroid lamina, *3* apex of right arytenoid, *4* vocal process of right arytenoid, *5* right vocal cord, *6* apex of left arytenoid, *7* mucosa of the left piriform recess, stretched forward, *8* left external perichondrium, everted back

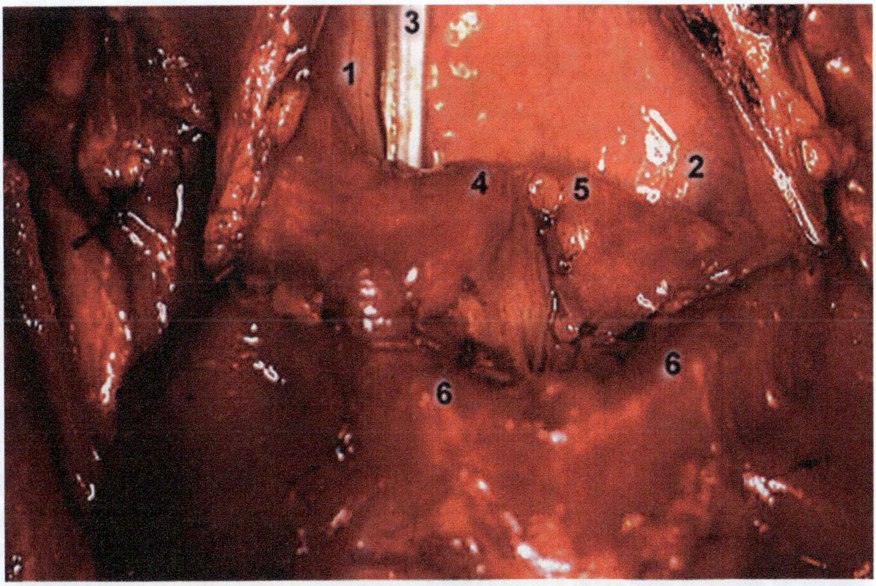

Fig. 14.17 *1* Entrance to the right piriform sinus, *2* entrance to the left piriform sinus, *3* feed tube, *4* apex of right arytenoid, *5* apex of left arytenoid, *6* thyroid lamina covered by external perichondrium

margin of the mucosa covering the vocal cord; we suture the posterior side of the perichondral triangle to the mucosa of the entrance to the piriform sinus. Two stitches restore the continuity of the mucosa on the anterior aspect the arytenoid cartilages (Fig. 14.16).

When reconstruction of the glottic plane is completed, the naso-oesophageal feed tube is inserted (Fig. 14.17).

The restoration of pharyngolaryngeal continuity is achieved by the pexy between the residual thyroid cartilage, hyoid bone, and tongue base. The main median suture surrounds at the centre the inferior edge of the thyroid cartilage piercing the crico-thyroid membrane and passes behind the anterior commissure; at the top it pierces, from the inside to the outside, the tongue base in the centre and surrounds the hyoid bone, if preserved. Some secondary lateral stitches are then applied; the most lateral of these perforates the mucosa of the lateral aspect of the piriform sinus and pierces at the top the mucosa of the gingivolingual fornix; the other stitches perforate the thyroid perichondrium at the bottom and the tongue base at the top then surround the greater cornu of the hyoid bone (Fig. 14.18).

After having flexed the patient's neck, all the stitches in this suture are tightened with several knots, with perfect adhesion between the two edges of the suture. We then make a second horizontal suture, with interrupted stitches, between the section margin of the subhyoid muscles and the suprahyoid cervical fascia. The central stitch is "U" shaped; the most lateral stitches of this suture are fixed at the top to the intermediate tendon of the digastric muscle, taking care not to damage the underlying hypoglossal nerve (Fig. 14.19).

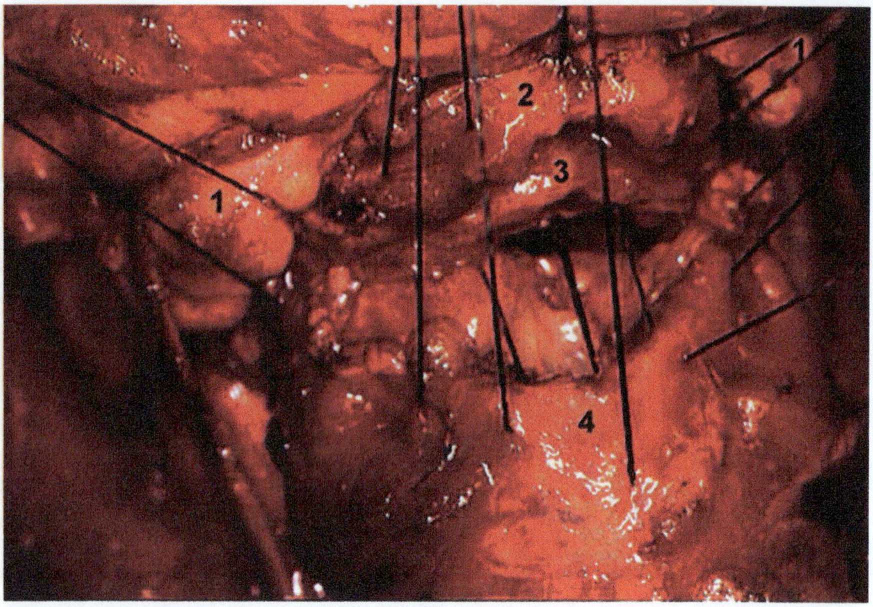

Fig. 14.18 *1* Submandibular gland, *2* hyoid bone, *3* tongue base, *4* thyroid lamina

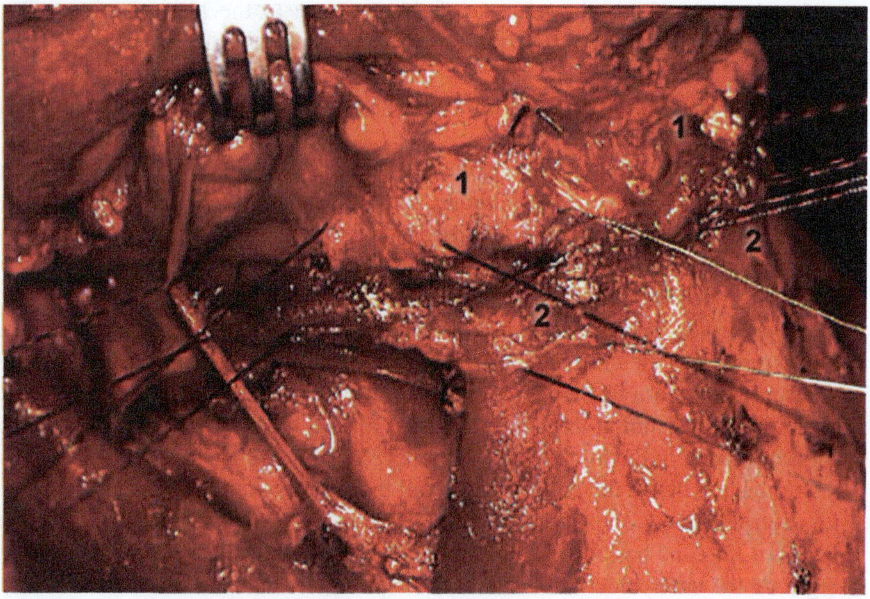

Fig. 14.19 *1* Suprahyoid fascia, *2* subhyoid muscles

At the end of the operation, the pars membranacea of the trachea is sutured to the myocutaneous flap in the centre and suction drains are placed in the laterocervical regions. The cutis and the subcutaneous tissue are sutured on two planes.

14.3 Supracricoid Laryngectomy

Performed for the first time by Foderl in 1896, resumed by Majer in 1959, and relaunched in the clinical field by Serafini [11], Labayle and Bismuth [12], and Piquet et al. [13], reconstructive laryngectomy, better known as supracricoid laryngectomy (SCL), became widely used in Europe only in the second half of the eighties. Nowadays, it is a routine operation in all the most qualified laryngeal oncology centres.

1 Principles of the Operation: SCL has numerous variants, improperly defined by the type of fixation carried out during the reconstructive phase of the operation.

(a) In glottic carcinomas, laryngectomy is performed preserving one or both arytenoids and the superior part of the epiglottis (crico-hyoido-epiglottopexy, CHEP, Mayer–Piquet); (b) in transglottic carcinomas, the operation is performed preserving generally one arytenoid and sacrificing the whole epiglottis with the content of the preepiglottic space (crico-hyoidopexy, CHP, Labayle); (c) in carcinomas with a subglottic extension or which infiltrate the inferoposterior paraglottic space and/or the cricoarytenoid articulation, supratracheal laryngectomy may be indicated, preserving a "cricoarytenoid unit" (tracheo-hyoido-epiglottopexy, THEP, Rizzotto–Serafini) [14] (Fig. 14.20a–d).

2 Indications: some T1b, some transglottic T2–T3–T4

3 Surgical Technique: For the operation limited to the larynx, we make a superficial "apron" incision with good exposure of the anterior cervical region from the hyoid bone to the jugulum. André's bilateral incision, with exposure also of the laterocervical regions, is preferable if lymph node dissection is expected. Surgery on the larynx is started at the bottom, exposing the thyroid gland with a section at the base of the neck and upward rotation of the subhyoid muscles (Figs. 14.21 and 14.22).

The thyroid gland is stretched upwards exposing the thyropericardial lamina: the inferior thyroid veins are ligated and sectioned, uncovering the cervical trachea. The recurrent nerves and the inferior thyroid arteries are preserved (Fig. 14.23).

A small tracheostomy is then made at the level of the fourth to fifth tracheal ring. Transferring the anaesthesia tube into the tracheostoma, the superior isolation of the larynx begins with the section of the superficial cervical fascia and of the subhyoid muscles along the inferior edge of the hyoid bone and the clearance of the contents of the preepiglottic space which will be removed with the specimen (Fig. 14.24).

The superior vascular pedicles are then ligated and sectioned, preserving the superior laryngeal nerves so as to maintain the sensitivity of the piriform sinuses (Fig. 14.25).

The lateral isolation of the larynx is achieved by sectioning the inferior constrictor muscles and the external perichondrium along the lateral margins of the thyroid cartilage (Fig. 14.26).

Laterally rotating the larynx, the lateral and anterior walls of the piriform sinuses are then separated with an internal subperichondrial approach (Fig. 14.27).

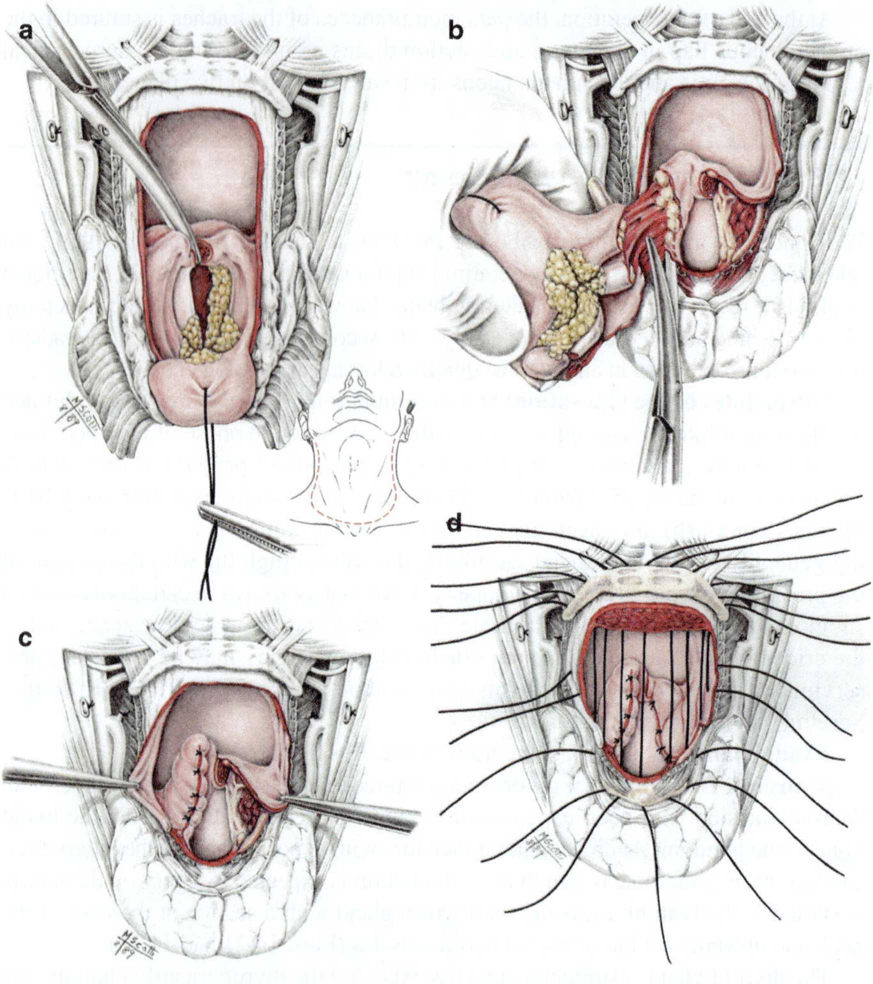

Fig. 14.20 Supracricoid Laryngectomy: surgical steps (**a**) (1) André's bilateral incision (if the neck dissections are performed) (2) section of subhyoid muscles at the base of the neck (3) small tracheostomy through the thyropericardial lamina (4) resection of subhyoid muscles along the inferior edge of the hyoid bone and clearance of the preepiglottic space (5) ligatures of superior vascular pedicles (6) section of the inferior constrictor muscle of hypopharynx and detaching of piriform sinuses from the thyroid lamina (7) dissection of the supraglottic larynx and of the posterior commissure (8) tracheostomy (**b**) caudal section along the superior margin of the cricoid cartilage, sparing the arytenoid cartilage on the side lesser involved by the tumour and the piriform sinus too (**c**) reconstruction of a neoglottis (**d**) (1) pexy between the cricoid ring, the tongue base at the top and the body of the hyoid bone (2) second layer between the superior edge of the thyroid gland and the suprahyoid muscles

The inferior thyroid cornua are sectioned at their base and preserved so as not to damage the underlying branches of the recurrent nerves (Fig. 14.28).

SCL may be performed with exeresis from top to bottom or from bottom to top. We prefer the first means of removal in cases with extensive carcinomas involving

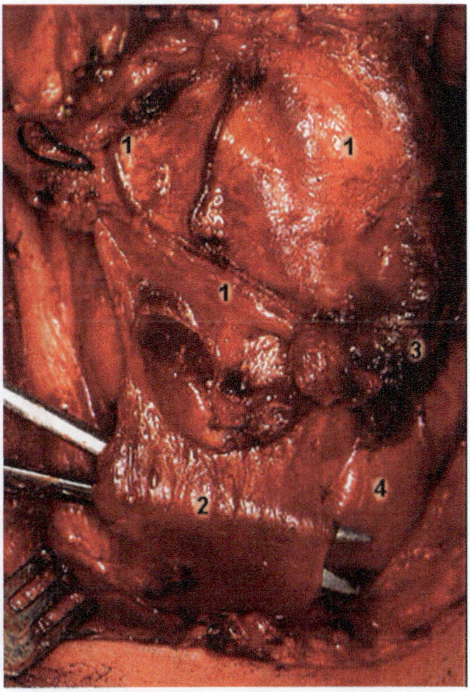

Fig. 14.21 *1* Superficial cervical fascia, *2* right subhyoid muscles, *3* left subhyoid muscles (sectioned), *4* thyroid gland

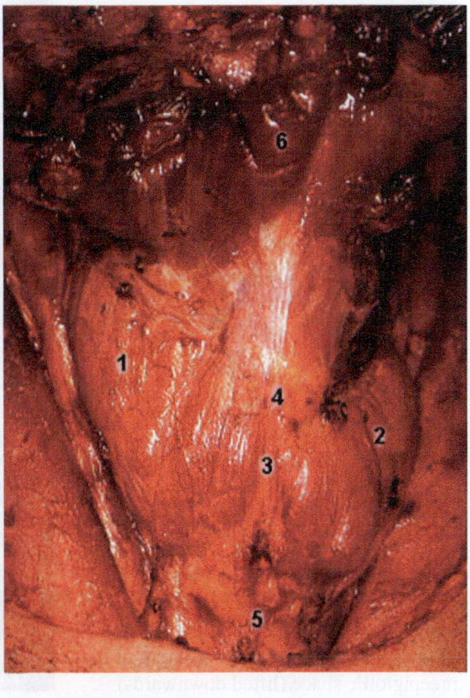

Fig. 14.22 *1* Right thyroid lobe, *2* left thyroid lobe, *3* thyroid isthmus, *4* cricoid cartilage, *5* thyropericardial lamina, *6* infrahyoid muscles

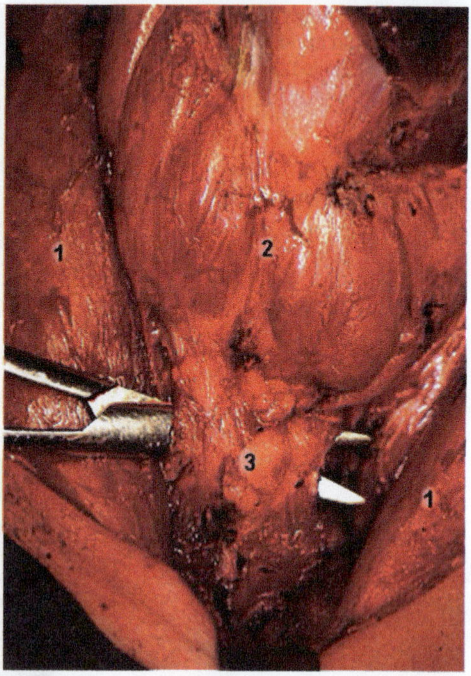

Fig. 14.23 *1* Sternocleidomastoid muscle, *2* thyroid isthmus, *3* thyropericardial lamina

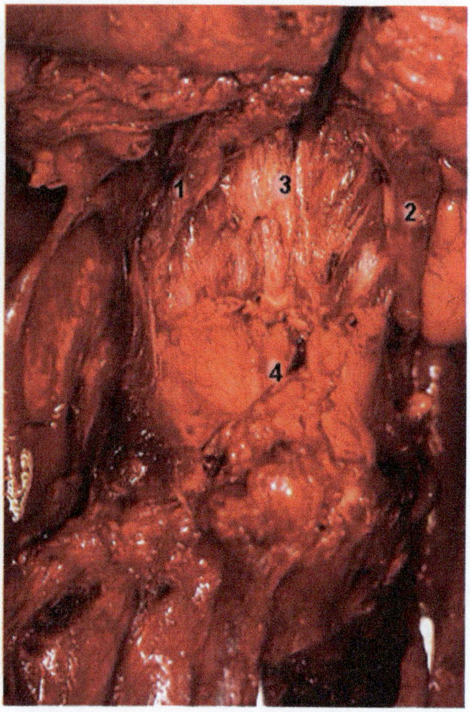

Fig. 14.24 *1* Right greater cornu of the hyoid bone, *2* left greater cornu of the hyoid bone, *3* submucosa of the glossoepiglottic vallecula, *4* content of the preepiglottic space (lifted downwards)

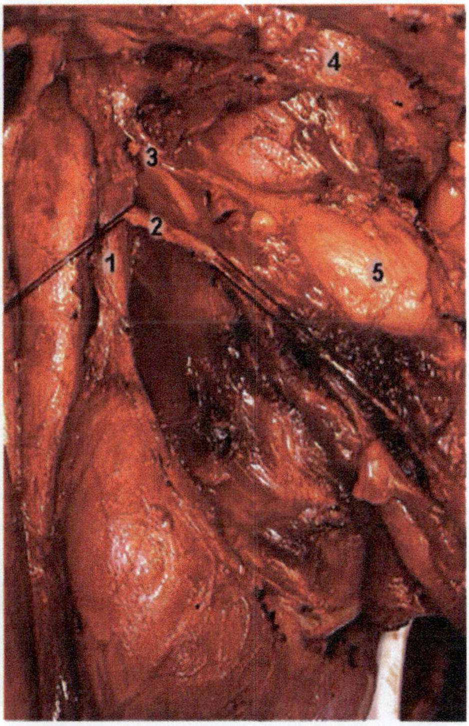

Fig. 14.25 *1* Superior thyroid artery, *2* superior laryngeal artery, *3* superior laryngeal nerve, *4* hyoid bone, 5 preepiglottic space

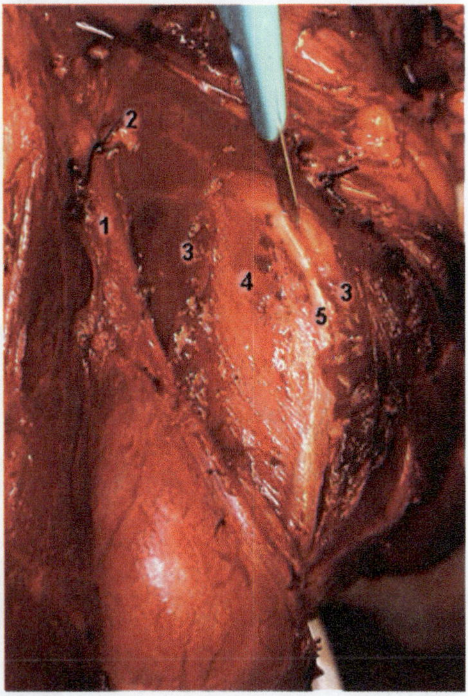

Fig. 14.26 *1* Superior thyroid artery, *2* superior laryngeal artery, *3* inferior constrictor muscle of the pharynx (sectioned), *4* submucosa of the right piriform sinus, *5* thyroid cartilage

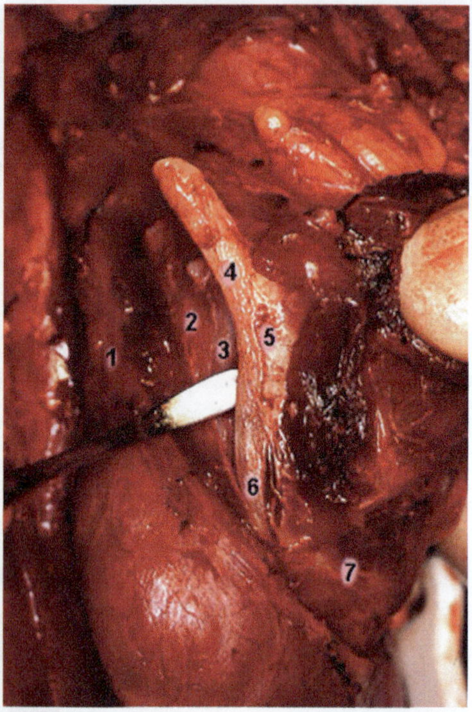

Fig. 14.27 *1* Superior thyroid artery, *2* submucosa of the right piriform recess, *3* internal perichondrium of the right thyroid lamina, *4* right superior cornu of the thyroid cartilage, *5* external perichondrium of the right thyroid lamina, *6* right inferior cornu of the thyroid cartilage, *7* cricoid cartilage

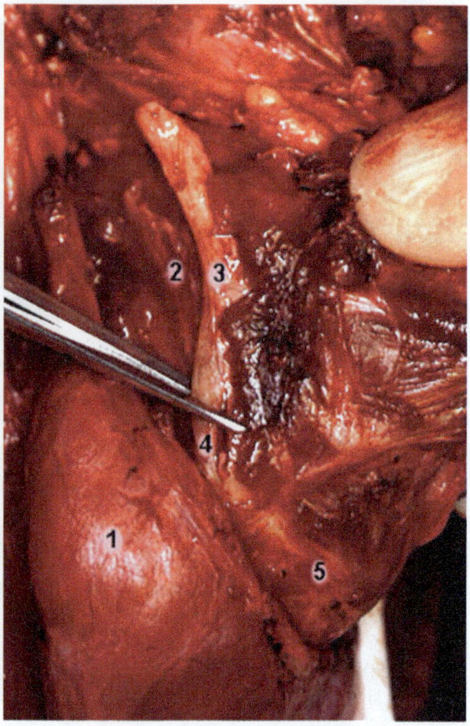

Fig. 14.28 *1* Right thyroid lobe, *2* right piriform recess, *3* thyroid cartilage, *4* right inferior cornu of the thyroid cartilage, *5* cricoid cartilage

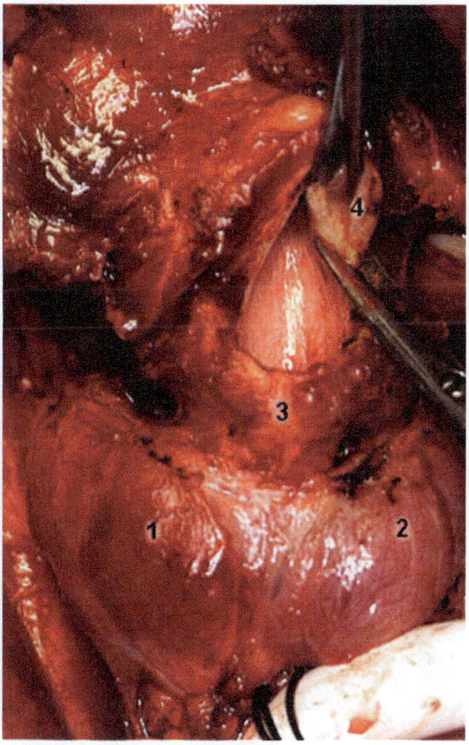

Fig. 14.29 *1* Right thyroid lobe, *2* left thyroid lobe, *3* cricoid cartilage, *4* left vocal cord

the supraglottic region, which require more extensive exeresis with good exposure of the tumour limits (reconstruction with crico-hyoidopexy), the second in cases of glottic carcinoma with the possibility of preserving the epiglottis (reconstruction with crico-hyoido-epiglottopexy).

4 Laryngectomy with Crico-Hyoidopexy (**CHP**): The demolitive phase begins with laryngectomy performed at the level of the superior margin of the cricoid (Fig. 14.29).

The section is prolonged horizontally as far as the vocal process of the side where the arytenoid cartilage is to be preserved then verticalised along the anterior wall of the same (Fig. 14.30).

The section involves the anterior mucosa of the entrance to the piriform sinuses and is horizontalised passing through the glossoepiglottic valleculae.

On the other side, which is the one more affected by the neoplasia, the horizontal section along the superior margin of the cricoid involves the cricoarytenoid articulation and arrives at the posterior commissure (Fig. 14.31).

From there, it goes up along the mucosa of the piriform sinus and joins the contralateral section at the top, thus completing the exeresis of the larynx (Fig. 14.32).

The continuity of the mucosa is reconstructed with absorbable interrupted stitches along the anterior face of the preserved arytenoid and the entrance to the piriform sinus on the same side by stretching the anterior mucosa forwards and fixing it laterally to the superior edge of the cricoid. Also, the anterior margin of the contralateral

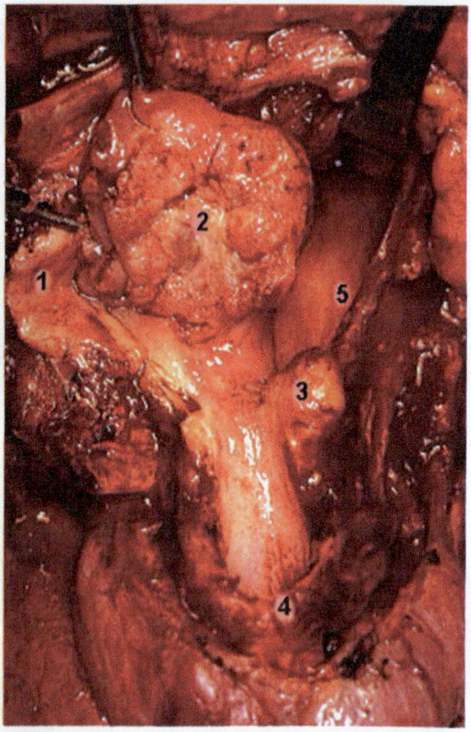

Fig. 14.30 *1* Epiglottis, *2* neoplasia, *3* left arytenoid, *4* cricoid, *5* entrance to the left piriform sinus

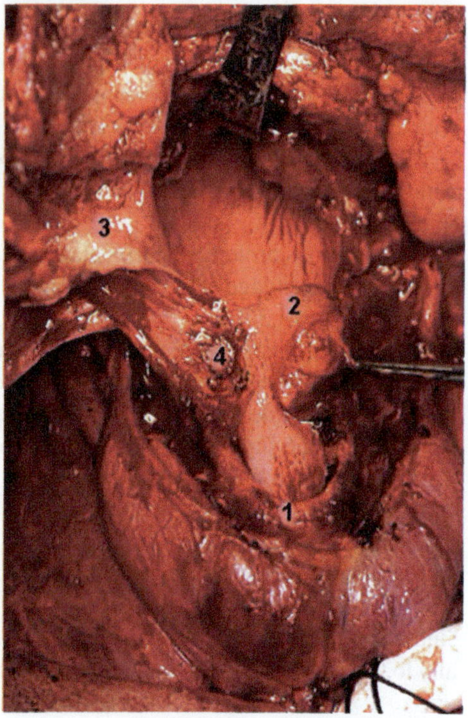

Fig. 14.31 *1* Cricoid cartilage, *2* left arytenoid, *3* right arytenoid, *4* right cricoarytenoid articulation

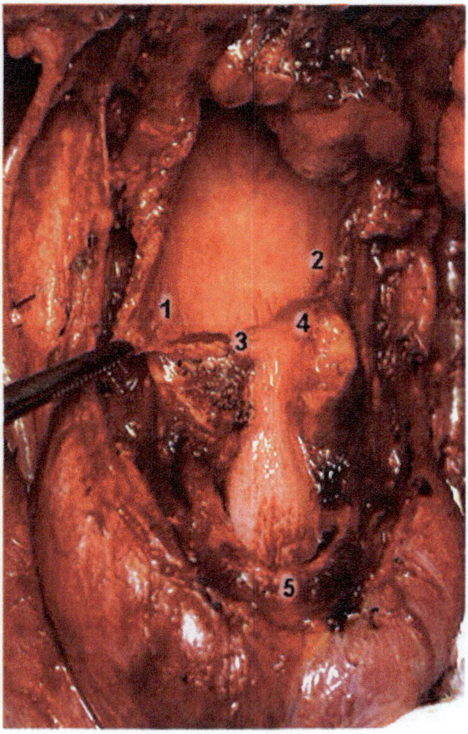

Fig. 14.32 *1* Aditus to the right piriform sinus, *2* aditus to the left piriform sinus, *3* posterior commissure, *4* left arytenoid, *5* cricoid cartilage

piriform sinus is stretched forwards and fixed to the superior edge of the cricoid cartilage (Fig. 14.33).

Once the reconstruction of the mucosa of the opening of the neolarynx has been completed, crico-hyoidopexy is started anteriorly with a central everted stitch of absorbable synthetic thread that surrounds the cricoid ring at the bottom, transfixes the tongue base at the top, and surrounds the body of the hyoid bone. Other sutures fix the cricoid ring to the tongue base and anteriorly to the body of the hyoid bone. The lateral mucosa of the entrance to the piriform sinuses is sutured posteriorly to the mucosa of the glosso-tonsillar sulci, taking care to include in the stitch also the greater cornu of the hyoid bone, which will have the function of keeping wide the entrance of the hypopharynx (Fig. 14.34).

5 Laryngectomy with Crico-Hyoido-Epiglottopexy (**CHEP**): SCL with crico-hyoido-epiglottopexy, indicated as mentioned above for glottic lesions, requires instead an approach from top to bottom, starting from the piriform sinus of the side less affected by the neoplasia and preserving the subhyoid epiglottis.

Crico-hyoido-epiglottopexy is performed with the same procedure as crico-hyoidopexy: the only difference is in the anterior central stitch which transfixes the epiglottis in the centre 5 mm above the inferior edge of the section before surrounding the body of the hyoid bone at the top. The two anterolateral stitches also transfix the epiglottis laterally and the two most lateral stitches join the anterior mucosa of the entrance to the piriform sinus to the pharyngoepiglottic fold.

Lastly, it must be stressed that the preservation of the piriform sinus in the demolitive phase often makes it superfluous to reconstruct the glottic plane before

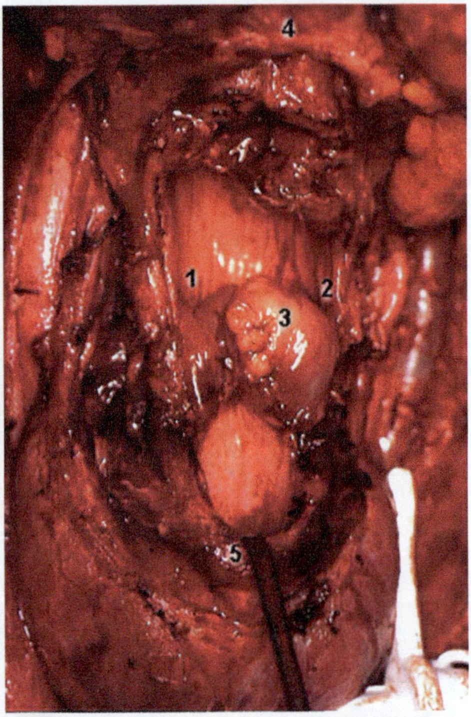

Fig. 14.33 *1* Entrance to the right piriform sinus, *2* entrance to the left piriform sinus, *3* left arytenoid, *4* hyoid, *5* cricoid cartilage

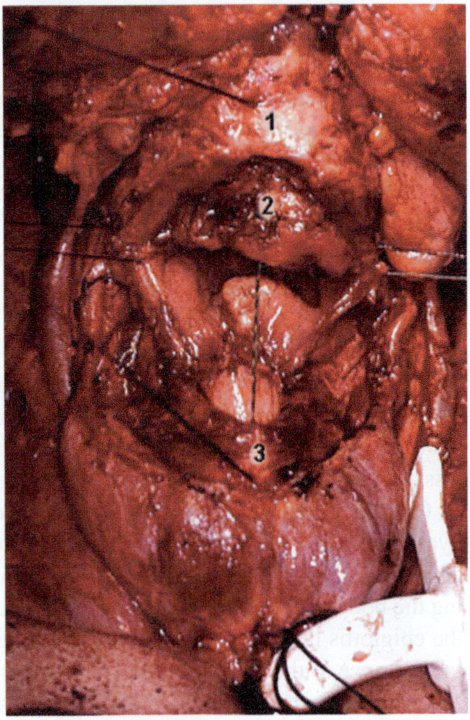

Fig. 14.34 *1* Hyoid, *2* tongue base, *3* cricoid cartilage

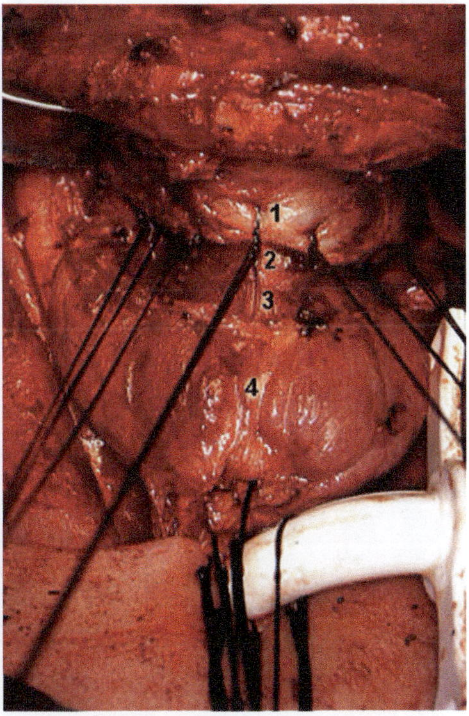

Fig. 14.35 *1* Hyoid, *2* cricoid, *3* trachea, *4* thyroid isthmus

plexy. It has also been demonstrated that the integrity of the piriform sinus appreciably favours the functional recovery of deglutition.

6 The subsequent reconstruction times are identical for both operations. Pexy is accomplished by ligating the sutures, with consequent raising of the cricoid with the cervical trachea/thyroid gland complex and lowering of the hyoid bone with the tongue base (Fig. 14.35).

A second suture is made with interrupted stitches of absorbable synthetic thread between the superior edge of the thyroid gland and the suprahyoid muscles. The thyroid gland thus covers and protects the region of the crico-hyoidopexy and its fixture to the suprahyoid muscles helps stabilise the neolarynx in the definitive position (Fig. 14.36).

7 Laryngectomy with Tracheo-Hyoido-Epiglottopexy (THEP) [14]: One of the contraindications for SCLs is the subglottic extension (selective neck dissection of the recurrential regions is mandatory, Fig. 14.37a) or the involvement of the cricoarytenoid joint.

In this case, the exeresis of a large part of the cricoid is then proposed and then the preservation of a portion of the cricoid lamina with the corresponding arytenoid, called "the cricoarytenoid unit" (Fig. 14.37b).

The neoglottis is reconstructed using the mucosa of the piriform sinus from the side without the arytenoid, which is stretched cranial, folded downwards, and sutured to the mucosa remaining after exeresis (Fig. 14.37c).

The next step is the tracheo-hyoido-epiglottopexy (THEP). The tracheostomy must be very low, because the trachea must be stretched very high up to allow anas-

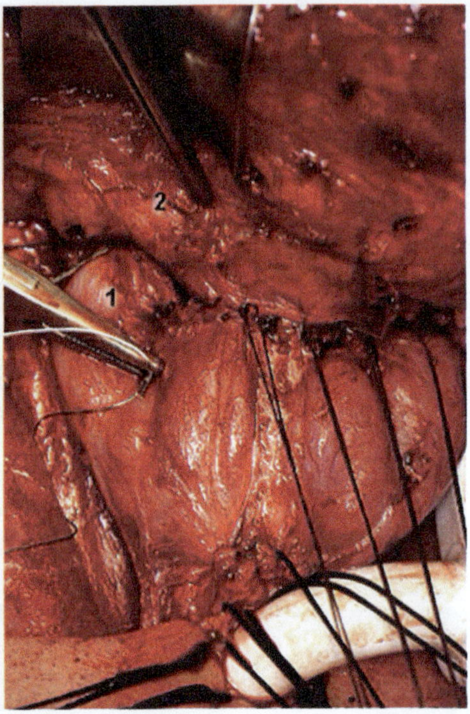

Fig. 14.36 *1* Right thyroid lobe, *2* suprahyoid cervical fascia

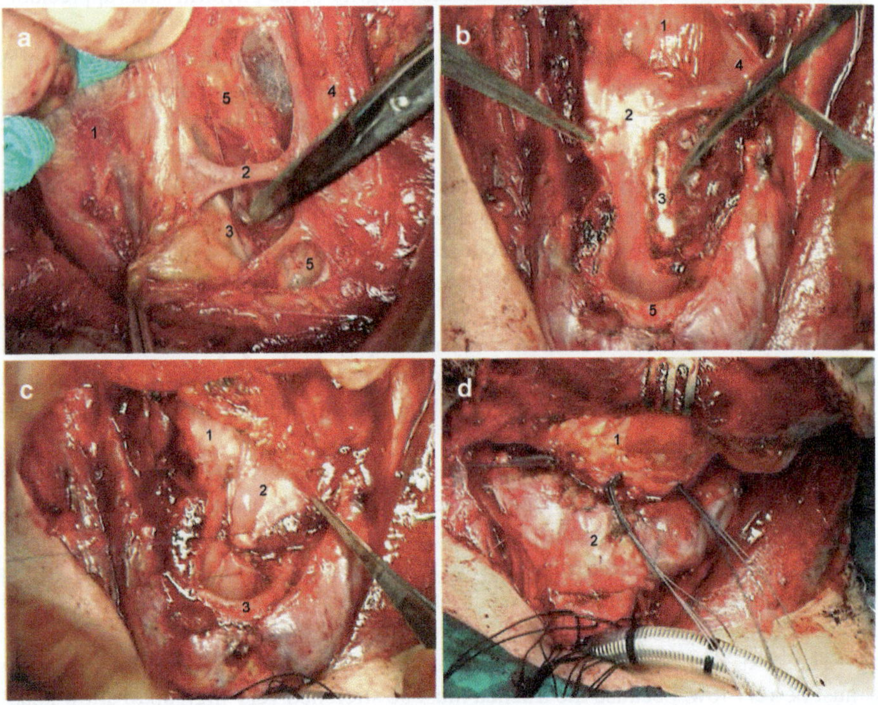

tomosis. In the reconstructive phase, it is necessary to flex the head on the neck (Fig. 14.37d).

14.4 Glottic–Subglottic Laryngectomy

The removal of the subglottic or glottic–subglottic segment of the larynx with reconstruction by means of terminoterminal anastomosis is the operation preferred in the treatment of severe subglottic or glottic–subglottic post-traumatic or iatrogenic stenosis. This operation was introduced in the oncological field by Bartual (1978) and it may be used in a number of extremely limited and selected cases.

1 Principles of the Operation: (a) horizontal removal of the inferior segment of the larynx, extending to the first tracheal rings, preserving the laryngeal vestibule and the arytenoids with the interarytenoid fold; (b) reconstruction with tracheolaryngeal continuity by means of terminoterminal anastomosis; and (c) provisional subisthmic tracheotomy (Fig. 14.38a–d).

2 Indications: The operation, which is vertically specular with respect to the supraglottic operation, can be used only for neoplasias confined below the glottic plane, with a modest subglottic extension and without signs of deep infiltration, that is, T1 and T2 (with preserved cord mobility).

N.B.: Subglottic primary carcinomas (fortunately very rare) and glottic–subglottic ones (more frequent) are an extremely severe pathology due to the lateness of the diagnosis and to the possibility of extralaryngeal and lymph nodal spread of the tumour. The treatment chosen is total laryngectomy, extended to the thyroid gland and to the first tracheal rings with contemporary recurrential lymph node clearance followed by radiotherapy on T and N. Only a very limited and carefully selected number of initial glottic–subglottic carcinomas (limited extension, low histological grading) can be treated with partial glottic–subglottic surgery followed or not by radiotherapy on the recurrential lymph node chains.

3 Surgical Technique: An "apron" incision is made, involving the cutis, subcutaneous tissue, and the platysma muscle, with elevation of the myocutaneous flap up to the level of the hyoid bone. The superficial and middle cervical fascia are separated together with the subhyoid muscles on the linea alba, with good exposure of the larynx and of the thyroid gland. The thyroid gland is elevated from the cricoid and from the first tracheal rings, then subisthmic tracheotomy[1] is performed. The

[1] In the case illustrated, the patient had already undergone transisthmic tracheotomy for acute dyspnoea.

Fig. 14.37 Tracheo-hyoido-epiglottopexy (**a**) Dissection of recurrent area *1* trachea, *2* inferior thyroid artery, *3* recurrent nerve, *4* common carotid artery, *5* recurrent lymph nodes. (**b**) Laryngectomy *1* hypopharynx posterior wall, *2* right arytenoid, *3* portion of cricoid cartilage, *4* mucosa of left piriform sinus, *5* first tracheal ring. (**c**) Neoglottic reconstruction *1* right arytenoid, *2* mucosa of left piriform sinus, *3* first tracheal ring. (**d**) Tracheo-hyoidopexy and second plane between thyroid gland and suprahyoid fascia *1* hyoid bone and suprahyoid fascia, *2* thyroid

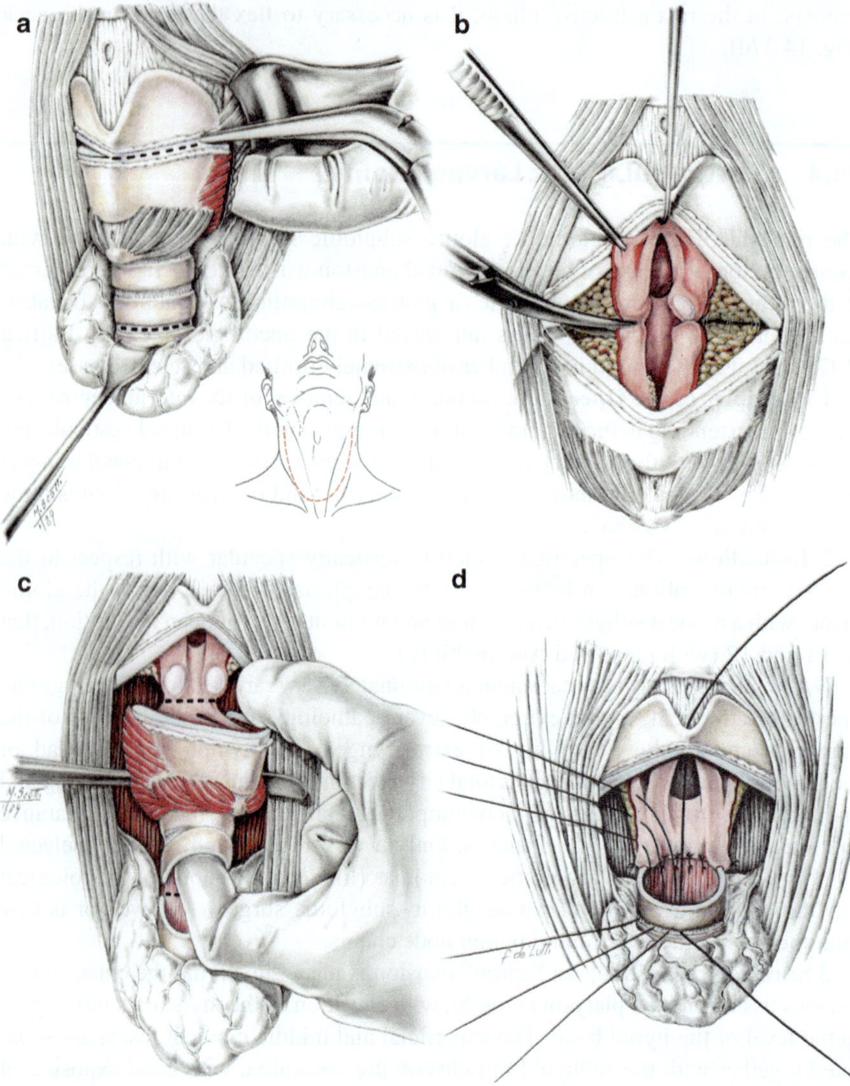

Fig. 14.38 Glottic–subglottic Laryngectomy: surgical steps (**a**) (1) "apron" incision (2) the subhyoid muscles are separated on the linea alba (3) subisthmic tracheostomy (4) the inferior constrictor muscles of the pharynx are sectioned (5) horizontal section of perichondrium and of the thyroid cartilage (6) section of the trachea between second and third ring (**b**) section of the anterior commissure and of Morgagni's ventricle passing through the cricoarytenoid joints (**c**) (1) cricoid cartilage and trachea are separated from hypopharynx and from oesophagus (2) section of subarytenoid posterior laryngeal mucosa with exeresis of laryngotracheal portion (**d**) (1) terminoterminal anastomosis between supraglottic larynx and trachea (2) second suture layer is made between the thyroid gland and the thyrohyoid membrane

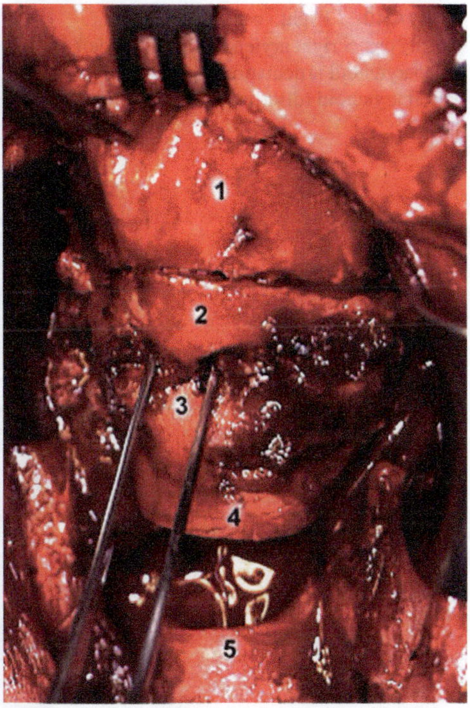

Fig. 14.39 *1* Thyroid lamina (preserved portion), *2* thyroid lamina (removed portion), *3* cricoid, *4* tracheal ring, *5* trachea

external thyroid perichondrium is cut along the line corresponding internally to the superior surface of the vocal cords.

The inferior constrictor muscles of the pharynx are sectioned on both sides along the posterior edges of the thyroid cartilage to be removed, and the anterior wall of the piriform recesses is elevated from the cartilage.

Horizontal supraglottic dissection is then performed along the line of the previous incision of the external perichondrium and the trachea is sectioned between the second and the third ring, anteriorly and laterally (Fig. 14.39).

After opening the larynx at the anterior supracommissural level and sectioning the mucosa along the lateral wall of the ventricle as far as the arytenoids, we proceed to disarticulate the arytenoids from the cricoid (Fig. 14.40).

At this point, the posterior wall of the cricoid and the wall of the superior segment of the trachea are separated from the anterior wall of the hypopharynx and from the cervical oesophagus. The exeresis of the laryngotracheal portion to be removed is now easy by sectioning the subarytenoid posterior laryngeal mucosa and the posterior tracheal wall between the second and the third ring (Fig. 14.41).

The reconstruction of laryngotracheal continuity begins posteriorly by fixing the bases of the arytenoids to the posterior ends of the superior tracheal ring (Fig. 14.42).

A suture is made between the edge of sectioning of the interarytenoid mucosa and the superior edge of the posterior wall of the trachea. The terminoterminal anastomosis is completed laterally and anteriorly by suturing the superior tracheal ring to the inferior

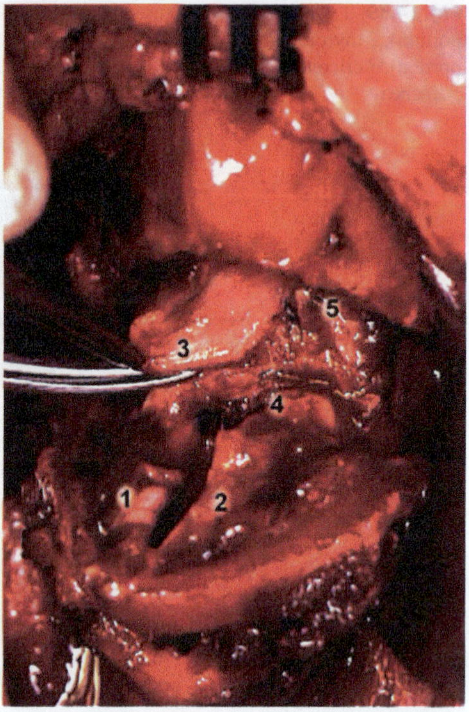

Fig. 14.40 *1* Right vocal cord, *2* left vocal cord, *3* right arytenoid mucosa, *4* left cricoarytenoid articulation, |*5* left arytenoid (disarticulated)

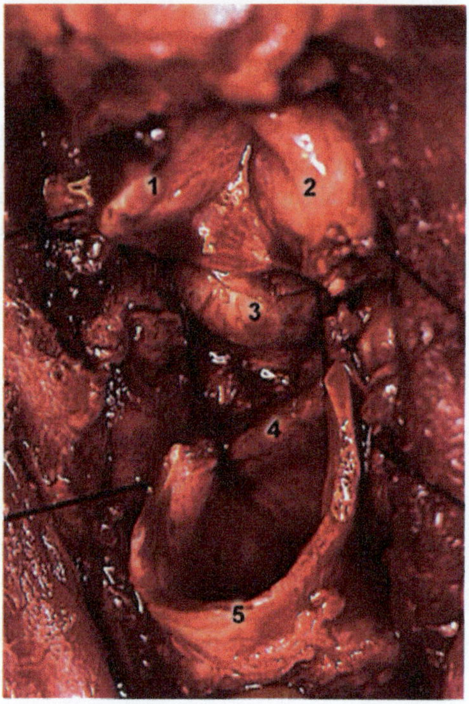

Fig. 14.41 *1* Right arytenoid, *2* left arytenoid, *3* hypopharynx, *4* pars membranacea of the trachea, *5* tracheal ring

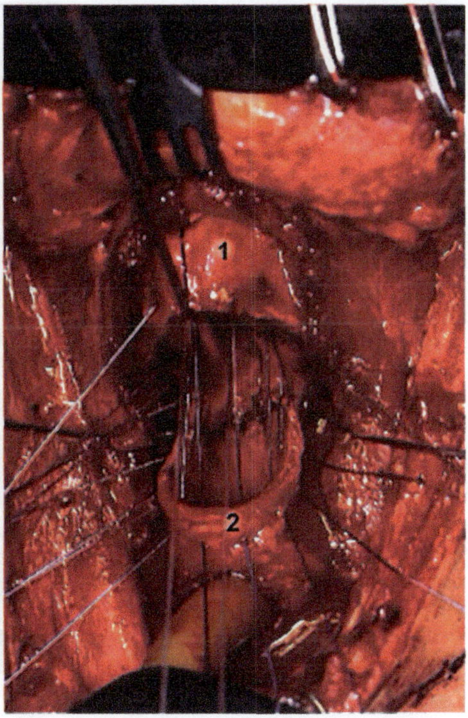

Fig. 14.42 *1* Thyroid lamina, *2* tracheal ring

edges of the lateral mucosa of the ventricles. An anterior anchoring suture surrounds the tracheal ring at the bottom and the thyroid cartilaginous isthmus at the top.

After completing laryngotracheal terminoterminal anastomosis (Fig. 14.43), a second suture is made between the superior edge of the thyroid gland and the thyrohyoid membrane, thus obtaining a good protection of the anastomotic line and a reinforcement ensuring that the union between the two structures is perfectly maintained.[2]

At this point, the operation is completed by suturing together the medial edges of the sternohyoid muscles, placing suction drains, superiorly finishing off the union of the tracheostoma to the cutis, and suturing the cutis on two levels.[3]

14.5 Total Laryngectomy

The conception of this surgical procedure is shared by P. Watson, who was probably the first to perform it (1866), and A.C. Billroth who performed it successfully and provided a complete description [15]. Despite the subsequent spread of various possibilities for

[2]In the case illustrated, since a large part of the thyroid gland is missing, the plane of reinforcement is constructed between the superior peritracheal tissue and the subhyoid cervical fascia.

[3]The neoplasia (in this case, an adenoid cystic carcinoma of the subglottis) can be clearly seen by observing the specimen from the tracheal side (Fig. 14.44).

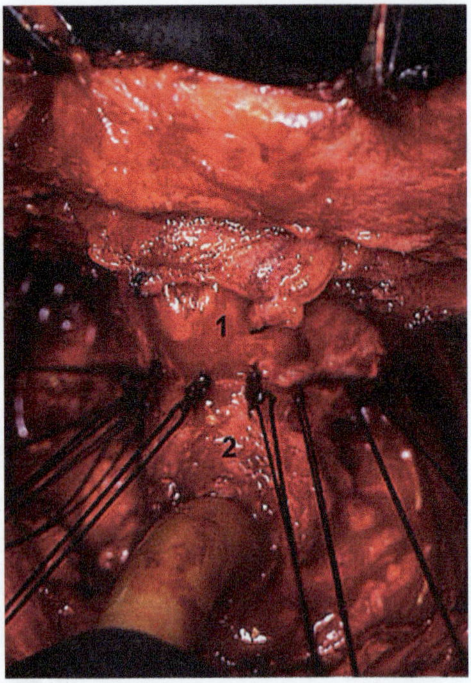

Fig. 14.43 *1* Thyroid cartilage, *2* trachea

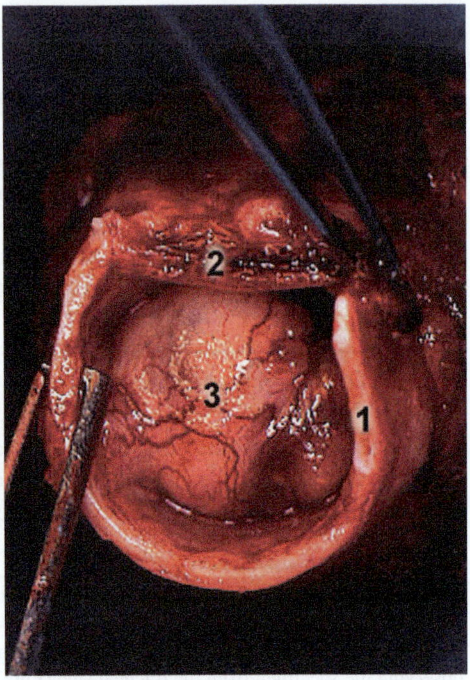

Fig. 14.44 *1* Tracheal ring, *2* pars membranacea of the trachea, *3* neoplasia

preservation surgery, even today, total laryngectomy is indicated in about a quarter of all laryngeal neoplasias, especially in advanced lesions.

1 Principles of the Operation: (a) ablation of the whole larynx from the first tracheal ring to the tongue base (this may be extended to the adjacent tissues, e.g. the piriform sinus, thyroid, trachea, tongue base), (b) reconstruction of the pharyngoesophageal alimentary canal, separating it from the airway, (c) anastomosis of the tracheal opening to the precervical cutis (definitive tracheostoma), and (d) possibility of creating a primary phonatory fistula (Fig. 14.45a–d).

2 Indications: some glottic–subglottic T2, some transglottic T3, some T4 (all cases in which the extension of the tumour or the general conditions of the patient advise against a more conservative treatment).

3 Surgical Technique: The incision of the skin preferred in total laryngectomy is André's incision, which involves the cutis, subcutaneous tissue, and the platysma muscle. The elevation of the myocutaneous flap with a suprafascial approach up to the lower edge of the mandible allows the ample exposure of the lateral and anterior regions of the neck (Fig. 14.46).

After drainage of the cervical lymph nodes, the subhyoid muscles are sectioned bilaterally at the bottom and rotated upwards, exposing the thyroid gland and the lower part of the larynx. The supramedial margins of the isthmus and lobes of the thyroid gland are elevated anteriorly and laterally from the cricoid (Fig. 14.47).

The greater cornu of the hyoid bone is freed laterally and then isolated with the scissors from their points of insertion in the ligaments, fascia, and muscles from the apex to the base; in performing this manoeuvre, the blades of the scissors must adhere to the bone to avoid damaging the lingual arteries, which lie immediately underneath. The hyoid bone is thus completely isolated laterally. Superiorly, the points of insertion in the suprahyoid muscles and the loose tissue beneath are sectioned, arriving at the submucosa of the glossoepiglottic valleculae (Fig. 14.48).

The superior laryngeal vasculonervous pedicle is isolated, ligated, and cut at the level of the thyrohyoid membrane on both sides (Fig. 14.49).

The larynx is now gripped and rotated laterally so as to expose the lateral margin of the thyroid cartilage, along which the inferior constrictor of the pharynx and the corresponding external perichondrium are sectioned; the inferior fibres of the constrictor going towards the cricoid are also sectioned (cricopharyngeus muscle) (Fig. 14.50).

With a separator, we then proceed to detach the anterior wall of the piriform sinus from the medial surface of the thyroid wings using an internal subperichondrial approach; this manoeuvre is also facilitated on both sides by the contralateral rotation of the larynx. Preserving the mucosa of the piriform sinuses, when this is made possible by the oncological situation, allows us to reduce the size of the pharyngostome resulting from the laryngectomy (Fig. 14.51).

We now section the full thickness of the anterior and lateral tracheal wall immediately above the first cartilaginous ring; the section at the same level of the rear tracheal wall, immediately under the lower margin of the cricoid lamina, must be performed taking care not to damage the anterior wall of the cervical oesophagus. During this incision, the inferior laryngeal pedicles are also cut, situated immediately

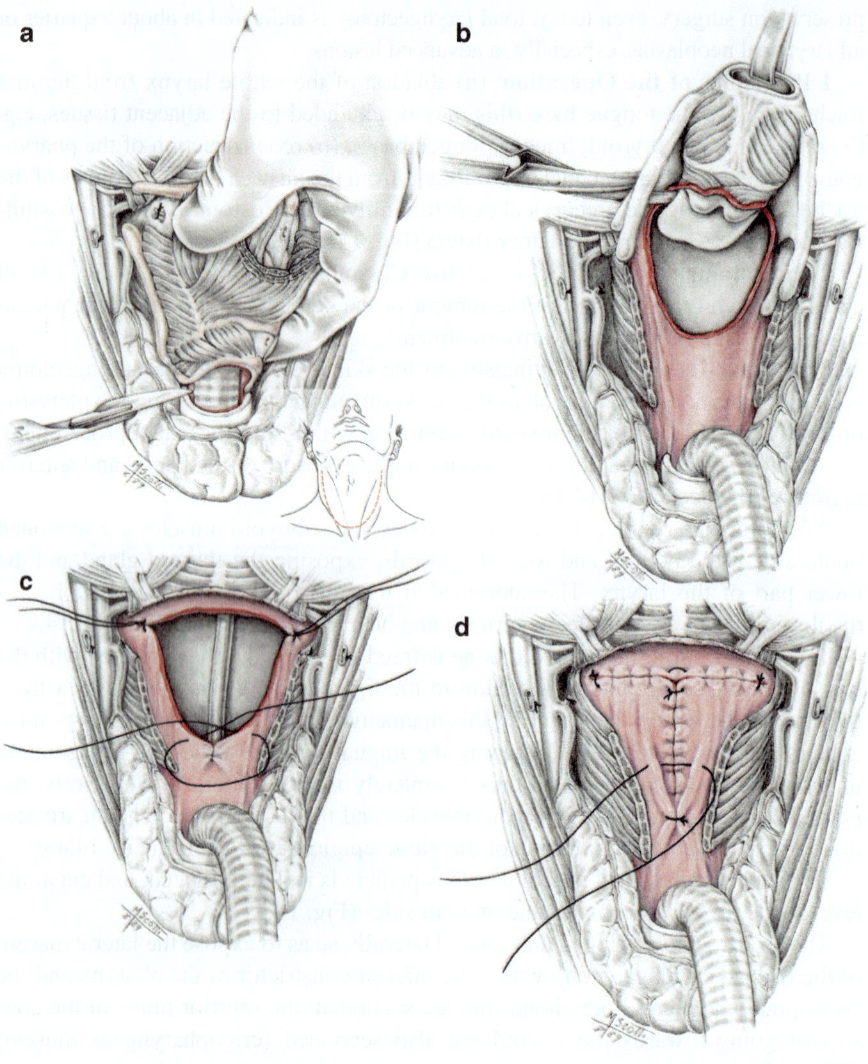

Fig. 14.45 Total Laryngectomy: surgical steps (**a**) (1) André's bilateral incision (if the neck dissections are performed) (2) section of subhyoid muscles at the base of the neck (3) resection of suprahyoid muscles along the superior edge of the hyoid bone (4) ligatures of superior vascular pedicles (5) section of the inferior constrictor muscle of hypopharynx and detaching of piriform sinuses from the thyroid lamina (6) caudal dissection above the first tracheal ring (**b**) detaching of larynx (rotated upwards) from hypopharynx preserving the piriform sinuses (**c**) (1) section passing through the valleculae mucosa and laryngectomy (2) T-shaped suture of the mucosal plane (**d**) (1) suture of the submucosal plane (2) suture of the muscles plane

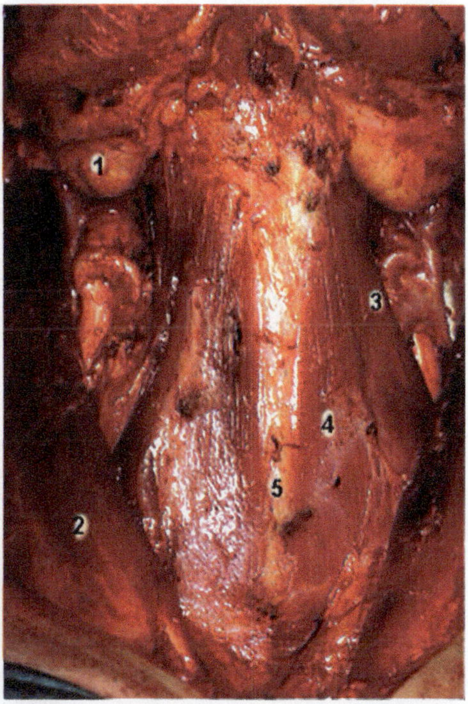

Fig. 14.46 *1* Submandibular gland, *2* sternocleidomastoid muscle, *3* omohyoid muscle, *4* sternohyoid muscle, *5* infrahyoid linea alba

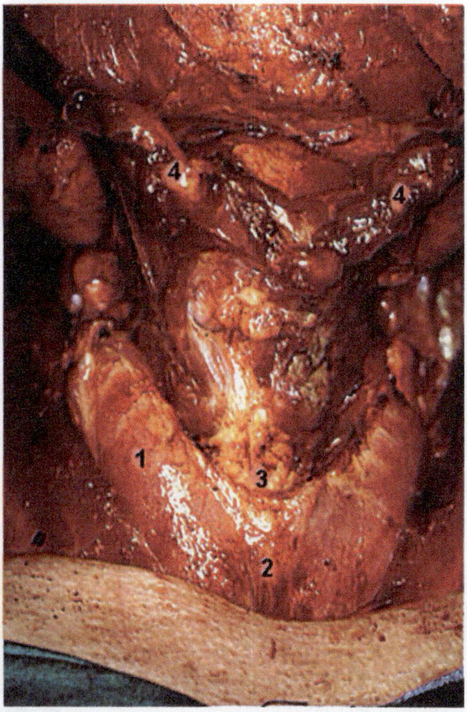

Fig. 14.47 *1* Right thyroid lobe, *2* thyroid isthmus, *3* cricoid cartilage, *4* subhyoid muscles (sectioned)

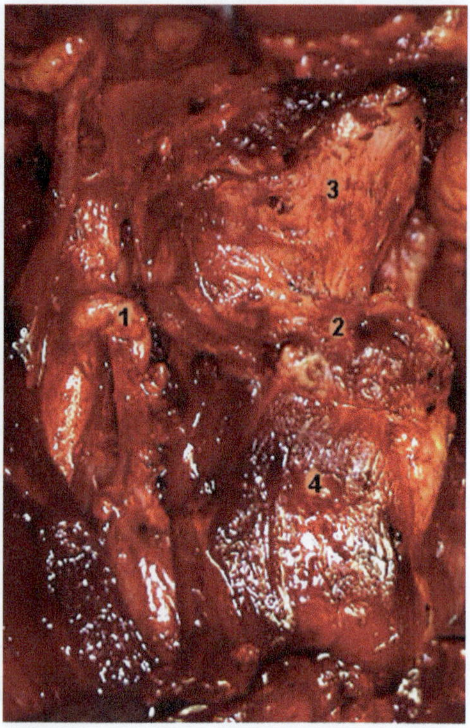

Fig. 14.48 *1* Superior thyroid artery, *2* hyoid bone, *3* preepiglottic space, *4* subhyoid muscles

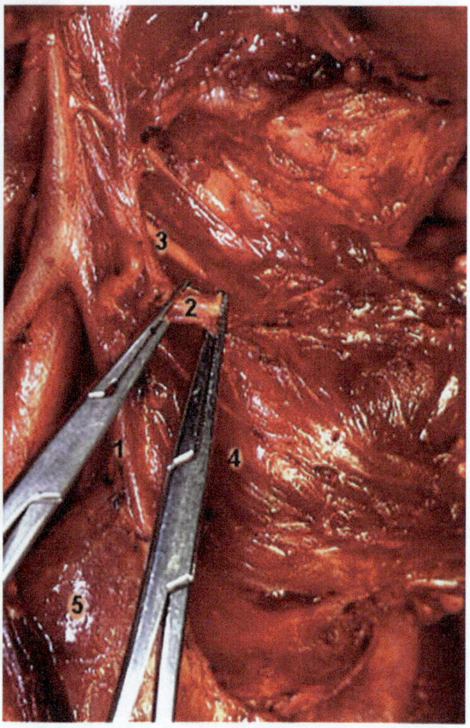

Fig. 14.49 *1* Superior thyroid artery, *2* superior laryngeal artery and vein, *3* superior laryngeal nerve, *4* inferior constrictor muscle of the pharynx

Fig. 14.50 *1* Superior thyroid artery, *2* right thyroid lobe, *3* inferior constrictor muscle of the pharynx, *4* thyroid cartilage

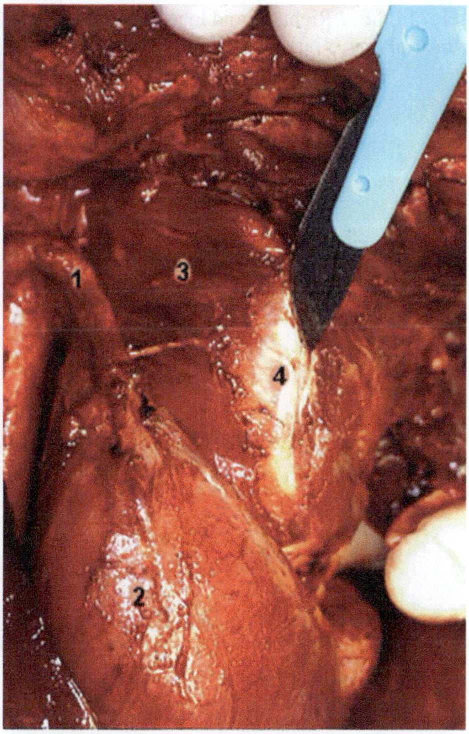

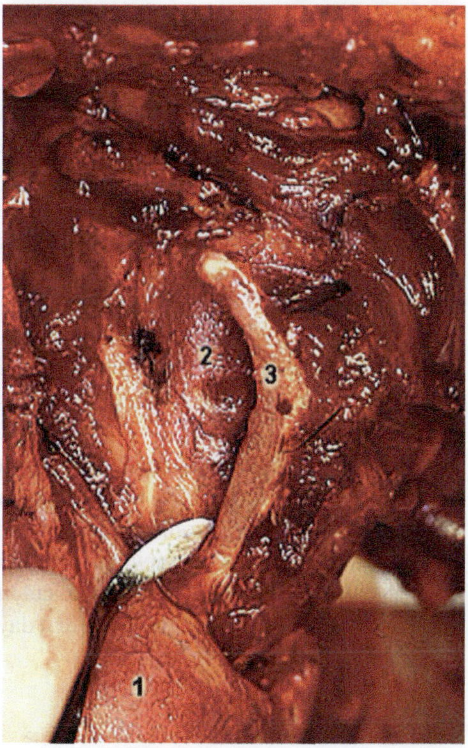

Fig. 14.51 *1* Right thyroid lobe, *2* right piriform sinus, *3* thyroid cartilage

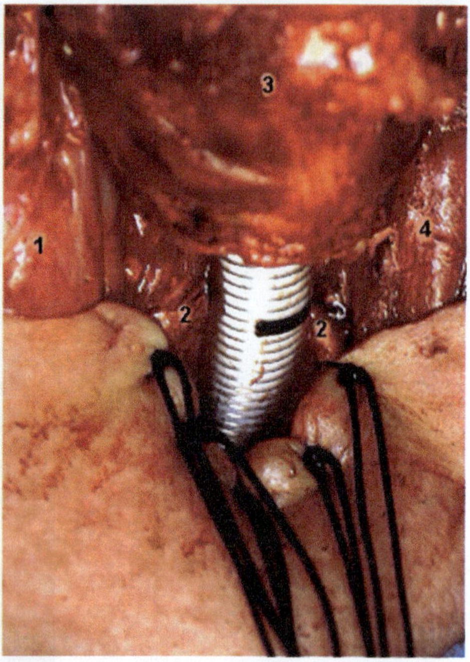

Fig. 14.52 *1* Right thyroid lobe, *2* tracheal ring, *3* larynx, *4* left thyroid lobe

lateral to the posterior ends of the first cartilaginous ring. The orotracheal anaesthesia tube is brought down and inserted in the opening in the trachea (Fig. 14.52).

The larynx is then rotated upwards, gradually detaching the anterior mucosa of the piriform sinuses and the retrocricoid mucosa inferosuperiorly with the scalpel from the posterior wall of the larynx as far as the base of the arytenoid cartilages (Fig. 14.53).

At this level, the mucosa is opened and sectioned along the contour of the laryngeal aditus (Fig. 14.54).

The complete detachment of the larynx is obtained by cutting the mucosa of the valleculae next to the tongue base with the scissors (Fig. 14.55).

We insert the naso-oesophageal feed tube. Two everted points of traction, fixed to the lateral ends of the inferior edge of the tongue base, cause the pharyngostome to assume a triangular shape with the base at the top (Fig. 14.56).

The pharyngostome is closed by means of a T-shaped suture of its edges, using interrupted and introflected stitches of fine absorbable thread, proceeding inferosuperiorly (Fig. 14.57).

Once vertical suturing of the mucosa has been completed inferosuperiorly and horizontal suturing latero-medially on both sides, a central everted "U" stitch completes the closure of the pharyngostome. A second suture, with everted interrupted stitches, is made on the outside of the previous one, through the submucosa. The closing of the pharyngostome must be carried out meticulously so as to ensure it is absolutely impermeable to the passage of saliva in the first days after the operation (Fig. 14.58).

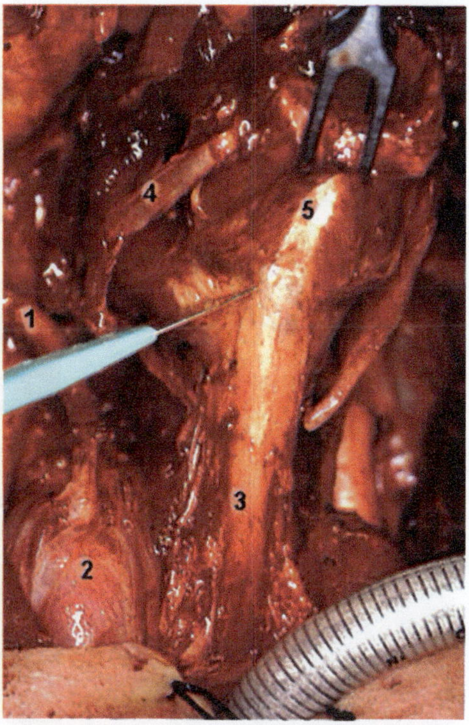

Fig. 14.53 *1* Superior thyroid artery, *2* right thyroid lobe, *3* hypopharynx, *4* thyroid cartilage, *5* cricoid cartilage

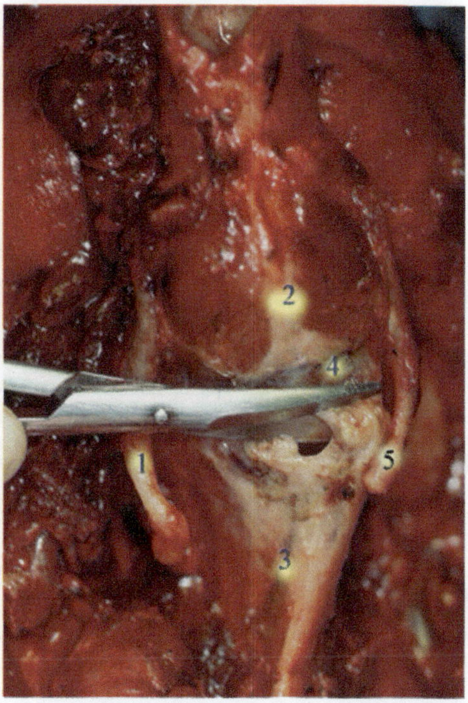

Fig. 14.54 *1* Right superior cornu of thyroid cartilage, *2* cricoid cartilage, *3* cervical oesophagus, *4* apex of left arytenoid cartilage, *5* right superior cornu of thyroid cartilage

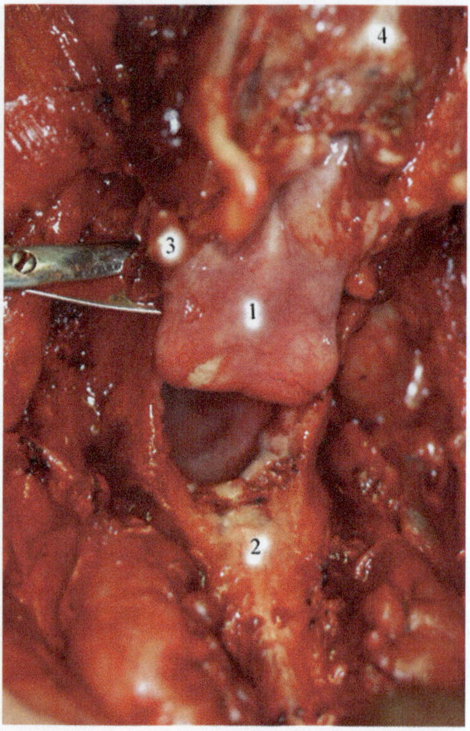

Fig. 14.55 *1* Suprahyoid epiglottis, *2* cervical oesophagus, *3* apex of right great hyoid cornu, *4* cricoid cartilage

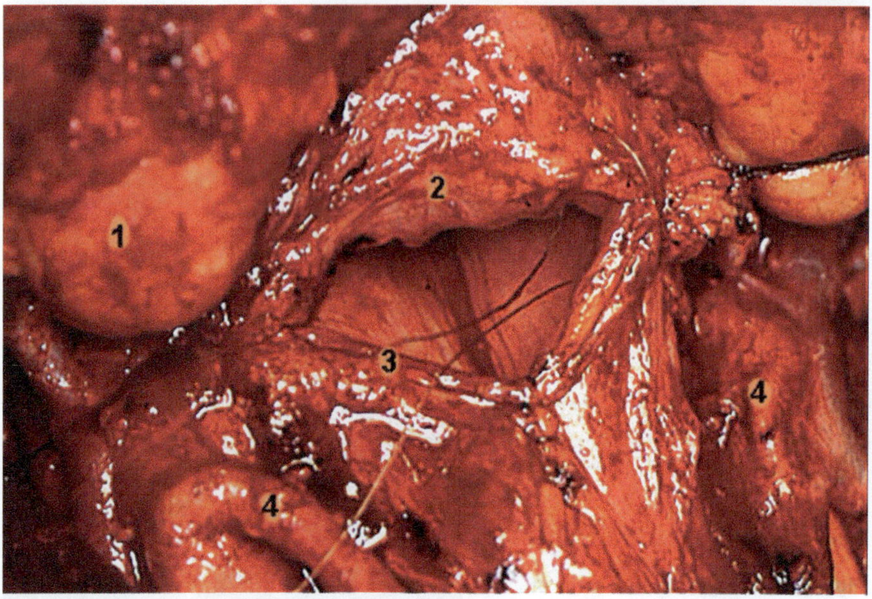

Fig. 14.56 *1* Submandibular gland, *2* tongue base, *3* hypopharyngeal mucosa, *4* superior thyroid artery

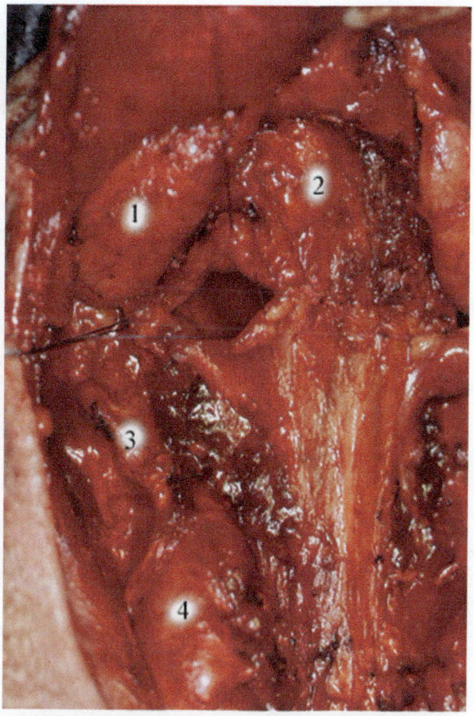

Fig. 14.57 *1* Right submandibular gland, *2* tongue base mucosa, *3* superior thyroid pedicle, *4* right thyroid lobe

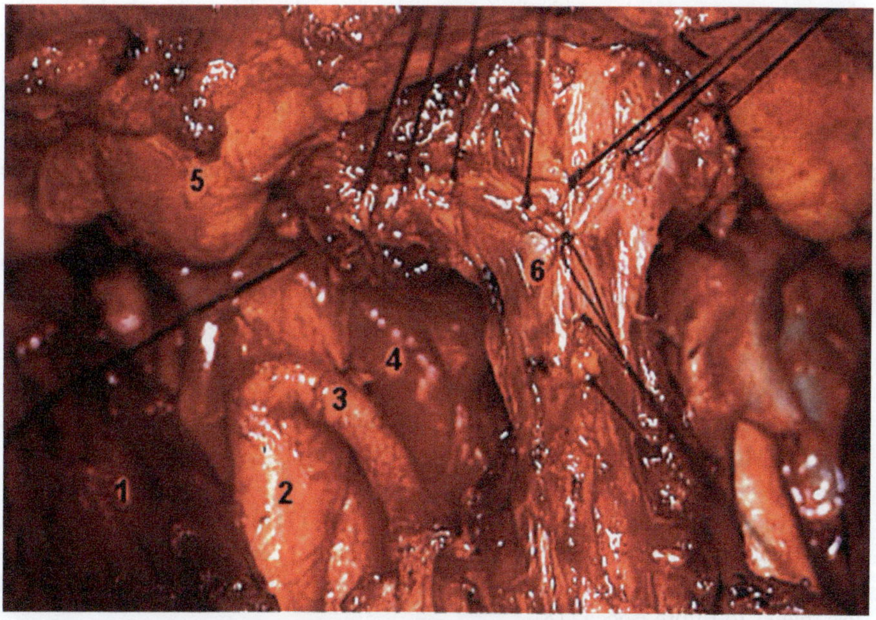

Fig. 14.58 *1* Sternocleidomastoid muscle, *2* external carotid artery, *3* superior thyroid artery, *4* inferior constrictor muscle of the pharynx, *5* submandibular gland, *6* hypopharyngeal submucosa

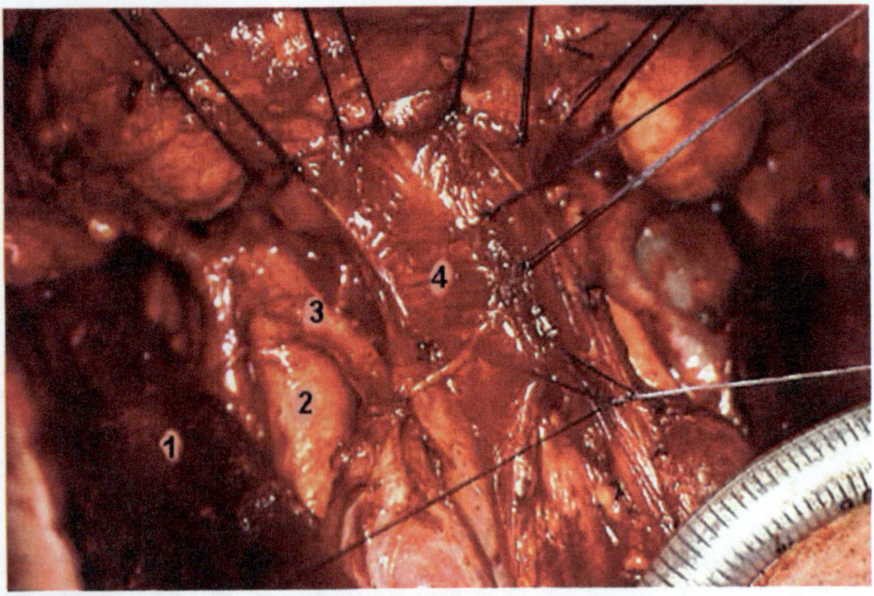

Fig. 14.59 *1* Sternocleidomastoid muscle, *2* external carotid artery, *3* superior thyroid artery, *4* inferior constrictor muscle of the pharynx

A third suture reconstitutes a muscle plane. The two section margins of the inferior constrictors, together with the two strips of perichondrium detached from the posterior edge of the thyroid cartilages and preserved in the demolitive phase of the operation, are sutured together vertically along the median line; the superior margin of the constrictors is horizontally sutured to the section edge of the suprahyoid muscles (Fig. 14.59).

The anastomosis of the tracheal opening to the cutis is obtained by suturing the anterolateral contour of the tracheostoma (corresponding to the cartilaginous ring) to the inferior cutaneous margin and the posterior edge of the tracheostoma to the cutaneous margin of the superior flap in the centre. The lateral traction exerted by the cutis on the two ends of the first tracheal ring helps keep the tracheostoma wide. The operation ends with the placement of suction drains and the suturing of the cutis on two levels.

14.6 Modified Radical Neck Dissection

Conceived in the early 1960s by Osvaldo Suarez [16, 17] in Montevideo, when from the early years of the century the standard treatment for cervical lymph node metastases had been radical dissection, it acquired its present importance and widespread use thanks to Ettore Bocca [18].

1 Principles of the Operation: (a) “En bloc” removal of the fasciae, the fatty tissue, and the cervical lymph nodes; (b) preservation of the sternocleidomastoid

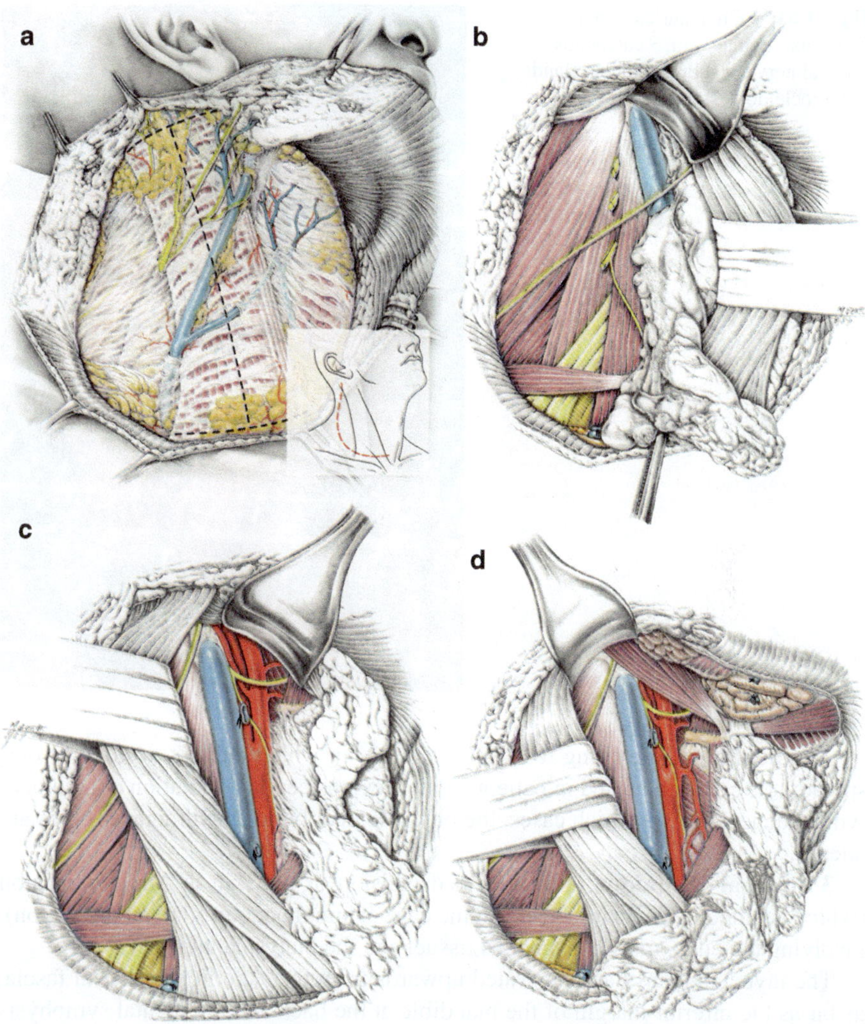

Fig. 14.60 Modified Radical Neck Dissection: surgical steps (**a**) (1) André's incision and detaching of platysma from the superficial cervical fascia (2) dissection of the superficial cervical fascia along the sternocleidomastoid muscle (**b**) dissection of supraclavicular space (VI Robbins level) spearing the spinal nerve, the phrenic nerve and the muscular portion of the cervical plexus (**c**) dissection of the jugulocarotid space (II–III–IV Robbins levels) spearing the internal jugular vein, the vagus nerve, the common trunk of spinal nerve and the hypoglossus nerve (**d**) the dissection arrives medially as far as the hyoid bone and the whole specimen must remain in continuity with the larynx

muscle, the internal jugular vein, and the spinal accessory nerve; (c) if the indications are correct, it has the same possibilities of regional control of the disease as radical dissection; (d) unlike radical dissection, it can be carried out simultaneously on both sides (Fig. 14.60a–d).

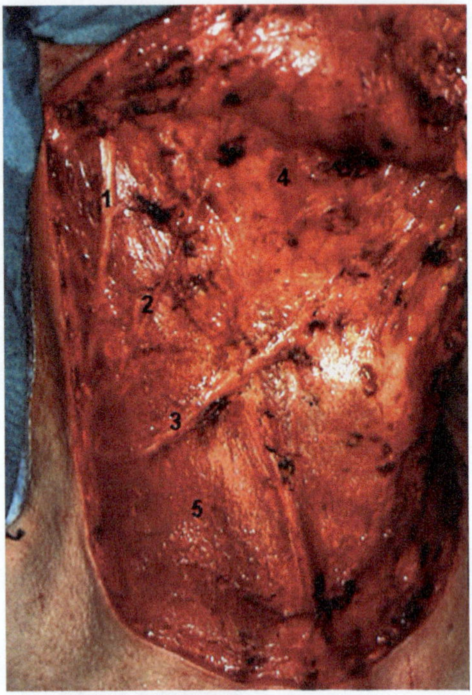

Fig. 14.61 *1* Great auricular nerve, *2* external jugular vein, *3* cutaneous cervical nerve, *4* submandibular gland, *5* sternocleidomastoid muscle

2 Indications: According to E. Bocca, all N in which there is no tumour outspread. It is contraindicated in patients with previous cervical radiation or results of lymphadenectomy, in which cases, the physiological lymphatic drainage routes are altered.

3 Surgical Technique: The preferred incision of the skin is André's incision (which is bilateral and symmetrical in the case of simultaneous bilateral dissection), involving the cutis, the subcutaneous tissue, and the platysma muscle.

The myocutaneous flap is elevated upwards from the superficial cervical fascia, as far as the inferior margin of the mandible at the back and the mental symphysis at the front; the rear flap as far as the anterior edge of the trapezius muscle; and the lower flap as far as the superior edge of the clavicle. In elevating the rear flap, care must be taken not to harm the peripheral branch of the spinal accessory nerve, the median section which lies very close to the subcutaneous tissue.

In this way, a large field of operation is obtained, traversed vertically in the centre by the sternocleidomastoid muscle, still covered by the superficial cervical fascia containing the system of the external jugular vein and some branches of the superficial cervical plexus (Fig. 14.61).

Dissection normally proceeds inferosuperiorly and posteroanteriorly. First, the anterior margin of the trapezius muscle is sought, then the peripheral branch of the spinal accessory nerve is identified and isolated, which we normally seek at its entrance into the trapezius muscle or where it comes out from the sternocleidomastoid muscle, about 1 cm above Erb's point (Fig. 14.62).

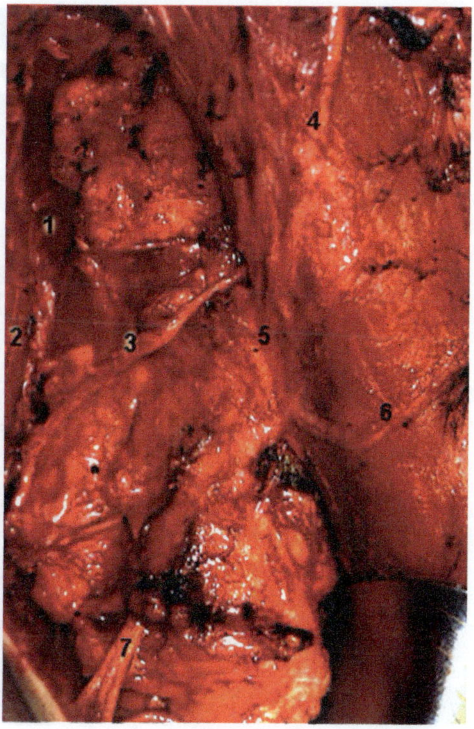

Fig. 14.62 *1* Levator scapulae muscle, *2* trapezius muscle, *3* spinal accessory nerve, *4* great auricular nerve, *5* Erb's point, *6* cutaneous cervical nerve, *7* cervical nerve for the trapezius muscle

Clearance of the fifth level does not normally include the resection of the muscular branches of the cervical plexus (Fig. 14.63); the phrenic nerve must be absolutely preserved.

The superficial cervical fascia is incised vertically along the anterior surface of the sternocleidomastoid muscle after ligation of the external jugular veins (Fig. 14.64).

The superficial cervical fascia is then elevated with the scalpel both on its external surface and on the internal surface of the sternocleidomastoid muscle, which is surrounded with a gauze swab and stretched anteriorly. At this point, removal of the fascial and fatty tissues of the supraclavicular fossa is begun latero-medially and inferosuperiorly, after ligation and section of the transverse pedicle of the neck and the branches for the trapezius muscle of the cervical plexus. A little lower down, the inferior belly of the omohyoid muscle is identified (Fig. 14.65).

The dissection of the supraclavicular fossa extends medially as far as the internal jugular vein, preserving the brachial plexus and the phrenic nerve and ligating and sectioning the external jugular vein. In the left side of the neck, it is necessary to recognise the thoracic duct and eventually interrupt it, after ligation.

Now we go back up. At this point, the retromandibular and external jugular veins are sectioned at the inferior pole of the parotid gland. A Farabeuf retractor pulls up the posterior belly of the digastric muscle and the stylohyoid muscle. Another retractor lateralises the sternocleidomastoid muscle. The upper landmark of dissection is the transverse process of the atlas, easily identifiable by palpation. The field of

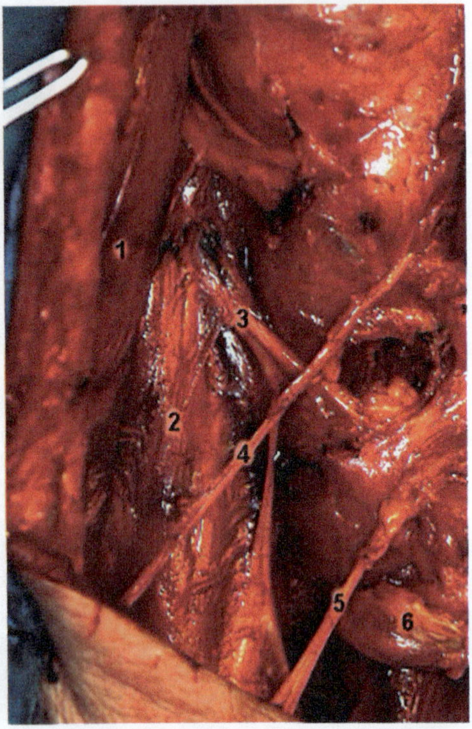

Fig. 14.63 *1* Trapezius muscle, *2* levator scapulae muscle, *3* branches of the cervical plexus, *4* spinal accessory nerve, *5* cervical branches for the trapezius muscle, *6* transverse pedicle of the neck

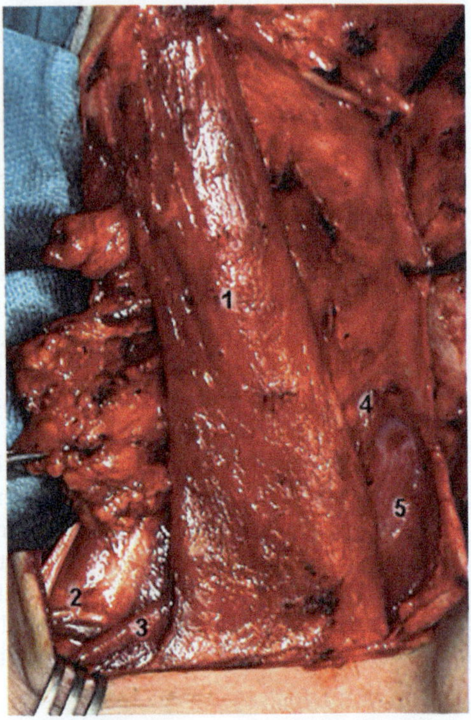

Fig. 14.64 *1* Sternocleidomastoid muscle, *2* traverse artery and vein of the neck, *3* inferior belly of the omohyoid muscle, *4* superior belly of the omohyoid muscle, *5* subhyoid muscles

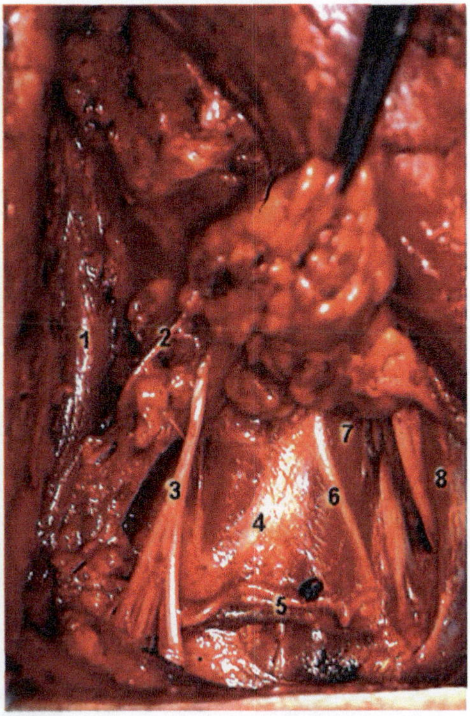

Fig. 14.65 *1* Trapezius muscle, *2* spinal accessory nerve, *3* cervical branch for the trapezius muscle, *4* brachial plexus, *5* traverse artery and vein of the neck, *6* phrenic nerve, *7* anterior scalene muscle, *8* internal jugular vein

operation is favourable for identification of the internal jugular vein with the spinal nerve that crosses it superficially (Fig. 14.66).

The loose tissue of level IIb is passed under the spinal nerve and rejoined to that of level IIa; this manoeuvre exposes the deep muscle plane, represented by the levator scapulae and splenius capitis muscles (Fig. 14.67).

We now go back down and proceed to complete dissection level III and IV in a latero-medial direction, taking the vagus nerve as reference (Fig. 14.68).

Also, the hypoglossal nerve and its descending branch are isolated from the fascial tissues (Fig. 14.69).

Using a subadventitial approach, the internal jugular vein is isolated (after ligation of the thyrolinguofacial trunk and of any other accessory branches) and then the carotid axis (Fig. 14.70).

Dissection is continued forwards in the parapharyngeal space, taking care to preserve the superior thyroid artery, and submandibular dissection, after ligation of the facial vein, preserving the salivary gland. During this surgical stage, it is necessary to preserve the marginal branch of the facial nerve, which must be identified beforehand and shifted cranially. Anterior clearance is completed as far as the median line with dissection of the submental region (interdigastric triangle) and of the superficial cervical fascia that covers the subhyoid muscles (Fig. 14.71).

Dissection is completed: all the lateral and anterior cavities, from the anterior edge of the trapezius muscle and from the splenius capitis muscle to the anterior median line in a horizontal direction and from the jugulodigastric space and from

Fig. 14.66 *1* Sternocleidomastoid muscle, *2* posterior belly of the digastric muscle, *3* common trunk of the spinal accessory nerve, *4* internal jugular vein, *5* Robbins level IIB, *6* right submandibular gland, *7* omohyoid muscle

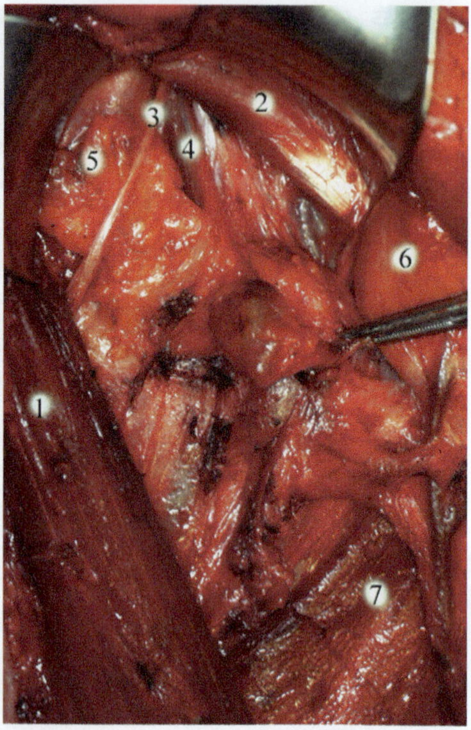

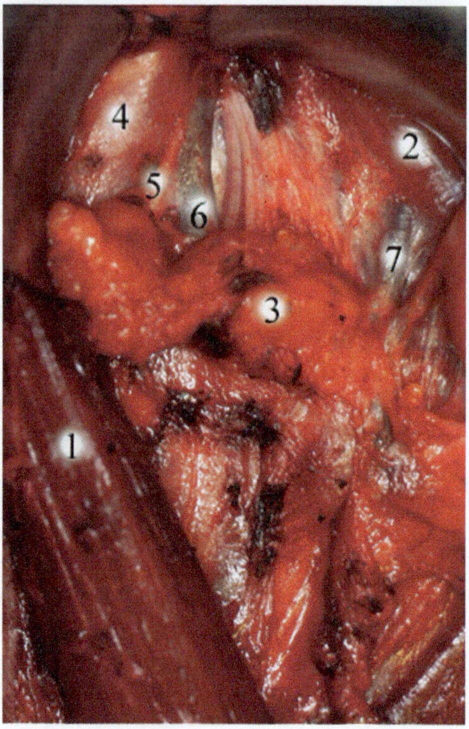

Fig. 14.67 *1* Sternocleidomastoid muscle, *2* posterior belly of the digastric muscle, *3* clearance specimen, *4* levator scapulae muscle, *5* common trunk of the spinal accessory nerve, *6* internal jugular vein, *7* lingual vein

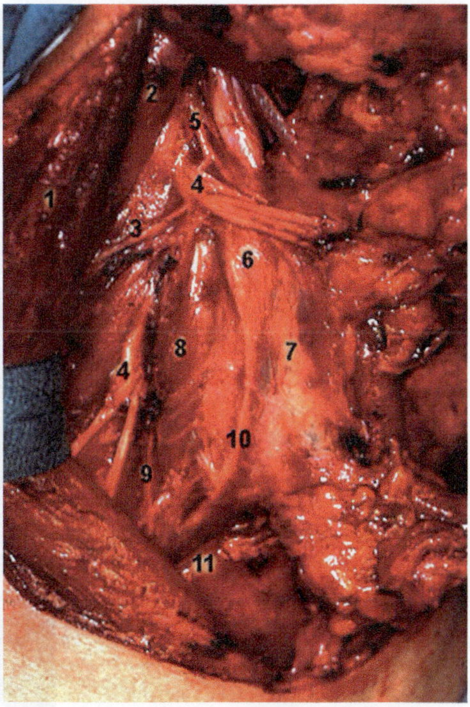

Fig. 14.68 *1* Sternocleidomastoid muscle, *2* levator scapulae muscle, *3* spinal accessory nerve, *4* cervical plexus, *5* hypoglossal nerve, *6* carotid bifurcation, *7* internal jugular vein, *8* anterior scalene muscle, *9* phrenic nerve, *10* descending branch of the hypoglossal nerve, *11* omohyoid muscle

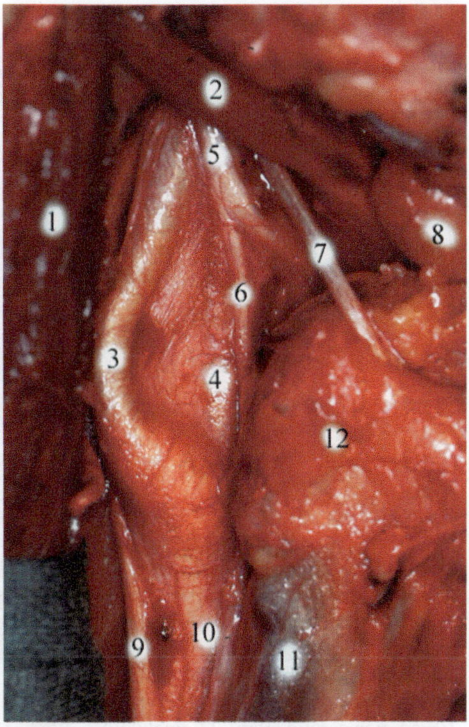

Fig. 14.69 *1* Sternocleidomastoid muscle, *2* posterior belly of the digastric muscle, *3* internal carotid artery, *4* external carotid artery, *5* hypoglossal nerve, *6* descending branch of the hypoglossal nerve, *7* lingual vein, *8* right submandibular gland, *9* vagus nerve, *10* common carotid artery, *11* internal jugular vein, *12* clearance specimen

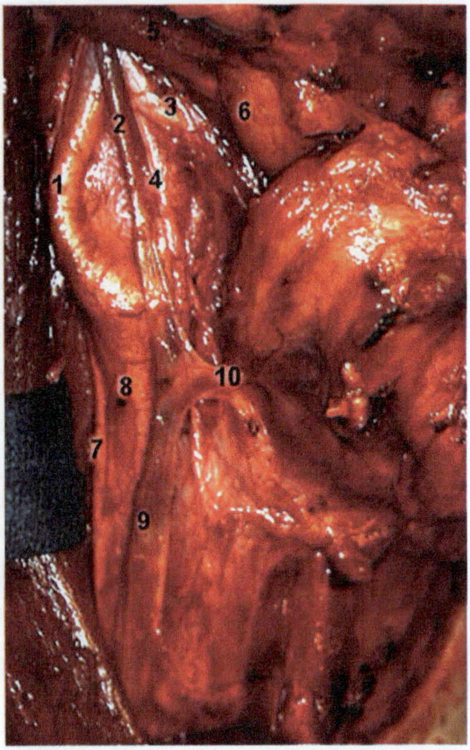

Fig. 14.70 *1* Internal carotid artery, *2* internal jugular vein, *3* hypoglossal nerve, *4* descending branch of the hypoglossal nerve, *5* digastric muscle, *6* submandibular gland, *7* vagus nerve, *8* common carotid artery, *10* thyrolinguofacial venous trunk

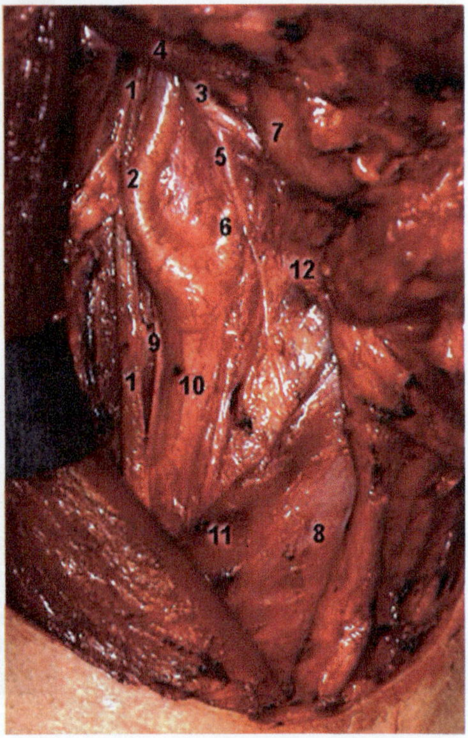

Fig. 14.71 *1* Internal jugular vein, *2* internal carotid artery, *3* hypoglossal nerve, *4* digastric muscle, *5* descending branch of the hypoglossal nerve, *6* external carotid artery, *7* submandibular gland, *8* sternohyoid muscle, *9* vagus nerve, *10* common carotid artery, *11* omohyoid muscle, *12* greater cornu of the hyoid bone

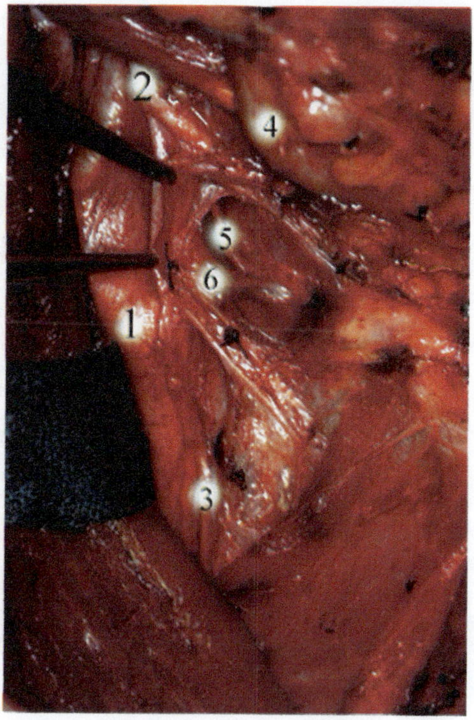

Fig. 14.72 *1* Carotid bifurcation, *2* hypoglossal nerve, *3* descending branch of the hypoglossal nerve, *4* submandibular gland, *5* superior laryngeal nerve, *6* superior laryngeal pedicle

the submandibular region to the superior margin of the clavicle in a vertical direction, must be radically cleared of the fasciae and of loose tissue. If the larynx is to be operated on immediately afterwards, at the end of dissection, the whole specimen must remain in continuity with the larynx, generally by means of the hyoid bone (Fig. 14.72).

4 Anatomical Variants: Those most observed are the following: (1) the peripheral branch of the spinal accessory nerve passes under the internal jugular instead of over it, or the same nerve runs completely outside the sternocleidomastoid muscle; (2) the branches of the internal jugular vein, sometimes also posterior, or the frequent presence of a median thyroid vain in addition to the thyrolinguofacial venous trunk; (3) the carotid bifurcation, which is usually at the level of the greater cornu of the hyoid bone, is much more caudal; and (4) kinking of the internal carotid artery just above the bifurcation (frequent) (Fig. 14.73) and carotid trifurcation (very rare) (Fig. 14.74).

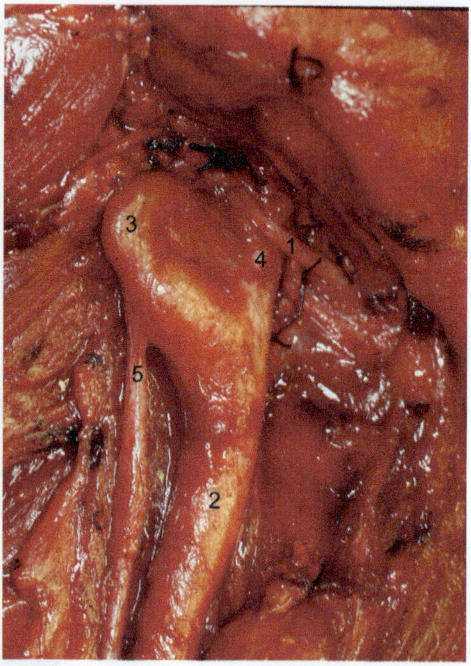

Fig. 14.73 *1* Hypoglossal nerve, *2* common carotid artery, *3* internal carotid artery, *4* esternal carotid artery, *5* vagus nerve

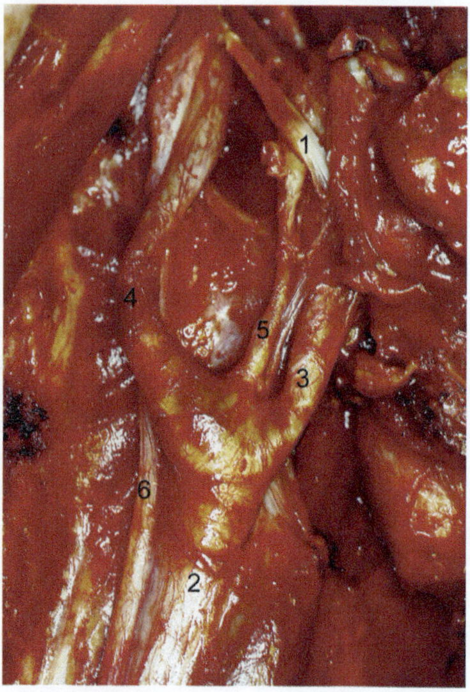

Fig. 14.74 *1* Hypoglossal nerve, *2* common carotid artery, *3* esternal carotid artery, *4* internal carotid artery, *5* abnormal facial artery, *6* vagus nerve

Take-Home Messages

- Endoscopic laser surgery has introduced new and revolutionary concepts into ENT surgery, such as the concept of tailored surgery as apposed to the invariable choice of the procedure of the cervicotomic approach. It has drastically reduced postoperative morbidity, hospitalisation, and consequently the costs of the treatments.
- Learning surgical techniques requires an initial theoretical time of absorbing pure knowledge for everyone. After that, the possibility of putting what has been learnt into practice is linked to countless variables, only some of which can be controlled. However, it is always true that knowing is better than not knowing. And then, things can always change.

References

1. Remacle M, Eckel HE, Antonelli A et al (2000) Endoscopic cordectomy:a proposal for a classification by the Working Committee, European Laryngological Society. Eur Arch Otorhinolaryngol 257:227–231
2. Remacle M, Van Haverbeke C, Eckel H, Bradley P, Chevalier D, Djukic V, de Vincentiis M, Friedrich G, Olofsson J, Peretti G, Quer M, Werner J (2007) Proposal for revision of the European Laryngological Society classification of endoscopic cordectomies. Eur Arch Otorhinolaryngol 264:499–504
3. Lucioni M, Marioni G (2008) Clinically based comments on the proposal for revision of the European Laryngological Society (ELS) classification of endoscopic cordectomies. Eur Arch Otorhinolaryngol 265:613–615
4. Remacle M (2008) Clinically based comments on the proposal for revision of the European Laryngological Society (ELS) classification of endoscopic cordectomies. Eur Arch Otorhinolaryngol 265:1143–1144, Reply to the letter to the editor by Lucioni and Marioni
5. Peretti G, Cappiello J, Nicolai P, Smussi C, Antonelli AR (1994) Endoscopic laser excisional biopsy for selected glottic carcinoma. Laryngoscope 104:1276–1279
6. Sjogren EV, vanRossum MA, Langeveld TP et al (2008) Voice outcome in T1a midcord glottic carcinoma: laser surgery vs radiotherapy. Arch Otolaryngol Head Neck Surg 134:965–972
7. Lucioni M, Bertolin A, D'Ascanio L, Rizzotto G (2012) Margin photocoagulation in laser surgery for early glottic cancer: impact on disease local control. Otolaryngol Head Neck Surg 146:600–605
8. Lucioni M, Marioni G, Bertolin A, Giacomelli L, Rizzotto G (2011) Glottic laser surgery: outcomes according to 2007 ELS classification. Eur Arch Otorhinolaryngol 268:1771–1778
9. Alonso JM (1947) Conservative surgery of cancer of the larynx. Trans Am Acad Ophthalmol Otolaryngol 51:633–642
10. Rouvière H (1938) Lymphatic system of the head and neck. (trans: Tobias M). Edwards Brothers, Ann Arbor
11. Serafini I (1988) Reconstructive surgery after total or partial laryngectomy. Acta Otorhinolaryngol Ital 8(Suppl 20):37–48
12. Labayle O, Bismuth R (1971) La laryngectomie totale avec reconstruction. Ann Otolaryngol 88:219–228
13. Piquet JJ, Desaulty A, Decroix G (1974) La crico-ioido-epiglotto-pexie. Tecnique operatoire et resultats fonctionels. Ann Otolaryngol Chir Cervicofac 91:681–690

14. Rizzotto G, Succo G, Lucioni M, Pazzaia T (2006) Subtotal laryngectomy with tracheohyoidopexy: a possible alternative to total laryngectomy. Laryngoscope 116(10):1907–1917
15. Schwartz AW (1978) Dr. Theodor Billroth and the first laryngectomy. Ann Plast Surg 1(5):513–516
16. Suarez O (1963) El problema de la metastasi linfatica y alejudus del cancer de laringe e hipofaringe Rev. Otorinolaring 23:83–99
17. Ferlito A, Rinaldo A (2004) Osvaldo Suarez: often-forgotten father of functional neck dissection in the non-Spanish literature. Laryngoscope 114(7):1177–1178
18. Bocca E, Pignataro O (1967) A conservation technique in radical neck dissection. Ann Otol Rhinol Laryngol 76:975–987

Index

M. Lucioni, *Practical Guide to Neck Dissection*,
DOI 10.1007/978-3-642-33977-6, © Springer-Verlag Berlin Heidelberg 2013

GPSR Compliance

The European Union's (EU) General Product Safety Regulation (GPSR) is a set of rules that requires consumer products to be safe and our obligations to ensure this.

If you have any concerns about our products, you can contact us on ProductSafety@springernature.com

In case Publisher is established outside the EU, the EU authorized representative is:

Springer Nature Customer Service Center GmbH
Europaplatz 3
69115 Heidelberg, Germany

Batch number: 10747773

Printed by Printforce, the Netherlands